Negotiating a River

The Nature | History | Society series is devoted to the publication of high-quality scholarship in environmental history and allied fields. Its broad compass is signalled by its title: nature because it takes the natural world seriously; history because it aims to foster work that has temporal depth; and society because its essential concern is with the interface between nature and society, broadly conceived. The series is avowedly interdisciplinary and is open to the work of anthropologists, ecologists, historians, geographers, literary scholars, political scientists, sociologists, and others whose interests resonate with its mandate. It offers a timely outlet for lively, innovative, and well-written work on the interaction of people and nature through time in North America.

General Editor: Graeme Wynn, University of British Columbia

A list of titles in the series appears at the end of the book.

Negotiating a River

Canada, the US, and the Creation of the St. Lawrence Seaway

DANIEL MACFARLANE

UBC Press • Vancouver • Toronto

© UBC Press 2014

All rights reserved. No part of this publication may be reproduced, stored in a retrieval system, or transmitted, in any form or by any means, without prior written permission of the publisher, or, in Canada, in the case of photocopying or other reprographic copying, a licence from Access Copyright, www.accesscopyright.ca.

21 20 19 18 17 16 15 14   5 4 3 2 1

Printed in Canada on FSC-certified ancient-forest-free paper
(100% post-consumer recycled) that is processed chlorine- and acid-free.

ISBN 978-0-7748-2643-3 (bound)
ISBN 978-0-7748-2644-0 (pbk.)
ISBN 978-0-7748-2645-7 (pdf)
ISBN 978-0-7748-2646-4 (e-pub)

Cataloguing-in-publication data for this book is available from Library and Archives Canada.

Canadä

UBC Press gratefully acknowledges the financial support for our publishing program of the Government of Canada (through the Canada Book Fund), the Canada Council for the Arts, and the British Columbia Arts Council.

This book has been published with the help of a grant from the Canadian Federation for the Humanities and Social Sciences, through the Awards to Scholarly Publications Program, using funds provided by the Social Sciences and Humanities Research Council of Canada.

Set in Garamond by Artegraphica Design Co. Ltd.
Copy editor and proofreader: Judy Phillips
Cartographer: Eric Leinberger

UBC Press
The University of British Columbia
2029 West Mall
Vancouver, BC V6T 1Z2
www.ubcpress.ca

For Jen, E.J., and Lucas

"Each generation exercises power over its successors, and each, in so far as it modifies the environment bequeathed to it and rebels against tradition, resists and limits the power of its predecessors."

– C.S. Lewis, *The Abolition of Man*

Contents

List of Illustrations / ix

Foreword: National Dreams / xiii
Graeme Wynn

Acknowledgments / xxvi

Abbreviations / xxviii

Introduction: River to Seaway / 3

Part 1: Negotiating

1 Accords and Discords / 21

2 Watershed Decisions / 48

3 Caught between Two Fires / 76

Part 2: Building

4 Fluid Relations / 111

5 Lost Villages / 139

6 Flowing Forward / 179

Conclusion: To the Heart of the Continent / 208

Notes / 232

Bibliography / 286

Index / 310

Illustrations

TABLES

2.1 Estimate of potential traffic on the canal systems of the St. Lawrence waterway / 61

6.1 Tolls on the St. Lawrence Seaway / 191

FIGURES

0.1 Great Lakes–St. Lawrence waterway / 4

0.2 St. Lawrence seaway / 4

0.3 Profile of Great Lakes–St. Lawrence waterway / 6

0.4 Welland Canal / 6

0.5 Contemporary aerial view of the submerged remains of the town of Aultsville / 8

0.6 Iroquois Lock and Control Dam during construction / 12

1.1 Historical canals on the St. Lawrence River / 22

1.2 Rapids in the St. Lawrence River with fourteen-foot canal in foreground / 26

2.1 Robert Saunders / 55

2.2 Citizens' Joint Action Committee / 68

3.1 Groundbreaking ceremony / 100

3.2 Signing ceremony / 104
4.1 Blueprint of the International Rapids section / 114
4.2 Robert Moses / 123
4.3 Vessel in completed seaway channel / 127
4.4 Construction of the Wiley-Dondero Canal and Moses-Saunders Power Dam / 129
4.5 Construction equipment at work on the St. Lawrence project / 131
4.6 Cofferdam / 132
4.7 Winter construction / 134
4.8 Workers / 134
4.9 Moses-Saunders Power Dam under construction / 136
5.1 Map of Lake St. Lawrence and the Lost Villages / 140
5.2 Moulinette / 144
5.3 Mille Roches / 145
5.4 Unhappy with HEPCO / 147
5.5 Relocated town of Iroquois / 151
5.6 House mover in action / 155
5.7 HEPCO's long-term planned layout of New Town 1 (Ingleside) / 157
5.8 HEPCO's initial layout of New Town 1 (Ingleside) / 157
5.9 HEPCO's long-term planned layout of New Town 2 (Long Sault) / 158
5.10 Sign in Iroquois / 160
5.11 New shopping centre / 166
5.12 View of Long Sault soon after inundation / 171
6.1 Opening ceremonies at Montreal featuring Queen Elizabeth II / 180
6.2 HEPCO model of the St. Lawrence Seaway and Power Project / 182
6.3 View from US side of the Moses-Saunders Power Dam under construction / 184
6.4 Power dam under construction / 185
6.5 View from the Canadian side of the power dam during construction / 185
6.6 Looking north from Welland Canal Lock 8, 2011 / 193

6.7   Vessel in lock / 194
6.8   Opening ceremonies at the Moses-Saunders Power Dam / 195
6.9   Snell Lock / 196
6.10 Inside of a seaway lock / 196
6.11 Eisenhower Lock / 197
6.12 Long Sault Dam and New York's Barnhart Island recreation area / 201
6.13 Saint-Lambert Lock and Victoria Bridge at Montreal / 203
C.1   Iroquois Lock, 2009 / 210
C.2   Remains of old fourteen-foot canal, Iroquois Control Dam in background / 210
C.3   Power dam from the American side, 2009 / 216
C.4   Long Sault Control Dam, 2009 / 216
C.5   Old Highway 2 disappearing into Lake St. Lawrence / 223
C.6   Old road and building foundations visible at the former site of Aultsville in Lake St. Lawrence, 2012 / 223

FOREWORD

National Dreams

Graeme Wynn

Canada has been a technological project. Harold Innis attributed the northern extent of the country to the fur trade and its dependence upon the canoe. The construction of the Canadian Pacific Railway was intended to knit together east and west and to secure the Prairies against encroachment from the south. In the age of iron and steam, the railroad was to form the backbone of the country; it was an instrument intended, in the memorable words of historian Pierre Berton, to realize the National Dream. When new forms of communication challenged the hegemony of the iron horse and the telegraph wires that marched alongside its rails, Canadians adopted and adapted them to the same end. Radio broadcasting began early in the twentieth century, and the Montreal station that became CFCF was among the first broadcasters in North America when it went on air in 1920. Within a decade, a Royal Commission on Radio Broadcasting had concluded that a country the size of Canada should have a publically funded radio broadcasting system. On its creation in 1932, Prime Minister R.B. Bennett harboured no doubts about its importance: the country, he said, needed "complete Canadian control of broadcasting from Canadian sources" to ensure that a "national consciousness may be fostered and sustained and national unity still further strengthened." Similar imperatives guided governance of television broadcasting when it began to expand in the 1950s.[1]

Late in the 1930s, the country's largest company, Canadian National Railway (a Crown corporation), underwrote the development of Trans-Canada Air Lines, to move the mail and to offer a new, quicker form of transport linking the country's far-flung regions. Reflecting the growing importance of automobiles as vehicles of choice among Canadians, construction of the Trans-Canada Highway began in 1950. Other technological projects – from the Alcan (Alaska) Highway begun in 1942 to the "extraordinary efforts at ecological re-engineering" entailed in the construction of massive hydroelectric-generating facilities across the boreal forest in the 1960s and 1970s – have served similarly to knit together the distant settlements of this vast territory by creating corridors for the movement of goods and people, or the energy to power a growing, increasingly integrated economy.[2] More generally, these late-twentieth-century developments allowed the forces of modernization to transform vast areas of the country and served (at least initially) to further the political, social, and cultural integration of Canadians.

Few of these twentieth-century developments have been as much remarked upon as the one at the centre of this book – the development, between 1954 and 1959, of the St. Lawrence Seaway as an artery into "the heart of the continent" and the concomitant harnessing of the river to produce electric power. The realization of powerful, brute-force technologies deployed when high modernist confidence in the value of transforming nature for human purposes met few challenges, this massive venture brought Canada and the United States together in a project of hitherto unrivalled magnitude. An American publication written a couple of years before the seaway opened captured the sense that this was not only a "matter of modern progress and international prosperity" but, in its turning of a mighty river to human ends, a project for the ages:

> It has taken 15,000 men to perform the labor, the basic digging, dredging, hauling and building that ended with the taming of the St. Lawrence River. They have used hundreds of machines worth sixty million dollars, including in round numbers 500 heavy trucks, 250 bulldozers, 150 of the biggest shovels and draglines, and 15 dredges. They have excavated 200,000,000 cubic yards of earth and rock, and impacted 10,000,000 cubic yards in the form of dikes, around 20 miles of them, some more than 50 feet high. They have built dozens of cofferdams … They have cut channels, removed islands, filled in points of access, laid down roads, set up bridges, [and] relocated everything from telegraph poles to towns.[3]

As the great project proceeded, embodying in its twin purposes – the improvement of transportation and the generation of hydroelectricity – the two great elixirs of the twentieth century, many Canadians came to see in it a reflection of Canadian aspirations. Print, radio, and television journalists documented the progress of what was widely regarded as "one of the most challenging engineering feats in history."[4] Others produced more extended accounts to document particular facets of the project, to celebrate its completion, to enhance public awareness of what had been achieved, or simply to record the magnitude of the accomplishment. Decades after the project was completed, the St. Lawrence Seaway Management Corporation produced a booklet intended to explain the seaway – "the tallest water staircase west of China" – to children; the great achievement of the 1950s found its place in general histories of the country; and, with the fiftieth anniversary of its opening, there was a resurgence of interest in what most agreed was among the top ten public works projects of the twentieth century.[5]

Perhaps inevitably, with an enterprise of this magnitude, there were people and places in the way of progress. All that digging, dredging, hauling, and building, and the rising water levels behind the big dams built to control the river's flow and generate electricity, in particular, meant displacement and relocation. This too was generally portrayed, at the time, as triumph rather than tragedy, or at least as a small price to pay for the benefits in store. New planned settlements would replace old villages; houses would have modern plumbing and efficient designs; for those who preferred treasured old dwellings, powerful machines would lift and relocate them above the rising waters. Some grumbled, others protested more vigorously, and many felt some sense of loss and disorientation, but the prevailing ethos was captured in a CBC television report on the relocation: "Outside the general store in Farran Point [sic], three youngsters are asked what they'll do when the town is flooded. One youth, unaware of the implications of the flooding, innocently answers, 'I don't know. Swim, I guess.'"[6]

Time and nostalgia have added poignancy to this harmless remark. Two decades after the waters crept into Farran's Point, Aultsville, Dickinson's Landing, Milles Roches, and other small riverside communities, the Lost Villages Historical Society was established "to inform the public, and specifically school children," about the settlements submerged by the St. Lawrence Seaway and Power Project (SLSPP). Although several buildings were removed from the Long Sault area in the 1950s to form the Upper Canada Village living history museum representing rural life in the colony on the eve of Confederation, the Lost Villages Society established its own museum in Ault Park to show what life was like before 1954. Websites, retrospective accounts of the flooding in regional newspapers, folk/pop songs, and art exhibitions memorializing the Lost Villages have followed.[7] So too have the displacements made their appearance in widely read works of fiction. Casablanca, the little cluster of government houses at "the end of the disappearing road," which runs down to a dock and begins again (beyond the submerged fence posts, crumpled steeple, and old foundations of a drowned settlement) "on one side of a small island and empties out into the water again on the other," is central to Johanna Skibsrud's *The Sentimentalist,* a novel "obsessed with feelings and events long submerged and only now half discerned."[8] And Anne Michaels's *The Winter Vault* brings the construction of Egypt's Aswan High Dam and the St. Lawrence Seaway and Power Project together in an extended, poetic meditation on memory and loss that is also a critique of progress and an expression of hope, however ambivalent, that the world, or parts of it, can be saved. Hence the efforts of Jean Shaw, the novel's female protagonist, to collect, record, and transplant the "particular generation" of plants about to be drowned by the rising waters of the St. Lawrence. And hence the enormous effort to mark, dismantle, and re-erect on ground above expanding Lake Nasser the ancient Nubian temple of Abu Simel – which provokes the rueful reflection that perhaps "the replica, which is meant to commemorate, achieves the opposite effect: It allows the original to be forgotten."[9]

All of this has focused attention on the great engineering project of the 1950s and its aftermath and allowed other dimensions of this story to be ignored, if not forgotten. Focusing on the second half of the twentieth century, prevailing Canadian interpretations of the St. Lawrence Seaway and Power Project see it variously as a high modernist, Cold War initiative that brought Canadians and Americans together to further common economic and strategic goals, that helped to knit the two countries together, and that had an immense impact upon society and environment in the Great Lakes–St. Lawrence corridor, even if some of these (the displacement of people, the introduction of invasive species) have since come to be regarded as part of a Faustian bargain.

These views are not wrong. But they are incomplete. It is Macfarlane's great achievement, in *Negotiating a River,* to complicate and clarify this story by focusing attention, in the first half of his book, on the "long submerged and ... [until] now only ... half discerned" political machinations that preceded the 1954 agreement to proceed with the SLSPP, and by offering, in the second, a detailed exegesis of the ways in which large-scale technological projects inevitably bring technology, politics, economics, and societal and environmental concerns into conflict and require compromise. Drawing from and speaking to many intellectual constituencies in this remarkably wide-ranging work, Macfarlane deepens our understanding of twentieth-century Canadian history even as he broadens the scope of Canadian environmental historical scholarship and leaves his readers with points to ponder about early-twenty-first-century debates over resource development and the illusory boundaries between technologies and environments. It is to a brief explication of these that we now turn.

IT WAS JACQUES CARTIER who gave the river its modern name, on 10 August, the feast day of Saint Lawrence, in 1535; elsewhere he described it as the "rivière du Canada." In the centuries since, this mighty stream has been a defining feature of northern North America, flowing across maps and through descriptions of the territory, blending itself with the sinews of those who worked its currents, and seeping into the narratives of historians, novelists, journalists, and others to shape Canadians' views of themselves and their country. Indeed, the St. Lawrence has been described as a force that made nations and moulded the lives of millions, and it has often served "as a synecdoche for Canada in general, and central Canada ... in particular" (see p. 14). The subject of Macfarlane's gaze is of such importance that some sense of its place in the interpretation of the country is vital to understanding the implications of his work.

On the single page depicting the new world in Ortelius's *Theatrum Orbis Terrarum* (1570), possibly the "first general atlas published after the revival of the sciences in Europe," the St. Lawrence was the only large river in eastern North America.[10] Early in the seventeenth century, it was commonly described, in echo of Cartier, as "the Great River of Canada" – although it made a fleeting appearance under the alias "de Groote Rivier van Niew Nederlandt," in John Ogilby's *America,* published early in the 1670s.[11] According to Cole Harris's retrospective account, those who settled along *le fleuve Saint-Laurent* conceived of it variously as both river and sea. In the seventeenth and early eighteenth centuries, it was an artery of the fur trade, linking the interior with the Atlantic and beyond to distant French ports; by the early nineteenth century, with commerce firmly in the hands of the English, and habitant society turning in upon itself, the St. Lawrence seemed, to French Canadians, to lead nowhere. They lived, observed Paul Veyret in a mid-twentieth festschrift for the French geographer Philippe Arbos, in isolation in a "region de passage"; in the minds of most of those who dwelled beside it, wrote Harris, the river had "become more completely an inland sea than ever before or since."[12]

Others, elsewhere, saw it differently. Surveyor and naval officer Joseph Bouchette, writing in 1832, thought the St. Lawrence "the most splendid river on the globe," and described it as an "indelible link formed by nature between the Canadas, and the source at once of the wealth, beauty and prosperity of both provinces."[13] For Henry David Thoreau, on a whirlwind northern excursion (travelling eleven hundred miles from Concord in little more than a week in 1850 at a cost of $12.75, "including two guidebooks and a map, which cost one dollar twelve and a half cents,") it was, simply, "the most interesting object in Canada."[14]

In the second quarter of the nineteenth century, the river-reach of the St. Lawrence was being extended and improved. Bouchette admired the soon-to-be-opened Welland Canal, a momentous work, forty-two miles in length, with thirty-seven locks capable of carrying vessels of 125-tons burden through the 330-foot difference in elevation between Lakes Erie and Ontario. He also noted the "stupendous magnitude and incalculable utility" of the Rideau Canal, with its forty-seven locks, all of five feet depth (matching those on the Lachine Canal immediately above Montreal), built both to ensure "unrestricted intercourse" between Kingston and the lower river should American hostility jeopardize passage along the shared section of the St. Lawrence below Lake Ontario, and to bypass the "physical embarrassments" to navigation on that stretch of the river. Indeed, Bouchette expressed "little doubt that when the whole line of canals from Kingston to Montreal will be completed, and it is now nearly so, the great thoroughfare of the Canadas will be transferred from the frontier to the Rideau route, until a canal shall have been opened along the St. Lawrence."[15] Indeed, the idea of doing just this along the north bank of the river had been raised in the Assembly of Upper Canada in 1826. By the time of Thoreau's visit, however, the fast-dawning railroad age had rendered most such plans moot. The iron horse would carry national dreams from ocean to ocean, clear across North America, far beyond the inland seas at the heart of the continent, though not without a deal of debate over whether the natural grain of the hemisphere ran north-south or east-west.

Early in the twentieth century, scholars began to reframe Bouchette's view of the river as a vital artery of Canadian development. The Scottish geographer Marion Newbigin may have initiated this trend when she made the St. Lawrence the subject of her 1926 book, *Canada: The Great River, the Lands and the Men.* Seeking to unravel "what Canada has meant and means" by retelling the story of the "great eastern river" and "looking at man and place together," she dwelled on the French presence along the St. Lawrence between Jacques Cartier's arrival and 1763. Barely a dozen of the book's three hundred pages were given to the "later history" of the country, but these were enough for her to conclude that early Canadien dependence on the birch bark canoe made it impossible for the French to sustain authority over the Ohio country and that it was the iron rail that had extended British authority "through the treeless prairies to the shore of the Western Sea of Champlain's dreams." Gushing superlatives at Newbigin's work in the "possibilist" geographical tradition, demonstrating that "man can modify to some extent the lands in which he dwells," but that "in other respects he must follow where nature leads," H.P. Biggar, chief archivist for Canada in Europe, described it as "probably the most brilliant book dealing with the history of Canada that has yet been written."[16]

Donald Creighton's *The Commercial Empire of the St. Lawrence, 1760-1850*, published approximately a decade later, won more lasting fame by incorporating political economist Harold Innis's insights into the course of development in new-settled countries with Newbigin's focus on the river to portray the St. Lawrence as integral to both the growth of staple trades and a market economy in early British North America. The Laurentian thesis, as Creighton's interpretation of the country came to be known, saw the river as a hinge between transatlantic political and economic links and the transcontinental contacts forged by the merchants and politicians who rose to prominence along its banks. In this telling, elaborated in Creighton's later biography of John A. Macdonald that portrayed Canada's first prime minister as the embodiment of a national will reflected in the construction of the Canadian Pacific Railway, the St. Lawrence and its connections "became the basis of an extensive communication system around which Canada itself took shape."[17]

At some level, the Laurentian interpretation of Canadian development was a reaction against "continentalist" arguments that drew inspiration from American progressive historians and gained strength as Canada developed what A.B. McKillop described as "a new sense of psychological distance from the British Empire" after the First World War.[18] As the American and Canadian economies became increasingly intertwined in the 1920s and 1930s, some began to think of Canada as a place shaped by many of the same cultural, economic, geographical, political, and social forces as the United States. Given substance in the work of A.R.M. Lower and Frank Underhill, among others, this school of thought was bolstered when the Carnegie Endowment for International Peace underwrote a twenty-five-volume series on Canadian-American relations. Canadian-born John Bartlet Brebner – trained at Oxford and Columbia (where he spent almost his entire professional career), and the author of the capstone volume in the Carnegie series, *North Atlantic Triangle* – summarized much of what this initiative was about when he wrote, in that volume, that his "primary aim was to get at, and to set forth, the interplay between the United States and Canada – the Siamese Twins of North America who cannot separate and live."[19]

Though it seems to matter less, in this age of globalization, than it once did, the tension between Laurentian (or critical nationalist) and continentalist traditions was central to twentieth-century interpretations of the Canadian past and to debates about the course of national development. Classic texts of the mid-century years tell the story in their titles: *Canada: An American Nation* (1935); *Lament for a Nation* (1965); *Close the 49th Parallel Etc.* (1970).[20] Macfarlane seeks the middle ground in this historiographic spat, attempting, as he says, to avoid "the excessive anti-Americanism and Canadian moral superiority" characteristic of the former, while steering clear of the tendency, evident in the latter, to assume inevitable benefit in closer integration between Canada and the United States. His approach is open to the possibilities of conflict as well as collaboration between the two countries at the political-diplomatic level, but recognizes that Canadian-American relations exhibit a basic stability and continuity because they rest, in the end, on the bedrock of everyday social, cultural, and economic interactions.

Macfarlane begins his story at the turn of the twentieth century, with appeals on both sides of the border for improved deep-water navigation on the St. Lawrence and discussions about the possibilities of hydroelectric power generation along the river, to show that the 1954 agreement to proceed with the SLSPP emerged from half a century of somewhat tangled and untidy negotiations that turned on an increasingly familiar handful of convictions and miscarried repeatedly on the same general issues. Among the convictions was the firm understanding that the river was something to be controlled for economic gain, and the growing sense, on the part of engineers at least, that they could manage both the flow of water and the buildup of ice along the mighty stream. Among the points at issue between the two governments were the route along which improvements were implemented – a sticking point shaped by the strong conviction of many Canadians that "our great river must be ours and ours alone" and American economic and security concerns about the implications of such a development – and the apportionment of costs and benefits between the two nations. The first half of the century saw a series of proposals mooted, discussions held, diplomatic notes exchanged, speeches made, and treaties and agreements drafted. Yet, gains were few as initiatives became ensnared in the United States "in domestic politics and in the conflicts between the regions and interest groups that had long plagued the project" (see p. 39), or were shunted aside by wartime exigencies. By carrying readers through these years of negotiation and at least temporary failure, Macfarlane adds a great deal to our understanding of the origins of the SLSPP. His discussion is fine-grained and complex, attentive to the challenges and practicalities of international diplomacy, and sensitive to the influence of context and personality (or what Donald Creighton called character and circumstance) on the course of negotiations. This is a history forged from the archives and written by a scholar with his sleeves rolled up. It does much to flesh out a too-long neglected Canadian perspective on discussions over a centrepiece of Canadian-American relations in the twentieth century.

For environmental historians, Macfarlane's work stands out as a substantial contribution to the literature on environmental diplomacy. Despite its oft-acknowledged promise, this subfield remains relatively untilled. Surveying one corner of it, the editors of a 2010 collection of essays titled *Environmental Histories of the Cold War* compared diplomatic and environmental historians to ships passing in the night, and in a similar vein American historian Kurkpatrick Dorsey wondered recently whether members of these two guilds have much to say to each other.[21] Dorsey's question was, of course, rhetorical. His 1998 book *The Dawn of Conservation Diplomacy*, analyzing three early-twentieth-century wildlife protection treaties between Canada and the United States is something of a touchstone of the field, and he has argued, often, that studies of the environment and diplomacy (or foreign/international relations) can be integrated in fruitful ways.[22] Still, the pickings, impressive individually, amount to a relatively thin harvest and most have appeared since 2000. In Dorsey's reckoning, they have concentrated on three broad topics: the oceans and the atmosphere; development as (environmental) diplomacy; and nature in the Vietnam War.[23] Although people, pollutants, rivers, resources, fish, mammals, and birds cross (and have crossed) the long boundary between Canada and the United States frequently, and although regulatory regimes and political interests are shaped by the abstract line that divides the two countries, even the fullest bibliographies of work in environmental diplomacy include relatively little specifically Canadian content.[24]

Acutely conscious that the St. Lawrence drainage basin forms an international bioregion, Macfarlane is also well aware that borders matter in environmental history, and that the story of the SLSPP is a story of environmental diplomacy through and through. This understanding is realized in the very structure of the book. The first part, "Negotiating," focuses on the national and international political machinations (the formal diplomacy) necessary to reach agreement on the project. The second, "Building," foregrounds the environmental transformations set in train by that agreement – the massive material adjustments constituting "the greatest construction show on earth," that "turned the St. Lawrence Valley into a hybrid waterscape that blended the mechanical and the organic" (see p. 111). This was the consequence, as New York politician Robert Moses had it, of "pitting against the rush of a mighty stream ... the vaulting ambitions of two democracies" (see p. 178), and Macfarlane's achievement lies in his consideration, in the pages that follow, of both the weight of water (calling into being dikes and dams, creating lakes, and flooding homes and fields) and the highs (and lows) of Canadian and American political and economic aspirations.

It is helpful here, in thinking about the contribution of *Negotiating a River*, to invoke an image conjured by the great American diplomat George Kennan and beloved of diplomatic historians. Reflecting on the evident passivity of a certain democracy under threat, Kennan wondered whether

> democracy is not uncomfortably similar to one of those prehistoric monsters with a body as long as this room and a brain the size of a pin: he lies there in his comfortable primeval mud and pays little attention to his environment; he is slow to wrath – in fact you practically have to whack his tail off to make him aware that his interests are being disturbed; but once he grasps this he lays about him with such blind determination that he not only destroys his adversary but largely wrecks his native habitat.[25]

In his Bernath Lecture to the Society for Historians of American Foreign Relations, Kurkpatrick Dorsey used this passage to urge his colleagues to move beyond their not unimportant preoccupations with the words and deeds of those "state actors who thrash about on stage" and "the pin-sized brain of Unclesamosaurus," to consider the implications of "laying about" the "comfortable muddy planet itself." There were, Dorsey suggested, actually two dinosaurs at play in Kennan's primeval ooze, and it was past time for scholars to think about "the one who fails to pay attention to its environment, fouls its own habitat ... and seems largely incapable of taking past lessons on the subject and applying them to impending problems."[26] Much diplomatic thrashing lay behind the SLSPP, but it also, as Macfarlane demonstrates, left its mark on the planet.

All of this leads us finally, to the third, implicit – but no less important – message of Macfarlane's book. Canadians reading *Negotiating a River* in the second decade of the twenty-first century might sense that there is a parable in these pages. Consider current debates about Canadian economic development and the need to link Canadian resources to markets. As the country's latest megaproject proceeds to "lay about" the oil sands of northern Alberta – and is frequently portrayed as a great catalyst for twenty-first-century nation building (see p. 75) – there is great debate about how best to get the bitumen (heavy oil) to market. Building a pipeline from the boreal forest to the Gulf of Mexico, the Keystone route, has been favoured in many circles: advocates find sound sense in using existing refining capacity in Texas rather than building expensive new facilities and tout the advantages of energy security (invoking the notion that "ethical oil" from Alberta will replace that from more dubious sources), and so on. Keystone is a multi-sectored project (much as was the St. Lawrence Seaway project in the 1950s), with some sections already operational. Building the fourth component of the system, the Keystone XL pipeline (which would increase capacity by over half a million barrels a day and shorten the route) has, however, proven a diplomatic as well as a technological and environmental challenge. To date, the project remains ensnared (just as plans for a seaway once were) in US domestic politics and conflicts between the regions and interest groups. Of course, there are differences between circumstances in the first half of the twentieth century and those prevailing today. Most striking perhaps is the contrast between the formal diplomacy that dominated discussions of the seaway and the high level of public participation and active engagement of NGOs and think tanks in debates about Keystone XL. But thinking about these two bilateral projects in parallel is useful, not least in providing a reminder of the complex contextual, political, diplomatic, and other considerations that lie behind the scenes of polarized and simplified public debates about such developments.

The comparison might be pressed even further. With the Keystone XL development (approved by Canada's National Energy Board in March 2010) hung up by environmental concerns and political manoeuvring in the United States, Prime Minister Stephen Harper probably unwittingly echoed Canadian Cabinet Minister CD Howe's excoriation of the Americans for their stance on the St. Lawrence in 1953 – "Why then, should your country withhold its cooperation and thus delay completion of this vital Canadian transportation outlet?" (see p. 89) – when he announced in September 2013 that Canada would refuse to take no for an answer on the Keystone XL project.[27] Anticipating the impasse, corporate and political interests in Canada had earlier began to advocate strongly for the development of an all-Canadian pipeline that would open new markets for Alberta oil in Asia and ensure that Canadians received a "fair" (i.e., greater than the prevailing, substantially discounted) price for their product. First mooted at about the same time as the Keystone option, the Northern Gateway proposal envisaged a pipeline between northern Alberta and a new shipping facility in Kitimat, British Columbia. A second proposal, to increase the capacity of the existing Trans Mountain pipeline from Alberta to Burnaby, British Columbia soon followed. But arguments for efficient, rational, and productive use of "resources" seemingly so persuasive in debates about the St. Lawrence project in the heyday of high modernist hubris no longer go unchallenged in societies well aware of the nest-fouling propensities of Dorsey's second dinosaur.

Both the Northern Gateway and Trans Mountain proposals have met vociferous opposition from Native peoples and environmental groups, and they have spawned interprovincial frictions as British Columbia politicians, responding to different local constituencies, have sought to secure both economic benefits from and environmental safeguards in the proposals. Undaunted, Alberta has begun to explore the prospect of moving its oil to market through the Arctic, and plans for a trans-Canada pipeline linking Alberta to refineries in New Brunswick are now considerably advanced. Readers of *Negotiating a River* might begin to wonder whether some of this activity represents an attempt to bluff or pressure the United States over the Keystone route, just as, some said, the all-Canadian seaway idea was an attempt to ensure American participation in the project. Time, and historians, will tell, but there is a final sobering straw – whether of hope or despair remains to be seen – to be taken from these pages by those concerned about the larger implications of expanding oil production and the consequences of global climate change. For all the technological progress represented by the seaway, Macfarlane notes, that great project was bordering on obsolescence when it was completed (see p. 198).

There is much to ponder here, and as ever, readers will find different insights and inspiration in the pages that follow. So let me conclude by turning once again to Anne Michael's evocative reflection on the transformation of the Great River. As "lost" villages were put to the torch, symbolically finalizing the displacement of people from their riverside homes, Michaels's protagonist Jean Shaw (who had come to know the area during childhood visits with her father), gazed across the broad expanse of water to reflect: "The St. Lawrence flowed as always. But already it was impossible to look at the river in the same way."[28] Daniel Macfarlane's wide-ranging, multi-faceted account of the St. Lawrence Seaway and Power Project provokes much the same response.

Acknowledgments

Since this book first took shape at the University of Ottawa, I need to extend my gratitude to Serge Durflinger for his continued support. Galen Perras was ready with help and tips on useful American sources. Others – Jeff Keshen, Eda Kranakis, and H.V. Nelles – pointed out a range of important improvements to an earlier version. Greg Donaghy not only originally suggested the St. Lawrence project as a topic but also offered constructive feedback on early drafts of the book manuscript. Norman Hillmer provided gracious advice and support – including walks and trips for ice cream – that was invaluable for completing this book and for navigating the waters of academia.

Others read portions of the book manuscript, including Tina Loo, Don Nerbas, Robert Passfield, and Frank Quinn. The NiCHE New Scholars Reading Group, participants in the "Environments of Mobility Workshop," and members of Quelques Arpents de Neige gave feedback on early concepts. A number of individuals provided encouragement, advice, or ideas: Cecil Chabot, Murray Clamen, Jim Clifford, Michèle Dagenais, Robert Englebert, Jameel Hampton, Lynne Heasley, Sean Kheraj, Maurice Labelle, Nancy Langston, Josh MacFadyen, Asa McKercher, Ron Stagg, Shirley Tillotson, Henry Trim, Phil Van Huizen, Andrew Watson, and Donald Wright. My original academic mentor, John Courtney, continued as a source of guidance. Many others assisted in some way, sometimes without knowing it. The footnotes to this work serve as a further testament to my many remaining intellectual debts.

Acknowledgments

NiCHE (Network in Canadian History and Environment) has helped create an incredible Canadian environmental history community, and I have benefited tremendously from its generosity, opportunities, and networks. Carleton University in general, and the History Department in particular, was an excellent place to hold a postdoctoral fellowship – I particularly want to thank Joanna Dean, Dominique Marshall, and John Walsh. Carleton's School of Canadian Studies subsequently granted me a visiting scholar position, for which I am very grateful. Students in my Carleton courses and seminars read and helped sharpen my work. St. Lawrence University, and Bob Thacker in particular, provided a wonderful academic home for a semester, and the students in FYP 187 read the draft manuscript and gave constructive feedback. Michigan State University enabled me to finish this book through a Fulbright Visiting Research Chair in Canadian Studies, and I am deeply appreciative of AnnMarie Schneider for that. The Program on Water Issues at the Munk School gave me an early forum for my findings. Louis Helbig graciously shared his fantastic photos.

A number of archivists made this research possible, and the staff at Library and Archives Canada, NARA II, Ontario Power Generation (formerly Ontario Hydro), the Government of Ontario, the International Joint Commission, the Eisenhower and Truman Libraries, and St. Lawrence University were very helpful. NYPS (formerly PASNY) provided some limited access to files. I cannot say enough about the folks at the Lost Villages Historical Society (especially Jim Brownell), not only for consenting to interviews and letting me use their resources, but for sharing their knowledge and passion with me. I also want to thank Ian Bowering and the Stormont, Dundas, and Glengarry Historical Society, as well as Marlene Beaudoin, Linda Halliday, and Rosemary Watson at Ontario Power Generation. A unique thanks goes to Canada Steamships Line, which let me ride the CSL *Niagara* through the Welland Canal to get a first-hand experience (a trip which provided the cover photograph for this book).

SSHRC, Fulbright, and OGS provided needed financial support. I also benefited from grants and subsidies from NiCHE and the Eisenhower Foundation. As an author navigating the unknown waters of his first book, I deeply appreciate the guidance of the UBC Press editorial team. Melissa Pitts was my starting point, and Randy Schmidt and Ann Macklem brought me to completion. They were a pleasure to work with, shepherding me through a new process, providing valuable advice, and putting up with my constant inquiries and requests. The copy editor, Judy Phillips, did a wonderful job of smoothing stilted prose and catching my many grammatical infelicities. Eric Leinberger provided wonderful maps. My profound thanks to the two anonymous readers, whose timely and judicious comments made this a far better book. The same is true of Graeme Wynn's penetrating insights, which helped save me from myself and the reader from a book that would have been more cumbersome than it needed to be. I am very honoured to be a part of the Nature | History | Society series. I shudder at the thought of what this book would have looked like without all the help, advice, and contributions I received from start to finish; of course, errors of fact, judgment, or just plain good sense remain my own.

I want to acknowledge my friends from my days in Saskatchewan, and all my friends at Celebration! Church in Ottawa for their supportive community. We have a responsibility to humbly steward creation rather than control and dominate it, to strive for a faith in truth rather than a misplaced faith in science, technology, and progress. My family has been wonderfully supportive: my parents Bill and Becky Macfarlane, brothers Tim and Eric and their families, my grandparents Art and Erna Reimer, and all my other relatives (particularly Bob and Marg Pepper for opening their place to us on many occasions). In addition to their loving support, my parents and brothers were also frequent proofreaders. My parents-in-law, Bob and Vivian Thomson, and their extended family have been supportive on so many levels.

Above all, I am indebted to my wonderful wife Jen and our children, Elizabeth and Lucas. I dedicate this book to them and their patient support during the copious hours I spent at the computer, in a book, or dragging them on trips along the St. Lawrence. But most of all for their love and faith – they are responsible for this book in more ways than they will ever know.

Abbreviations

FPC	Federal Power Commission
HEPCO	Hydro-Electric Power Commission of Ontario
IJC	International Joint Commission
IRS	International Rapids section
PASNY	Power Authority of the State of New York
SLSA	St. Lawrence Seaway Authority (Canadian)
SLSDC	Saint Lawrence Seaway Development Corporation (US)

Negotiating a River

INTRODUCTION

River to Seaway

The St. Lawrence "is more than a river, more even than a system of waters. It has made nations. It has been the moulder of the lives of millions."¹ So wrote noted author Hugh MacLennan in 1961. Previously, in his quintessentially Canadian novel, *Two Solitudes*, published at the close of the Second World War, MacLennan had perceived hydroelectric development on the St. Lawrence River as either the key to the future or the death knell for the sleepy Quebec parish of Saint-Marc-des-Érables. These writings nicely bookend the period on which this book focuses and, in doing so, highlight the perceived centrality of the St. Lawrence to the history of Canada. Moreover, MacLennan's portrayals of the river and hydroelectric development reflect the dominant perspective of the period, an attitude that saw the river as something to be controlled and harnessed through science and technology for the progress of the nation and humanity.

Prominent academics of the time, such as Donald Creighton and Harold Innis, were equally enraptured by the St. Lawrence. They found inspiration in the idea that the river determined Canada's historical development – enough so that this notion became one of the great metatheories or narratives in the annals of Canadian history: the Laurentian thesis. Although this thesis is now dated, it cannot be denied that the St. Lawrence River has exerted a major influence on Canada, serving as the cradle and lifeblood of the country's economy and development. From the First Nations groups sustained by its waters to the early European explorers and settlers – the habitants and Loyalists who populated its environs – to the location of many major communities and the majority of the country's population, much of Canadian history has played out along the banks of the St. Lawrence.

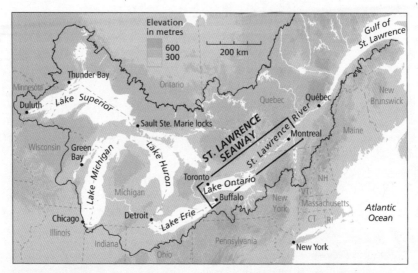

FIGURE 0.1   Great Lakes–St. Lawrence waterway. *Cartography by Eric Leinberger*
FIGURE 0.2   St. Lawrence Seaway. *Cartography by Eric Leinberger*

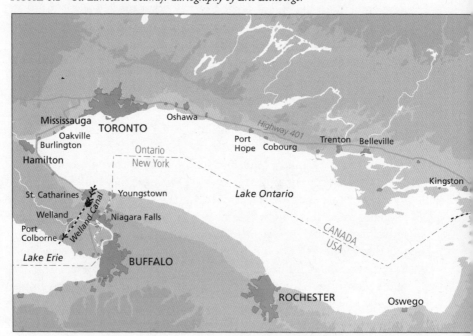

The St. Lawrence Seaway and Power Project was built between 1954 and 1959 by Canada and the United States, after decades of cooperative efforts to create the combined navigation and hydroelectric project. Technically, the seaway is a series of navigation works (channels, dams, canals) that runs 181 miles (291 kilometres) from Montreal to Lake Erie. It includes the earlier-constructed Beauharnois and Welland Canals, and has a continual minimum depth of twenty-seven feet, four large dams (two of which generate hydroelectricity), and fifteen locks with a depth of thirty feet each. The larger Great Lakes–St. Lawrence water route system, which includes connecting links in the St. Marys River, the Straits of Mackinac, the St. Clair River, and the Detroit River, provides a network of deep canals, channels, and locks that stretches some 2,300 miles from the western end of Lake Superior, a little over 602 feet above sea level, to the Atlantic Ocean.

The St. Lawrence Seaway and Power Project was superimposed on the majestic St. Lawrence River, which drains a vast basin of about 800,000 square miles, including the Great Lakes, the largest combined body of fresh water in the world. The third longest river in North America, the St. Lawrence proper has a length of about 745 miles and passes through a range of physiographical features. The upper St. Lawrence is flanked by lowlands, though punctuated by the protruding rock of the Frontenac

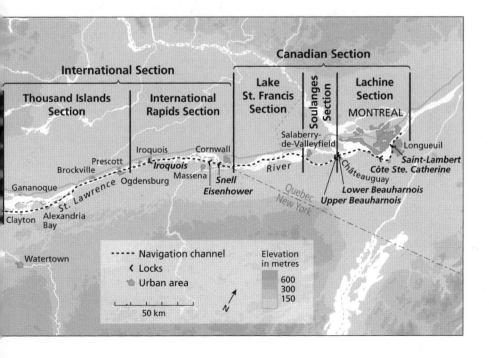

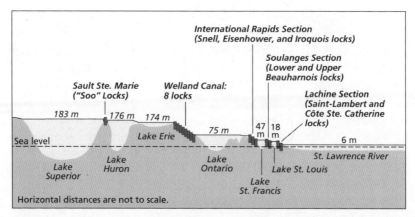

FIGURE 0.3 Profile of Great Lakes–St. Lawrence waterway. *Cartography by Eric Leinberger*

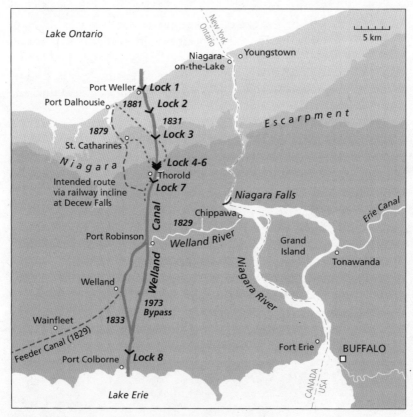

FIGURE 0.4 Welland Canal. *Cartography by Eric Leinberger*

Axis that create the vaunted Thousand Islands between Brockville and Kingston. There are higher banks near Quebec City, and further east an estuary zone where the river mixes into the Atlantic Ocean. Here the historic mean annual flow is 16,800 cubic metres per second, more than double the flow rate at the river's starting point, with the increase attributable to the many tributaries, the Ottawa River being the largest. Unlike many other rivers with high flow volumes, the St. Lawrence River is known for its regular flow levels (i.e., the amount of water does not fluctuate much during different seasons). Between Kingston and Montreal, where most of the St. Lawrence Seaway and Power Project construction took place, the river was divided into five sections: Thousand Islands, International Rapids, Lake St. Francis, Soulanges, and Lachine.[2] The latter three downriver sections are solely in Canada or, more precisely, in Quebec. The upper two sections, the Thousand Islands and International Rapids, form the border between Canada and the United States (Ontario and New York) from the foot of Lake Ontario to the Ontario-Quebec border; any change to the river levels in these sections is thus a Canadian-American, as well as a federal-provincial and federal-state, issue.

The St. Lawrence undertaking was a megaproject. It involved construction of the massive Robert Moses–Robert H. Saunders Power Dam between Cornwall, Ontario, and Massena, New York. The two halves of the dam are bisected by the international border. Hydroelectrical production was a prime factor for building the entire project. The resulting Lake St. Lawrence flooded out much of the surrounding area, dramatically pushing back the shoreline into land that had formerly been farmers' fields. Numerous Canadian communities between Iroquois and Cornwall, including the nine submerged "Lost Villages," were affected by the raised water level, which flooded approximately 20,000 acres and 6,500 people in the Province of Ontario, and about 18,000 acres and 1,100 people in the sparsely populated riverine part of the State of New York. Another 1,500 permanent residents east of Cornwall, chiefly in Quebec, were moved because of the seaway. This was the largest rehabilitation project in Canadian history, with engineers and planners relishing the opportunity to employ modern planning principles to redesign the rehabilitated areas. The St. Lawrence Seaway and Power Project would prove to be one of the great engineering achievements of the twentieth century, and with a generating capacity of 1,880 megawatts, the power project was at the time of its completion second only in North America to the Grand Coulee development on the Columbia River (1,954 megawatts).

Figure 0.5 Contemporary aerial view of the submerged remains of the town of Aultsville. © *Louis Helbig, sunkenvillages.ca*

Despite the remarkable rapidity with which construction of the St. Lawrence Seaway and Power Project was completed, it took over half of the twentieth century – spanning two world wars, the Great Depression, and the formative years of the Cold War – and multiple failed negotiations and agreements for Canada and the United States to commence the seaway and power project. According to Canadian political scientist James Eayrs, writing in 1961, the St. Lawrence matter was one of the "most difficult and most momentous" Canadian foreign policy issues.[3] On the American side, it was the longest continually running issue in US congressional history. The authors of a study of Canadian-American relations assert that "nothing represents the bilateral [North American] relationship during the cold war better than that seaway."[4] The completed waterway was, in the words of another historian, comparable to a gigantic "zipper" pulling together Canada and the United States and accelerating the economic, trade, and defence integration of the two North American countries.[5] The resulting hydroelectricity allowed for the industrial and economic expansion of central Canada, and deep-draught inland navigation permitted the flow of foreign goods and the movement of iron ore to the Great Lakes region while simultaneously allowing for the increased export of the products of manufacturing, industry, and western agriculture.

This book is divided into two sections: negotiating and building. The first three chapters proceed chronologically and cover negotiations over the St. Lawrence project up to 1954; the next three chapters examine its construction and are organized along both chronological and thematic lines. This dual structure requires engaging a range of academic disciplines mostly from, but not limited to, different fields of history: international, political, environmental, nationalist, cultural, state building, water, transnational, borderlands, and technology. Many of these are prominent throughout; however, the first section out of necessity puts more emphasis on hydropolitics and thus on political history, nationalism, Canadian-American relations, and environmental diplomacy.[6] The construction of the joint project had a profound impact on St. Lawrence land- and waterscapes, and the second section therefore relies more heavily on environmental, technological, and borderlands approaches.

In the first section, I argue that in the late 1940s the Liberal government of Louis St. Laurent began to consider a unilateral Canadian waterway and a bilateral Canadian-American hydro development. The idea of "going it alone" struck a responsive nationalist chord in Canada. Between 1949 and 1951, both the St. Laurent government and the Canadian public progressively embraced the concept of an all-Canadian seaway, and it became the preferred policy. However, the US government, and specific American regional and economic interests, considered an all-Canadian route to be an economic and national security threat and used various means to stop the Canadian plan and secure American participation. Out of concern for the impact on the broader Canadian-American relationship, the St. Laurent government reluctantly acquiesced in a joint seaway project in 1954. Canada technically had the right under the Boundary Waters Treaty of 1909 to construct a deep canal system entirely on its own side of the border if such a system would not change the water levels of the St. Lawrence. However, a hydroelectric power dam, which raises the level of the river and then uses the ensuing drop of water to spin turbines that generate electricity, was necessary to make a seaway feasible: the raised water levels of the headpond made achieving deeper navigation channels much easier, and the cost involved in constructing a canal and lock system of sufficient depth to accommodate deep-draught shipping was seen as prohibitive without the dammed river. Many proponents of a St. Lawrence project had viewed the waterway and power project in tandem since at least the First World War. But because a power dam would raise the water level in the international section of the St. Lawrence, it needed the concurrence of both the Canadian and American governments via the International Joint Commission, the bilateral body established to rectify Canadian-American border environmental issues. As a result, a unilateral Canadian waterway was indirectly subject to American assent to the power dam.

My reconceptualization of the diplomacy of the St. Lawrence project contributes to our understanding of Canadian-American relations in general, and the St. Laurent years in particular. I argue that there was a unique – terming it "special" would be going too far – Canadian-American relationship during the early Cold War.[7] The relationship was unique in the sense that Ottawa considered the United States to be its primary friend and ally and, accordingly, the main aim of Canadian foreign policy in this period was to ensure smooth relations with the United States. For its part, Washington was often willing to tolerate, accommodate, or humour Canadian policies and sensitivities. More specifically, the US Department of State, or at least the section responsible for relations with Canada, often accommodated Ottawa. This study seeks a middle ground between the continentalist and critical nationalist traditions in the Canadian historiography on the northern North American relationship, avoiding the excessive anti-Americanism and Canadian moral superiority characteristic of the latter while eschewing the tendency of the former to see Canada as inevitably benefiting from increased integration with the United States. I generally align my approach with the North American school of Canadian-American relations, which points to the importance of shared continental outlooks and tendencies, highlights cooperation – without obscuring conflict – between the two nations, and sees the bilateral relationship as constituted by everyday social, cultural, and economic interactions – which generally provided the relationship with an inbuilt momentum, balance, and continuity – as much as by negotiations at the elite and executive government levels.[8]

In the long history of the two nations oscillating between conflict and cooperation, the decade after 1945 was mostly characterized by the latter; the St. Lawrence Seaway and Power Project, however, was a key exception.[9] As many recent borderlands, transnational, and regional studies have demonstrated, there is a strong upper North American interrelationship forged by many years and forms of regional, social, and personal transborder contacts.[10] However, such perspectives can potentially obscure the policy conflicts that did exist and exaggerate the impact of informal cross-border networks and shared cultural affinities on formal governmental policies, which generally played a more important role in determining the nature and tenor of the Canadian-American relationship. Each nation's political, economic, and security policies were predicated on self-interest, and cultural notions of Canadian-American kinship often went out the window when these national interests did not align.

In much the same way as recent works have identified the importance of culture and race in shaping Canadian foreign relations, I point to cultural conceptions of nature, water, and technology as important determinants of Canadian foreign policy.[11] Although this book could be accused of perpetuating the old idea of a North Atlantic triangle, linking Canada, Britain, and the United States, the fact remains that in the years after the Second World War these two countries were the main allies and key concerns of Canadian international policy. Indeed, it is difficult to overestimate the importance of the Ottawa-Washington relationship in the eyes of the St. Laurent government, as well as the desire to balance that relationship through an Atlanticist policy based on multilateral alliances and institutions. Some scholars claim that the Canada-US relationship in the 1950s avoided the use of "linkage" – a diplomatic approach in which one side attempts to put pressure on the other by tying together unrelated policy issues – but I contend that linkage attempts were prominent in the St. Lawrence dispute.[12] Realizing the diplomatic limitations inherent in aligning itself squarely with the United States, Canada sought to maximize its freedom of manoeuvre and protect its sovereignty to the fullest possible extent while simultaneously contributing to, and benefiting from, the spreading American economic empire and security umbrella. The United States hoped to bring Canada, and its resources, more tightly into the American orbit while at the same time protecting America's northern flank from Soviet encroachment and was willing to override Canadian sovereignty or desires when American security and important national interests were at stake. I believe that although the St. Laurent government certainly furthered Canadian-American integration, it did so with some reluctance and in order to advance what it perceived to be Canada's best interests.[13]

Most of the main political and diplomatic accounts of the St. Lawrence project – for example, those authored by Theo L. Hills, Lionel Chevrier, William Willoughby, and Carleton Mabee – date from the late 1950s and early 1960s.[14] The latter two are the strongest works, but they do not adequately explain the Canadian decision-making process or the nature

FIGURE 0.6   Iroquois Lock and Control Dam during construction (the old lock and canal can be seen to the left). © *Dumas Seaway Photograph Collection, Mss. coll. 124, Special Collections, St. Lawrence University Libraries*

of Canadian nationalism, nor the environmental and technological history. The publication of several articles on the topic over the last decade suggests there has been a growing interest in the seaway, and the fiftieth anniversary of the opening of the St. Lawrence Seaway and Power Project in 2009 resulted in a spate of publications on the subject.[15] Considerable attention within the United States has been paid to elements of the political history of the St. Lawrence project, particularly the ability of special interests to block the project, its consideration in Congress over many decades, and what this process discloses about the American system of government and separation of powers.[16] However, this literature substantially neglects the Canadian perspective. The history of the seaway is meanwhile reduced to a minor side issue in broader works on Canadian-American relations and studies of modern Canadian politics, defence, and external affairs, instead of being properly treated as one of the major joint disputes between Canada and the United States.[17]

The literature on Canadian decision makers involved in the development of the seaway – such as R.B. Bennett, William Lyon Mackenzie King, Louis St. Laurent, Lester Pearson, C.D. Howe, and A.G.L. McNaughton – has not added a great deal to our understanding of the genesis of the St. Lawrence project. Even when scholars have paid attention to Ottawa's attempt at an all-Canadian seaway, they have generally argued that the St. Laurent government wanted the Americans involved all along and was only trying to cajole the United States into a cooperative project. Some, such as the venerable John Holmes, assert that Canada outmanoeuvred the Americans by bluffing about an all-Canadian seaway in order to induce American participation.[18] Yet, Donald Creighton, for example, charged that American involvement in the seaway was "on the ungenerous terms of its own choosing" and contended that the manner in which this participation took place was as a serious blow to Canadian sovereignty and national identity.[19] On this particular occasion, Creighton was much closer to the mark.

The river – and, by extension, the seaway – offered the potential, as William Kilbourn phrased it, to "fulfill that age-old dream at the heart of Canadian history, the Empire of the St. Lawrence."[20] The Laurentian thesis, most prominently forwarded by Creighton in his 1937 *The Commercial Empire of the St. Lawrence, 1760-1850* (and rereleased in 1956 as *The Empire of the St. Lawrence*), holds that "Canadian economic and national development derived fundamentally from the gradual exploitation of key staple products – fur, timber, and wheat – by colonial merchants in the major metropolitan centres along the St. Lawrence River system," which "provided the means by which both a transatlantic and a transcontinental market economy could be created."[21] This east-west axis was further enhanced by the St. Lawrence's connection to the Great Lakes, and the "empire" extended west by railway after Confederation to the western interior and Pacific Ocean.[22] Although the grander aims of this empire may have failed, the extended attempts to bring it to fruition did serve to geographically and psychologically carve out the country of Canada, resist the pull of the United States, and forge the various colonies and English- and French-speaking peoples together. According to Creighton, "The impulse towards unity in the interests of strength and expansion is one of the oldest and most powerful tendencies in the history of the Empire of the St. Lawrence."[23]

The St. Lawrence River holds an iconic place in the Canadian national imaginary. I am interested in how the manipulation of the St. Lawrence basin was shaped by culture, identity, region, and environment. This is in keeping with a global, even ancient, tradition of viewing rivers, and water control projects, as the bloodstream of nations on which nationalist obsessions were projected as reflections or repositories of cultural or national character.[24] The role of the St. Lawrence River (and of rivers in general) in the development of Canada is central, perhaps even unsurpassed, in the paradigm of national development. From the early explorers who travelled up the St. Lawrence and dreamed of bypassing its rapids to the settlers who populated the riverine basin in subsequent centuries, it served as the crucible of Canadian settlement and development. Canals were central to this evolution, and though they may have seemed in some ways an anachronistic technology by the mid-twentieth century, the seaway as a deep canal system (joined with hydroelectric development) could simultaneously link romantic nationalist associations and modern transportation and industrialization goals.

I argue that the link between identity and riverine environments has unique manifestations in the Canadian context. Given the importance of rivers and water to Canadian identity, it is no surprise that the St. Lawrence, the greatest of all of Canada's rivers, is the leading protagonist in historical writings that personify geographic factors in the nation's historical development, often acting as a synecdoche for Canada in general, and central Canada (Ontario and Quebec) in particular. Indeed, geographically determinist explanations of Canadian history animated many prominent historical texts of the day. These metahistorical and nationalist interpretations include the staples and metropolitan theses, which are part of – or contribute to, depending on one's perspective – the Laurentian thesis.[25] Creighton built on the work of Harold Innis, elevating Innis's exalted view of the St. Lawrence into "new poetic realms"[26] in which "the dream of the commercial empire of the St. Lawrence runs like an obsession through the whole of Canadian history ... The river was not only a great actuality; it was the central truth of a religion."[27] A range of prominent historians, although taking issue with unabashed Laurentianism and its geographically determinist and inherently anti-American stance, nonetheless accepted that the St. Lawrence had played a pivotal role in Canada's historical development.[28] Popular histories from the era forwarded similar narratives.[29]

Stéphane Castonguay and Darin Kinsey point out the tautological nature of the Laurentian thesis, for, in tandem with the linked staples and metropolitan-hinterland theses, "one is led to believe that cod, beaver, grain, and other staples are a part of the elements of the triumphant environmentalism responsible for the 'neo-Wagnerian myth-symbolism complex Canadian nationalists have woven around the St. Lawrence Valley.'"[30] Yet, according to historian Janice Cavell, "No other interpretation of history has ever been so widely and whole-heartedly accepted [in Canada] as Laurentianism" was at the height of its popularity from the 1940s to the 1960s.[31] The linked but not identical metropolitan thesis, for its part, had achieved "near-doctrinal status" by the 1960s and, along with the staples thesis, found some resonance in the following decades, particularly among political economists.[32] Though some of the concepts underlying these grand theories are implicit to varying extents in recent studies that foreground the importance of natural factors in order to explain Canada's past, the growth of environmental history approaches in Canada since the 1990s has not resulted in direct attempts to rehabilitate these metatheories; nonetheless, contemporary Canadian historiography would benefit from a reappraisal of the metropolitan, staples, and Laurentian theses.[33]

It was no coincidence that the peak of the Laurentian thesis's influence coincided with the conclusive stretch of negotiations for the St. Lawrence project, and then its construction. It is apparent that the Laurentian thesis helped sustain the conception of the St. Lawrence watershed as the defining and fundamental aspect of Canadian history and identity and, in turn, infused the notion of an all-Canadian seaway with nationalist importance and symbolism. The hydroelectric development of the St. Lawrence was equally a repository of hydraulic and technological nationalist associations, for generating stations represented modern Canada's ability to control the exploitation of its natural resources and reap the benefits.

For many, the St. Lawrence served as both a bridge and a barrier between English and French Canada, and between Canada and the United States. The river could unite, but it could also divide. Joseph Bouchette, writing in 1831, preferred the former interpretation: "The St. Lawrence, originally called the Great River of Canada, or the Great River, to mark its preeminence, is the indelible link formed by nature between the Canadas, and the source at once of the wealth, beauty, and prosperity of both countries."[34] A good deal of French-language literature on the St. Lawrence sees the river and valley as fundamentally intertwined with Quebec's identity, history, and nationalism.[35] This study does not explicitly aim to differentiate between French- and English-Canadian nationalism concerning the St. Lawrence, but it is safe to say that in both central Canadian provinces, and maybe even more so in Quebec, the St. Lawrence was viewed as a

Canadian – or *canadien* – river, rather than an American one.³⁶ Although the St. Lawrence River is profitably viewed as a bioregion that eschews man-made boundaries, the history of the St. Lawrence Seaway and Power Project also underlines some of the ways that borders do matter in environmental history.

Not all Canadians conceptualized the development of the seaway as a zero-sum game in which either Canada or the United States would be victorious, or conceived of the St. Lawrence canals as symbolic nation-building devices. Many were primarily concerned with gauging the narrow economic benefits the St. Lawrence offered to themselves and their immediate community. Those who thought about the seaway in terms of nation building did so in different ways, and it is necessary to avoid the assumption that ideologies and ideas were more coherent or more widely shared than was actually the case. Nevertheless, for all that, many Canadians, particularly those with the ability to shape public opinion or make governmental decisions, addressed the nationalist, technological, and state-building implications of the St. Lawrence project commonly and consistently enough to be worthy of identification, consideration, and generalization. Moreover, although I often refer to and make generalizations about "the state," I attempt to balance this by recognizing and detailing the various divisions and diversities within the different levels of government that make up what Christopher Armstrong and H.V. Nelles have called the "internal pluralism of the state."³⁷

I interrogate water as both a natural element and as a culturally constructed object.³⁸ Since a central contribution of this study is the detailed examination of the process by which governments and engineers transformed a river system into a power-producing lake and system of canals and locks, it contributes to the large corpus of work on hydroelectricity in Canada.³⁹ In doing so, the subject at hand also crosses into the growing field of envirotech history.⁴⁰ Conceptual tools that are clearly applicable to the scale and nature of the St. Lawrence project include Paul Josephson's description of "brute force technologies" and David Nye's portrayal of the "technological sublime" (though Nye attaches the label only to the United States).⁴¹ But there is an even more compelling concept that better encapsulates the St. Lawrence project: what James C. Scott pejoratively identifies as "high modernism."⁴² According to Scott, since states inherently need to reduce complexity in order to realize their social engineering aims, increase efficiency, and centralize their control, they employ a narrow and selective way of viewing – tunnel vision rather than a synoptic view – that privileges technocratic scientific expertise, excluding local and vernacular knowledge, to order both nature and society.[43] This type of state-led effort to make landscapes and societies "legible" relies on simplification, abstraction, and standardization, and was practised by states across the political spectrum. Although modernist forerunners are obvious in western Europe throughout the nineteenth century, Scott argues that "high" modernism can be dated to German mobilization in the First World War, with the high modernist wave cresting during the Great Depression and Second World War.

I argue that high modernism is a useful concept for understanding and characterizing the organizing logic and imperatives that drove plans for the St. Lawrence project. As numerous authors have shown, since hydrological resources are so fundamental to human life, state control of water is a means of controlling society as well. Noted environmental historian Donald Worster refers to "imperial water" and "hydraulic society" while other scholars have used terms such as "hydraulic bureaucracies and "water wizards."[44] Because of the St. Lawrence Seaway and Power Project's strategic value, it represented a physical defence against the growing threat of the Soviet Union. In a 1951 speech, Canadian Minister of Transport Lionel Chevrier emphasized that Communist forces threatened the Western democracies to the point that the "survival of our civilization now depends above all on our scientific and technical superiority" and specifically cited the engineers working on the St. Lawrence project as a prime example.[45] Indeed, the hubristic domination of nature inherent in the seaway megaproject was intimately intertwined with Cold War symbolism and Canadian and American attempts to assert the more progressive, modern, and powerful nature of capitalist democracies as compared with Communist nations. Moreover, as is discussed in the Conclusion, there was not only a unique North American variant of high modernism but also a particularly Canadian version that emerged from the differing national views of the links between technology, environment, and culture.

The end result was a technological and engineering marvel, albeit one that had an immense and mixed range of social, environmental, and political impacts. The St. Lawrence Seaway and Power Project is the integral part of the Great Lakes–St. Lawrence waterway, the longest navigable inland waterway in the world, stands as one of the biggest borderlands project ever undertaken jointly by two countries, and is perhaps the largest construction project in Canadian history. It is a vitally important part of the environmental, technological, and state-building history of North

America in general, and Canada specifically, as well as one of the defining issues of the modern Canadian-American political and transnational relationship. This critical event in North American history has been neglected, and providing the complete story of both the negotiation and construction of the St. Lawrence Seaway and Power Project, particularly from the Canadian perspective, is the central aim of the following pages.[46]

PART I
Negotiating

1
Accords and Discords

The St. Lawrence flows through the annals of Canadian history with a momentum perhaps equivalent to the river itself; no natural feature has played a more fundamental role in the development of Canadian identity than the great river. However, the early European explorers to the territory through which the river passes were initially dismissive of its potential, particularly once they realized that the river was not the coveted trade route to Asia. They regarded the St. Lawrence as an obstacle-filled water highway into the heart of an undiscovered continent. As European settlement advanced, the St. Lawrence was central to the development of the country that would become Canada, with defence and economic considerations at the forefront of schemes to modify the St. Lawrence and connecting waters.[1] The canal mania that seized the eastern part of North America during the first half of the nineteenth century fostered a transportation revolution that had direct imperial and nationalist motivations. The American Erie Canal was quickly countered by the Canadian Welland Canal, which was soon joined by Canadian canals and locks in the St. Lawrence at Lachine, Soulanges, Cornwall, and Williamsburg, as well as navigation works by both countries at Sault Ste. Marie.[2] These canals collectively provided a navigable channel with a depth of nine feet from Lake Erie to the Atlantic; by 1905, there was a toll-free navigable channel with a minimum depth of fourteen feet from Lake Superior to the Atlantic Ocean.

Amidst the flurry of canal proposals and developments in the last decades of the nineteenth century, the first serious conceptions of an unbroken

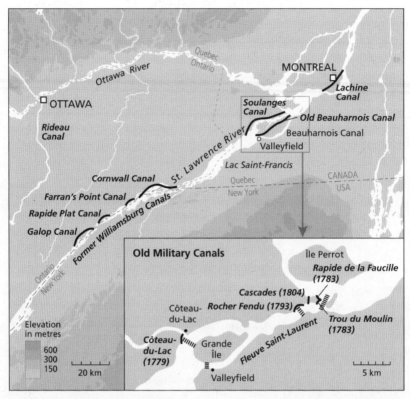

FIGURE 1.1　Historic canals on the St. Lawrence River. *Cartography by Eric Leinberger*

deep St. Lawrence waterway allowing access to the interior of the continent were articulated. In 1892, motions in both the Canadian Parliament and the US House of Representatives appealed for improved navigation on the Great Lakes–St. Lawrence route. Then, at the 1894 International Deep Waterways Convention in Toronto, Canadian and American delegates called for the cooperative improvement of the existing St. Lawrence canals. In response, the two governments formed the International Deep Waterways Commission, made up of three members from each country, and charged it with studying the feasibility of constructing waterways that would allow ocean-going vessels to pass between the Great Lakes and the Atlantic Ocean. Each national group released a report in 1897. The American contingent suggested linking the Great Lakes to the sea through American territory, such as down the Hudson River; the Canadian report favoured the St. Lawrence.[3]

Thus, a tension between competing national routes characterized even the initial discussions of a deep waterway. Later in 1897, a three-person American board of engineers evaluating different routes from the Great Lakes to the eastern seaboard promoted a depth of twenty-one rather than thirty feet and voiced concerns that the St. Lawrence option would be situated partially in a foreign country. Considering the tensions then prevalent in Canadian-British-American relations, such as those emerging from the Alaskan boundary dispute, it is not surprising that the Americans were wary of relying on the St. Lawrence as a water highway.[4] Nor would the Canadian public have been too willing to share the St. Lawrence, as they took a possessive view of the river and so were "psychologically unprepared for cooperative development and use of the waters of the St. Lawrence."[5]

Competing Diversions

Calls for an international waterway, however, were not easily stilled. In 1909, the United States and Britain, the latter acting for Canada, agreed to the Boundary Waters Treaty, which created the International Joint Commission (IJC). The result of improved Anglo-American and Canadian-American relations, established amidst a flurry of treaties concerning natural resources such as fish, seals, and migratory birds, the IJC grew out of the International Deep Waterways Commission. The IJC had jurisdiction to deal with Canadian-American boundary water issues, as well as any other matters affecting the borderlands region (such as air pollution).[6] Although the reputation of the IJC as a model of international cooperation, dispute resolution, and environmental management might be exaggerated at times, it was a political coup for Canada, since it gave Canada a voice within the commission equal to the United States. The commission was given only a circumscribed range of jurisdiction and powers, which made its creation politically possible in the first place, but its flexible structure and mandate allowed the IJC to anticipate and address many future pollution issues.

Interest in canal development was piqued by the potential to harness hydroelectricity. Just as people had, in the first half of the nineteenth century, hoped that canals could fulfill their every ambition, so too did they hold a pervasive belief, in the late nineteenth and early twentieth centuries, that electricity would solve every problem and usher in a grand, new age. The efficacy of hydro power and its distribution over long distances had been proven at Niagara Falls, and hydroelectric projects soon proliferated across North America. For example, the installed hydroelectric capability in Canada as of 1910 was about 1 million horsepower (1 horsepower is equivalent to about 745 watts), with more than half in Ontario and a number of generating stations in the area of the St. Lawrence River eventually subsumed by the seaway and power development. By 1960, the Canadian installed capacity had increased twenty-five-fold over 1910.

On the New York side, the Aluminum Company of America (Alcoa) had built at the tail end of the nineteenth century a short diversion canal from the St. Lawrence to supply power at its plant near Massena, and its subsidiaries soon began surveying power dam possibilities in the river at Barnhart Island. Alcoa applied for permission to build a dam, but that was unsuccessful, in part because of Canadian opposition. In 1906, the Government of Ontario created the Hydro-Electric Power Commission of Ontario (HEPCO), a publicly owned power utility that was often referred to simply as Ontario Hydro (that name was officially adopted from 1974 until the utility's dismantlement in 1999), and it immediately began pushing for hydro development on the St. Lawrence.[7]

The following year, the Long Sault Development Corporation sought authorization for a power dam development near Cornwall and Massena in the International Rapids section (IRS) of the St. Lawrence, the turbulent stretch of the river shared by Ontario and New York between the Thousand Islands section to the west and the Quebec border to the east. Bills to that effect were introduced into the US Congress several times in the first decade of the twentieth century. The Long Sault Development Corporation was a combination of primarily American commercial and financial interests, most prominently Alcoa and the St. Lawrence Securities Company; it proposed a network of power developments (four powerhouses and two dams) in the vicinity of Barnhart and Long Sault Islands, with most of the power developed on the American side and for American use.[8] It also included a navigation channel and lock in US territory, but this was more of an afterthought, since the company's dominant concern was the development of hydroelectricity. Representatives from the Ontario riverfront communities of Cornwall, Prescott, and Brockville appeared before the International Waterways Commission in support of the scheme, contending that their communities needed access to cheap power.[9] Studies by HEPCO were drawing the same conclusion.[10] Other St. Lawrence towns opposed the proposition, and Canadian interests complained about the drawbacks of the plan, including the injury to navigation on the St. Lawrence and the placement of navigation facilities in US territory. Echoing the detractors, a 1913 study by the Canadian Commission of Conservation objected that the majority of the hydroelectricity developed would be controlled by private US interests at the expense of Canada, that the proposed development would render navigation almost impossible, and that the engineering plans were flawed.[11] Other interests, particularly those connected to tourism, also complained that the development would spoil the natural beauty of the renowned Long Sault Rapids, though some argued that the way to conserve natural resources was to utilize them. In the end, most commentators, whether pro or con, viewed the river as something to be controlled for economic gain: what was chiefly at stake was who would benefit from its exploitation.

The Canadian report adumbrated later scientific debates about ice control while using language that would contrast dramatically with engineering rhetoric in subsequent decades about taming the river. Contending that the corporation's plans drastically underestimated the threat of ice (including frazil ice, sometimes referred to as "slush ice," which is fine, spicular, crystal-like ice that forms in open channels that flow too quickly for surface/sheet ice to form), the report in its discussion of ice buildup and flooding demonstrated regard for nature's power and the limits of engineering expertise: "While the calculations and opinions of engineers are to be respected, nevertheless, when they are expressed by way of *minimizing* the possible disastrous effects of damming the St. Lawrence, it is well to discount such opinions." Indeed, "the magnitude of the St. Lawrence River, and the tremendous forces latent in it, may act in ways as yet undiscerned by members of the engineering profession."[12] Ultimately, the Long Sault Development Corporation's plans fell through when the US Congress failed to pass the requisite legislation; Canada's necessary concurrence was not forthcoming, in any event. But in broaching the ice issue, bringing up public versus private hydro development concerns, revealing the engineers' belief in their ability to control the river, highlighting the Canadian view of the St. Lawrence as a national resource, and focusing on the Barnhart Island region of the International Rapids section as the site of development, these plans helped establish the parameters for future discussions on developing the St. Lawrence.

Ottawa looked at other waterway transportation improvements it could undertake alone. The Sir Wilfrid Laurier government established a royal commission to investigate a comprehensive Canadian waterway system, but it was abandoned in favour of the Grand Trunk Railway. A number of improvements to Canadian inland navigation were made, such as the

FIGURE 1.2   Rapids in the St. Lawrence River with fourteen-foot canal in foreground. *Photograph by Eleanor L. (Sis) Dumas. © Dumas Seaway Photograph Collection, Mss. coll. 124, Special Collections, St. Lawrence University Libraries*

Trent Canal, upgrades to the Lachine Canal, and the deepening and enhancement of Montreal's port facilities. Laurier then made the Georgian Bay Ship Canal one of his campaign promises during the 1911 election. This route would employ a series of locks and waterways to connect the Ottawa River to Georgian Bay via Lake Nipissing. But Laurier's Liberal Party was defeated and the new government, led by Robert Borden, opted instead to build a new alignment of the Welland Canal.

In the years immediately before the First World War, the United States and Canada remained ambivalent about a cooperative waterway, and wary of international commitments in general. Yet, several pro-seaway organizations were formed (such as the Canadian Deep Waterways and Power Association) and there were further attempts in Congress and Parliament to legislate a St. Lawrence project. In early 1914, the American government approached the Canadians with a diplomatic note suggesting that the IJC be allowed to investigate the possibility of a jointly developed St. Lawrence waterway. Prime Minister Borden, however, was concerned about railway expenses and the political consequences of funding a project that would be perceived as primarily benefiting Ontario.[13] The cataclysm of the First World War soon crashed down, and the Canadians never replied to the note. Nevertheless, prompted by the St. Lawrence River Power Company's application to the IJC for a submerged weir in the St. Lawrence to divert water for power development, which was allowed on a temporary basis despite Canadian objections, the Canadian cabinet authorized two of its members to travel to Washington to propose a joint project. The timing was not propitious, in part because a report by the US Army Corps of Engineers had recently advised against any improvement in the International Rapids section until Canada had sufficiently deepened the connecting links in the St. Lawrence. The Canadian proposal did not elicit a response from the American government.

The lines of opposition to the St. Lawrence project were firmly set by the 1920s. The St. Lawrence was a relatively minor trade route for the United States compared with its importance to Canada, and by the 1920s the American exports trade was generally channelled to east coast ports via rail, either directly or by trans-shipment from the Great Lakes, or on barges down the Mississippi River to ports on the Gulf of Mexico. The St. Lawrence route drew strong opposition from these established routes and their supporters. Hostility in the United States also came from Atlantic port cities, New England, and downstate New York. Even though New York State stood to benefit from the river's development, powerful port interests were worried that a seaway would draw away business from the City of New York.[14] The state legislature was also divided, as Republicans favoured private development of the St. Lawrence waters. States around the Gulf of Mexico and along the Mississippi River opposed the St. Lawrence route too, desiring to retain their shipping business. Railway interests and unions were among the strongest detractors of a deep waterway, and private power concerns were antagonistic to hydroelectric development on the St. Lawrence by state or federal governments. Some American isolationists and Anglophobes joined the ranks of the seaway critics, arguing that it would open up the interior of the continent to foreign vessels.[15] The umbrella organization uniting the various forms of opposition was the National St. Lawrence Project Conference, which by 1930 boasted 250 member organizations, with the Association of American Railroads the most influential.

Members of Congress who were favourably disposed toward the concept of the St. Lawrence project often felt pressured to vote against it, depending on the power of port and railway unions in their constituencies.

In addition, since the St. Lawrence issue stretched over many decades and burdened different administrations, Democrats were generally reluctant to vote for a St. Lawrence agreement negotiated by a Republican administration and vice versa. Various individuals and organizations weighed in as well, on both sides of the argument, with unsolicited statistic-laden tracts and studies. Indeed, the long history of the St. Lawrence waterway brings into sharp relief the differences between the Canadian and American systems of government, and the extent to which the American separation of powers enabled certain viewpoints and sectors to block in Congress a project that had strong support from consecutive administrations.

Canadian opposition too seemed to follow regional patterns. Some of the same types of special interests, although less powerful than their American counterparts, opposed a Great Lakes–St. Lawrence waterway. Geographer Theo Hills suggested that Canadian feelings toward a St. Lawrence project were marked by a lack of enthusiasm rather than by effective opposition.[16] Whatever the case, there was considerable resistance during the interwar period, though changes after the Second World War saw any disapproval virtually evaporate in most areas of the country. The prairie provinces had favoured a Hudson Bay railroad as an outlet for their produce, but they shifted their support to the Laurentian route when that option failed to meet expectations. Ontario provided solid support, though this was at times circumscribed by premiers who had personal reasons for opposing St. Lawrence development. Until the early 1950s, Quebec generally resisted the project out of concern for the port and trans-shipment business in Montreal and Quebec City, and a reluctance to share the river with the Americans. The Atlantic provinces saw improvements to the St. Lawrence as potential competition for Atlantic shipping.

The First World War brought about myriad political, economic, and technological changes that significantly influenced the fate of the St. Lawrence project. In particular, the wartime growth of North American industrial and agricultural production, the strain that the great conflict placed on railroad systems, the need for protected shipbuilding, and shortages of electrical power gave the St. Lawrence idea momentum in the post-1918 years. Although the Canadian state may have historically been slightly more disposed toward subsidization of large-scale transportation routes connected to nation-building goals (e.g., railways), this was as much out of necessity as ideology, and the genesis of the seaway indicates that American transportation policy too was shaped by both state interests and private capital. Instead of attempting to avoid rapids and obstacles by going around them, post-1918 engineering plans called for the obstacles to be removed by means such as flooding them out. Nature would not be circumvented; it would be tamed, dominated, and made to fit humankind's needs. Reconfiguring the environment implicitly also carried with it ideas about reshaping social and economic structures, as agricultural land would be converted to what the North American governments considered to be more modern purposes: creating the sufficient head of water to produce hydro power and allow deep-draught navigation.

Late in the war, HEPCO had concluded that a power dam at Morrisburg, Ontario, would produce approximately 400,000 horsepower, help regulate the water levels of Lake Ontario, and reduce dangerous ice problems by replacing rapids with slack water.[17] This dam could work in conjunction with another near Cornwall. After the war drew to a close, Sir Adam Beck, the chairman of HEPCO, publicly boosted hydro development on the St. Lawrence, arguing that power production would pay for a seaway. Sensing that Canada might prove receptive, the US Congress again requested that the International Joint Commission be permitted to examine the St. Lawrence case. After careful consideration, the Canadian government consented and the question was referred to the IJC, which began an extensive series of hearings.[18] In 1920, Washington and Ottawa created their own joint two-man engineering board to study the issue. In 1921 the board produced the Wooten-Bowden Report, named after the two presiding engineers, which favoured a general deepening of the waterway from Lake Ontario to Montreal.[19] It suggested that the St. Lawrence River be canalized to a depth of twenty-five feet, that nine locks and two hydroelectrical developments capable of producing 1,464,000 horsepower be built in the International Rapids section, and that the two countries share the estimated $252,728,000 cost. At the end of 1921, the IJC endorsed the St. Lawrence route and urged the two national governments to forge a treaty authorizing the project and to include the Welland Canal as part of a wider waterway system.

After some delays, the United States recommended a treaty pledging the two governments to undertake the St. Lawrence scheme. American agricultural interests, especially in the Midwest, were keen to acquire outlets for their products and pushed Congress and the Harding, Coolidge, and Hoover Republican administrations for improvement of the St. Lawrence. Like the opponents of the waterway, proponents had specific economic and transportation interests that they framed as national benefits. However, according to historian C.P. Stacey, William Lyon Mackenzie

King, the new prime minister, "shot this project down without even the courteous pretence of careful consideration."[20] With a minority government, King, who was cautious under any circumstances, was reluctant to push forward on an issue involving the Americans that seemed to have little support in Quebec. King appears to have been ambivalent about the seaway throughout his long tenure as prime minister, and his constant concern that relations with the United States could cause domestic discord made it especially unpalatable. In November 1923, the US government again asked whether Canada was willing to enter into talks, but to no avail.[21]

Washington's inquiries caught the attention of Canadian nationalists, who insisted that the Americans should not be allowed to encroach on the St. Lawrence, for it was a "Canadian" river. The Montreal Board of Trade, the Montreal Chamber of Commerce, the Montreal Harbour Commission, and the Shipping Federation of Canada strongly questioned the viability of developing the St. Lawrence, raising doubts about whether ocean-going ships would use a deep waterway, and whether there would be sufficient cargoes to justify the huge expense that the waterway represented.[22] The *Montreal Gazette* opined that a joint project would amount to a loss of Canadian sovereignty, while Senator J.P. Casgrain labelled the potential project as an American "Trojan horse," a view that found echo in the Legislative Assembly of Quebec.[23] Anglican canon F.G. Scott, who was born in Montreal but had resided in Quebec City since the late nineteenth century, declared that American involvement on the St. Lawrence would be "equivalent to moving the United States boundary line north to the St. Lawrence River, and Canada might find herself in a hopeless national position at any time." "The St. Lawrence is a great Canadian asset; it is also an Imperial asset, and must be wholly under [Canadian] control," Scott continued; "it is the very special marrow of Canada, and on its shores and the shores of its tributaries lie the cities or villages of a large part of Quebec and Ontario ... Our great river must be ours and ours alone."[24]

Scott and Casgrain caught the feelings of many Canadians, particularly in Quebec, where resistance to a cooperative waterway development also stemmed from fears that the province's ports would be bypassed. For Quebec Premier Louis-Alexandre Taschereau, the waterway might result in Montreal losing cargo business and American manufacturing investment. It made sense to align with Ontario against the federal government's claims that navigation rights on rivers superseded provincial power possibilities.[25] In 1923, Ontario had elected a new government, and HEPCO accelerated its attempts to obtain federal approval for a power development, separate from navigation works, near Morrisburg.²⁶ The following year, HEPCO applied for permission from the Province of Ontario to build a powerhouse in Canadian territory. Since the British North American Act, the base constitutional document that had established the country of Canada, was unclear on the division of jurisdiction in regard to water rights, this opened up the constitutional question of whether the federal government or the provinces had jurisdiction over power development. The act gives the provinces control over most issues related to the management of water resources, subject to the federal government's specific powers over navigation and fisheries, and broad powers in the areas of trade and commerce, interprovincial-international works (e.g., transborder canals), and the peace, order, and good government clause. In no small part because of the demand for power in Ontario, which increased at a rate of about 10 percent annually in the 1920s – by 1929, 69 percent of all Canadian industry depended on electricity – this debate persisted into the 1930s and was referenced to the Supreme Court.²⁷

Canadian constitutional issues were of little concern to American companies. In 1922, the International Joint Commission approved the construction of a submerged weir in the St. Lawrence River to divert up to 25,000 cubic feet per second (cfs) into the Massena Power Canal. This weir had initially been approved by the IJC in 1918 on a temporary basis, despite Canadian objections that private development would hinder the ongoing plans of the two federal governments.²⁸ In 1923, the American Superpower Company applied to the US Federal Power Commission (FPC) for the right to dam the whole of the St. Lawrence River, proposing to market two-thirds of the approximately 1 million horsepower in the United States. Canadian opposition would have denied IJC approval for this proposal, but the FPC did so first. Canada was open to other options, however, and in 1924, an enlarged Joint Board of Engineers, with three members from each country, was created to work with the IJC. Meanwhile, the United States assembled a St. Lawrence Commission, and Canada instituted a similar body, the National Advisory Committee, though King's appointment of this committee was merely a gesture meant to buy time, particularly as the Americans were more interested in the seaway than in the power.²⁹

Both the St. Lawrence Commission and the National Advisory Committee favoured the St. Lawrence scheme. In its 1926 report, the Joint Board of Engineers called for twenty-five-foot-deep canals and five locks, though its membership differed in respect to their opinions about the power development. The Americans favoured a single-stage project with one dam at the foot of Barnhart Island near Cornwall. The Canadians preferred a dual-stage project with two powerhouses, the largest of which would be in the vicinity of Barnhart Island, the other on the south shore near Ogden Island. The single-stage would cost less than the dual-stage – $235,110,000 versus $274,742,000 – and require one fewer lock.[30] The Canadians argued that construction of a dual-stage project could be carried out in two parts, meaning that power could be produced before the entire project was completed, allowing for better control of the river flow and reducing the amount of flooded land from twenty-eight thousand acres to about eighteen thousand acres.

There were, however, several valid reasons for King to delay. Power shortages were predicted for Ontario, and the province's manufacturing strength derived from access to cheap power.[31] In addition to seeking the diversion of water from the Albany River system into Lake Superior, which could then be used for hydroelectric production at Niagara, Premier Howard Ferguson began exploring electricity exports from Quebec. Between 1926 and 1931, Ontario signed a series of contracts with different Quebec power companies to furnish Ontario with electricity.[32] As a result, both the Quebec and Ontario governments were uninterested in developing hydroelectric power from the St. Lawrence as long as these contracts remained in effect. There were rumours that both the Canadian Pacific Railway and the British government were also exerting pressure on the King government to thwart a St. Lawrence development. Historian Gordon Stewart further identifies a range of reasons for King's trepidation about moving ahead on the St. Lawrence:

> (1) Such an agreement with the United States would sacrifice Canadian sovereign rights over the St. Lawrence; (2) special interests in the United States wanted cheap power from the St. Lawrence; (3) the Americans must agree to restrictions on the Chicago drainage scheme; (4) Canada's political autonomy would be undermined; (5) the scheme would not be supported in the Conservative-controlled Senate even if the government were to go ahead; and, finally, (6) the St. Lawrence was Canada's most valuable asset, and she should drive a hard bargain by demanding significant US tariff reductions in return for Canadian cooperation on the seaway project.[33]

In 1927, Canada opened its first legation in Washington, as did the United States in Ottawa. The St. Lawrence topic was the US Department of State's top Canadian priority, and it again asked, in light of the recent reports, whether the Canadian government was prepared to enter into discussions. The Canadian prime minister hoped to avoid the prickly federal-provincial problems that a St. Lawrence project would surely bring up and subtly stalled negotiations, advocating further study.[34] The King government was concerned too about the cost and entertained hopes that the US government might be willing to shoulder the expenses. Indeed, there had been considerable advocacy of an entirely American deep waterway in the United States throughout the 1920s, though such proposals generally focused on joining with the Hudson River.

The steady stream of official reports recommending the project continued. The Canadian National Advisory Committee suggested that Canada develop the parts of the prospective waterway and power project that were exclusively in Canadian territory, and the United States develop the international section of the St. Lawrence as well as the connecting channels between Lake Erie and Lake Superior. This led to some movement from the Canadian government, which in January 1928 proposed the committee's idea to the United States, and included giving Canada credit for the cost of the new Welland Canal (which would prove to be about $130 million) that it was in the process of constructing. The United States was willing to accept this as a basis for negotiations, but King again balked, and his apprehension led to US threats to resurrect the all-American waterway possibility.[35]

Discussions on other water control schemes connected to the St. Lawrence met with varying levels of success. In January 1929, Canada and the United States signed a Convention and Protocol for the Preservation of Niagara Falls, allowing for an additional 10,000 cfs diversion above the falls for power generation. However, this agreement was not able to make it through the US Senate.[36] After considerable back and forth, the two countries agreed to dredge deeper channels in the Thousand Islands. These developments corresponded with the creation of the Beauharnois Canal and hydro dam, which were constructed along the St. Lawrence just west of Montreal between 1929 and 1932.[37] The opening of the Beauharnois project coincided with that of the new Welland Canal, which would also prove to be a monumental achievement. Work had begun in 1913 but, because of the economic impact of the First World War, progress was halting. This new iteration of the Welland, the fourth, reduced the number of locks from twenty-six to seven, and included a guard lock; its dimensions measured 859 feet long and 80 feet wide, with a controlling depth of 25 feet and 30 feet of water over the sills.[38]

1932 Great Lakes Waterway Treaty

There had been support for a St. Lawrence project from the Harding and Coolidge administrations during the 1920s, but advocacy for the development was most pronounced under President Herbert Hoover, who was in office from March 1929 to March 1933. Among other reasons for the president's support for a seaway was his desire to reduce water transportation rates for agricultural products and to placate midwestern farmers alienated by Republican farm policies.[39] Indeed, the spate of multipurpose dam building and comprehensive water developments that characterized the New Deal era later in the 1930s was already underway during Hoover's presidency.[40] Hoover was a conservationist of Progressive era vintage and a strong associationalist who approved of government financing for water navigation and dam development, but he was philosophically opposed to complete public ownership of electricity distribution and marketing.[41]

As debates about federal-provincial jurisdiction over power development continued, so too did American pressure for a St. Lawrence treaty, including threats to delay the ratification of accords concerning Canada, namely Pacific salmon fisheries and cooperative action to divert water from Niagara Falls for hydro power while simultaneously preserving its beauty. But these did not achieve any results. King briefly toyed with the idea of exchanging a seaway agreement for American tariff lenience but then abruptly retracted the offer once this potential linkage was leaked to the press.[42] The majority of King's ministers considered a St. Lawrence undertaking desirable from numerous perspectives, though the economic feasibility depended on the proportion of the cost that the United States was willing to assume.[43] The possibility of an all-Canadian seaway was raised by both Canadian Undersecretary of State for External Affairs O.D. Skelton and Ontario Premier Howard Ferguson.[44]

However, a Canadian federal election was imminent, with the result that the Canadian prime minister remained unwilling to embark on any risky moves, particularly because he was sensitive to accusations of catering to the Americans.[45] Nonetheless, both King and Conservative leader R.B. Bennett promised action on the St. Lawrence, which was a major issue in the campaign. The latter pledged to "blast a way" into foreign markets, particularly those closed by the recent US Smoot-Hawley tariffs, while playing down his party's adoption of an all-Canadian seaway platform at its 1927 convention. King attempted to blame the obstructionist tactics of the Ontario government for the seaway impasse.[46] Voters selected Bennett and the Conservatives over King and the Liberals. The Conservative

Party remained in power in Ontario, but George Henry succeeded Ferguson in the premier's office. Henry was more inclined to constructively discuss the seaway project than was his predecessor. Bennett was willing to ignore Quebec's opposition, since he was not as dependent on that province for electoral support as King had been, and the changed provincial-federal conditions put the Canadian government in a position to sign a treaty with the United States, despite ongoing constitutional questions about jurisdiction over water rights.[47]

Canadian and American leaders met in Washington in January 1931. Bennett, who disliked Hoover and was apprehensive about the president's unpopularity, rejected Hoover's strong overtures to move ahead on a St. Lawrence treaty. Many of the same factors that had caused King to move cautiously now gave Bennett pause. As scholar William Willoughby stated, "In view of the many unfavorable factors, the surprising thing is not that the prime minister delayed for more than a year the opening of negotiations but that he had the temerity to agree to discussions when he did."[48] Bennett kept sending mixed signals.[49] The Depression was just beginning, canal and railroad usage had decreased, and there were concerns about undertaking a project of the magnitude of the St. Lawrence development. Issues such as tariffs and the ongoing Illinois diversion through the Chicago Sanitary and Ship Canal, which sent water out of the Great Lakes watershed to the Mississippi River, continued to aggravate Canadian officials.[50] Nevertheless, by the autumn of 1931, serious discussions were underway. The economic and employment benefits of the seaway project, including the ability to take full advantage of the recently opened Welland Canal by the removal of the St. Lawrence bottleneck between it and the Atlantic Ocean, were hard to ignore. The possibility that the United States would proceed with its own national waterway, as recommended by a 1930 US Army Corps of Engineers report and subsequently approved by the House of Representatives Rivers and Harbors Committee, further motivated the Canadian prime minister.[51] Perhaps indications that Hoover, who was strongly in favour of developing the St. Lawrence, would not win the upcoming election were instrumental in bringing Bennett to the table.

Talks slowed when the Canadian prime minister departed for an extended trip to Britain, and then gathered momentum heading into 1932.[52] The negotiations between the Canadian minister and American Secretary of State Henry L. Stimson were fairly straightforward; according to an unidentified American governmental source, the resulting treaty was largely the result of the personal commitment of the powerful William Herridge, the new Canadian ambassador to the United States as well as the prime minister's brother-in-law, to the seaway. Interestingly, Herridge told the Americans that an all-Canadian St. Lawrence project would be needlessly costly and impracticable, and that suggestions to secure it were "nothing more than a sop to the nationalists in Canada who might criticize the Conservative Government for ignoring the [1927 Conservative all-Canadian] resolution."[53] A.G.L. McNaughton, who would play a major role in seaway negotiations decades later and be a major advocate of an all-Canadian seaway, served as a key advisor to Herridge and helped draw up drafts of the treaty.

The Canadian and Ontario governments had agreed on a dual-stage dam, which became a point of binational contention, as each country's engineering boards had made different recommendations.[54] Because of objections to the higher cost, the Americans inquired about Canada's inclination to revisit a single-stage dam, but the Bennett government was unwilling to consider this for domestic political reasons, the impact on water levels downriver in Montreal, and the flooding in Ontario that would result from damming the river. Canada's disinclination to budge on this issue was aided by the April 1932 Joint Board of Engineers report, which recommended a dual-stage hydro plan for the IRS and, in the end, the Americans acquiesced. To make the dual-stage dam program more palatable, the State Department negotiators requested that Canada be more amenable to the levels of the Chicago diversion, and a compromise was arrived at by creating a tribunal. In turn, Canada asked for the ratification of the salmon treaty that had been stalled in the US Senate, and for more favourable terms on cattle imports.[55]

Potential complications arose from different quarters. Just as federal-provincial control over water power had created problems in Canada, so too were there disputes in the United States over which level of government had rights to the electricity harvested from the St. Lawrence. The governor of New York, Franklin D. Roosevelt, who was most concerned about the hydroelectric power and wanted it controlled by the state government, was causing difficulties for Hoover, who was more concerned about the navigation aspect of the project and looked unfavourably on New York's intrusion into territory that the president saw as federal prerogative.[56] Premier Taschereau of Quebec was a similar nuisance for Bennett. In response to Roosevelt's concerns, the New York legislature, acting on the recommendation of the St. Lawrence Power Development Commission, created the Power Authority of the State of New York (PASNY) to deal with hydroelectric development on behalf of the public interest. The creation of this authority was part of Roosevelt's conservationist approach in which publicly owned and distributed hydroelectricity was to be distributed as cheaply as possible to aid rural rehabilitation; such policies served as a precursor to the projects he would initiate as president.[57] The formation of HEPCO a quarter century earlier served as a model.

Discussions between Ontario and Ottawa were ongoing during the treaty talks. The pivotal issue was the cost-sharing arrangement for the joint works (benefiting both navigation and power) in the proposed treaty. A division of costs agreement was reached in July 1932 under which Ontario would receive the rights to the Canadian share of the St. Lawrence waters for hydroelectric development. Queen's Park would pay for power development and 70 percent of the cost of joint power-navigation facilities, whereas Ottawa was responsible for the navigation elements.

The Great Lakes Waterway Treaty, signed on July 18, 1932, generally satisfied the governments of both countries. It provided for a twenty-seven-foot-deep waterway from the head of the Great Lakes to Montreal and addressed other water issues in the Great Lakes basin. The United States was responsible for any navigational improvements above Lake Erie; Canada agreed to furnish the appropriate navigation works in the Quebec part of the St. Lawrence. Both countries would cooperatively build in the International Rapids section, with each providing a lock and canal on its side – the Canadians at Crysler Island and the Americans at Barnhart Island – and perform other channel improvements. The dual-stage development provided for dams at Crysler and Barnhart Islands, with powerhouses on both sides of the international border at each location. This scheme would protect downstream interests in Quebec concerned about water levels. The hydroelectricity produced would be an estimated 2,200,000 horsepower. Each country would get half of this power, though the treaty deliberately left open-ended the precise manner in which the power was to be distributed, given the jurisdictional federal-state and federal-provincial issues in each country.

Estimates put the cost of the entire project at $543,429,000, split as $270,976,000 for the Canadians and $272,453,000 to the Americans. However, Canada received credit for the Welland Canal and work already done in the Thousands Islands, though not for efforts previously undertaken in Quebec. Canada was therefore accountable in actual expenditures for only $142,204,000 of the costs, with the United States handling the difference. The United States assumed the cost of most of the works on the Canadian side of the IRS, except for the lock and canal, even though

Canada would perform the actual construction. The treaty also established a joint commission composed of five members from each country to supervise the whole endeavour and other water diversions featured in the treaty. Canada and the United States affirmed that they would build compensation works at Niagara Falls and Lake St. Clair to offset waters diverted for hydroelectricity production at Niagara and the Chicago diversion.[58]

The treaty signing was accompanied by fanfare throughout the Great Lakes region, and initial assessments of public opinion indicated that the St. Lawrence treaty was warmly received in all areas of Canada. Hoover was ecstatic, calling it "the greatest internal improvement yet undertaken on the North American continent."[59] Nevertheless, the treaty still needed to be approved in both countries, and Bennett had decided not to present the treaty to the Canadian Parliament for approval until it had first been ratified in the United States. If the treaty had been submitted to the US Senate immediately, it might well have been rapidly ratified, considering the sentiment and momentum in its favour.[60] However, Congress had already adjourned for the summer, and there was not time in the remaining portion of Hoover's term for both houses of Congress to sanction the compact. A presidential election was in the offing. Moreover, the Democratic nominee was Franklin D. Roosevelt, the thorn in Hoover's side over New York's rights to the hydroelectric power of the St. Lawrence. After Congress resumed in autumn 1932, the Senate Committee on Foreign Relations favourably reported the treaty, and PASNY and the US Army Corps of Engineers arrived at an arrangement under which the former would develop the power aspect of the St. Lawrence project and the latter would construct the navigation elements. Shortly thereafter, Roosevelt was elected president.

Roosevelt was a relatively new convert to the seaway concept, as he had always emphasized the hydro power aspect. Pinning down Roosevelt's approach to enabling the treaty is not easy, in part because of his general decision-making manner. The new president favoured a St. Lawrence agreement with Canada and eventually put the 1932 treaty to the Senate. Although Roosevelt did not formally introduce the treaty until January 1934, over the course of 1933 it was debated and the Senate requested several studies on the St. Lawrence project. But since the treaty had been Hoover's accomplishment, Roosevelt neither advocated strongly for it nor expended his political capital to aid the legislation. Many Democrats were unwilling to support the treaty's passage because it had been negotiated by Hoover, a Republican, whereas many Republicans opposed it because the Democrats were trying to take credit for the treaty.[61] The seaway treaty was also briefly tied to the ongoing transboundary dispute over the smelter in Trail, British Columbia.

The treaty legislation received congressional consideration for several months, and the president even suggested that Canada might go it alone if Congress did not act.[62] When it came time to vote in March 1934, the majority of the Senate approved the resolution by a division of 46-42; however, a two-thirds majority was required for ratification of a treaty, and the Great Lakes Waterway Treaty went down to defeat. According to John D. Hickerson, an official in the State Department who had served in Canada, those who voted against the treaty tended to have general objections to the whole project, or specifically to the Chicago diversion aspects.[63] The treaty had become ensnared in domestic politics and in the conflicts between the regions and interest groups that had long plagued the project in the United States.

For proponents on both sides of the border, it was a crushing outcome. At the same time, after the treaty was signed, support in Canada for the seaway and power project had apparently dwindled because of the Depression. In fact, the Canadian government quietly informed Washington that it was not displeased that the treaty had failed to pass.[64] Subsequent inquiries from Washington about the Bennett government's willingness to introduce the treaty into Parliament or approve of changes to the American St. Lawrence bill met with either promises of consideration or refusal, and ultimately no action was taken by Canada. Mitchell Hepburn, the newly elected Liberal premier of Ontario, quickly repudiated his predecessor's unratified cost-allocation arrangement with the federal government, which would have seen the province pay $67,202,500 for its share of the costs.

The tempestuous Hepburn had been a backbencher in the King government during the 1920s and had previously supported the seaway concept. He became a dedicated opponent for a range of reasons: the belief that the seaway would never justify the enormous expenditures it would require, a desire to publicize his province's surplus of power, and the wish to protect private railroad or power interests.[65] There was also speculation that US power companies, which were more interested in hydroelectric development than transportation when it came to the St. Lawrence, had assisted Hepburn's campaign and he was beholden to them as a result.[66] Despite Hepburn's opposition to St. Lawrence development, he sought some of the provisions of the 1932 treaty, such as generating power through additional diversions at Niagara Falls, which would be aided by extra water from the Ogoki and Long Lac (Long Lake) diversions from the Hudson's Bay basin into Lake Superior. The Ogoki diversion alone would allow for the generation of a further 100,000 horsepower at Niagara. However, since these diversions changed boundary water levels, they needed American concurrence, which was not forthcoming as long as Hepburn proved intransigent about the St. Lawrence.

Despite the St. Lawrence treaty defeat, Roosevelt was committed to achieving a St. Lawrence project and was said to have remarked that a seaway would someday be built "as sure as God made little green apples."[67] Additional American inquiries, extending into 1935, about Canada's willingness to entertain even minor changes to the 1932 St. Lawrence treaty to make it more attractive to Congress met with resistance from Prime Minister Bennett and Skelton.[68] In a conversation with the US ambassador, Bennett said that the treaty was "dead" in Canada and US ratification offered the only hopes of reviving it.[69]

1941 Great Lakes–St. Lawrence Basin Agreement

King returned to office in October 1935. Roosevelt, who was more knowledgeable about Canada than any previous president, got along well with King; FDR was a relentless flatterer and charmer, and the Canadian leader lapped it up. They immediately came to a major trade agreement.[70] The two national leaders talked about revising the Bennett-Hoover treaty, but instead started to explore the possibility of a new and wider agreement encompassing issues such as the diversions at Niagara and Ogoki–Long Lac and the export of power. In late 1936, the Roosevelt administration approached Canada in connection with a new treaty. Prime Minister King seemed reluctant, citing the opposition of Hepburn and Maurice Duplessis, the recently elected Union Nationale premier of Quebec who had joined Hepburn in resisting the federal government and the St. Lawrence project. For a brief time it appeared that Hepburn was willing to approve a St. Lawrence treaty in exchange for US consent to the Ogoki diversion and the ability to stagger construction of St. Lawrence hydroelectric facilities. However, complications stemming from Ontario's power contracts with Quebec, among other reasons, led the premier to abandon interest in a St. Lawrence treaty and instead call an election in which he focused on the possibility of Niagara power.

Disputes over water rights remained tied into wider constitutional questions and federal-provincial issues, which were addressed by the creation of the Royal Commission on Dominion-Provincial Relations (the Rowell-Sirois Commission) in 1937. The acrimony between the federal government and Hepburn – HEPCO and the majority of Queen's Park appeared to be in favour of restarting St. Lawrence negotiations – was only exacerbated by Ottawa's unwillingness to allow Ontario to export hydro power to the United States or undertake other water diversion schemes. The Ontario premier charged King with conspiring with Roosevelt to force Ontario to accept the seaway; Roosevelt was in fact using the push for a comprehensive program as a means of making Hepburn more amenable to a St. Lawrence treaty. The Ontario premier therefore "launched a diplomatic offensive of his own early in 1938 to persuade Roosevelt to adopt a piecemeal approach to waterways problems."[71]

The United States transmitted another draft treaty in May 1938, outlining a program by which the United States would build *all* works in the IRS except the Canadian powerhouse. This engendered a good deal of debate in the Canadian government; nevertheless, Ottawa failed to formally reply. It seemed that Hepburn might be agreeable to this plan, but he was apparently put off by remarks in August 1938 that Roosevelt made promoting the seaway while he dedicated the new bridge at Ivy Lea in the Thousand Islands.[72] Although the media generally focused on the president's statement that "the United States will not stand idly by if domination of Canadian soil is threatened" – a threat as much as it was a promise – the *New York Times* report paid almost as much attention to the president's St. Lawrence remarks.[73] The Ontario premier showed occasional signs of willingness to cooperate in the following months, but his obdurate opposition and epistolary attacks, along with Premier Duplessis's hostility, meant that matters remained at an impasse during the last half of 1938 and into 1939.[74]

The Second World War broke open the Canadian domestic St. Lawrence stalemate. Duplessis was defeated in the October 1939 Quebec election by Liberal Adélard Godbout, who was more willing to cooperate with Ottawa. Meanwhile, in order to make common cause in the war effort, and convinced by HEPCO that the St. Lawrence project would be necessary to meet the province's power needs in the event of a protracted conflict, Hepburn performed a volte-face and suddenly became open to initiating St. Lawrence negotiations, provided Ontario could export power and proceed with the water diversions into Lake Superior.[75]

Canadians re-elected King's Liberals in the March 1940 federal election. Although there was still opposition from port interests in the Atlantic provinces and Quebec, as well as from power companies in Quebec, in the wake of its election victory the King government was willing to step up the treaty discussions. But Roosevelt was now the more reluctant partner, as he too had to face the electorate in 1940. The president was unsure whether he had the requisite votes to put a waterway treaty through the Senate, and that made him hesitant to force the issue.[76] To help compensate for his delay, the Roosevelt administration permitted Ontario to divert extra water for hydro production at its Sir Adam Beck plant at Niagara Falls and ordered preliminary engineering investigations of the prospective St. Lawrence works. After initial reluctance, motivated in part by Quebec private power interests emitting propaganda against the St. Lawrence project, Canada asked to participate in this work.

In October 1940, the Canadian government created a Temporary Great Lakes–St. Lawrence Basin Committee to work with the American St. Lawrence Advisory Committee, formed the previous week, and the two countries briefly looked at developing only the power works. The committees reported in early 1941, recommending a single-stage project under which there would be a control dam and a power dam rather than the two power dams envisioned in previous dual-stage proposals.[77] Over the course of the previous year, Canada had agreed to the single-stage project, and the American preference for this type of project was a powerful inducement. The US Army Corps of Engineers recommended in 1942 a "combined" single-stage plan, a compromise between the single-stage and dual-stage in which there would be multiple dams but only one that produced power in the international section. This combined plan envisioned joined powerhouses at the foot of Barnhart Island, a Long Sault spillway dam at the southern head of Barnhart Island, a control dam in the vicinity of Iroquois, and three US locks (two near Barnhart Island and another across from Iroquois, at Point Rockaway). Even though a dual-stage scheme would flood only 5,900 acres, compared with the 15,600 inundated by a combined single-stage project, the latter offered considerable financial savings.[78]

The case for improving the great river was further buttressed by the *St. Lawrence Survey*, an extensive seven-volume study completed between 1940 and 1941 by the US Department of Commerce.[79] This study advocated the construction of the seaway and power project on both economic and national defence grounds. It considered an enormous range of economic factors connected to numerous industries and sectors, and compared shipping/railroad distances and times, then used this information to conclude that the project was not only feasible, but extremely desirable. A mountain of information and statistics was heaped into the Department of Commerce's report; it is possible here to provide a summary only of the most salient figures. Between 1910 and 1938, traffic on the existing St.

Lawrence canals had risen consistently, to a high of 8,285,167 tons of cargo in 1938. That same year, the United States had exported 58,418 tons of cargo via the canals and imported 56,952 tons. Considering anticipated economic and population growth, a deep waterway would allow for significantly increased traffic, along with a wide range of attendant benefits to the American economy, all without seriously injuring railroad and eastern port interests. A seaway would be 30 percent cheaper than rail for certain types of freight, and up to 340 percent cheaper ($9.24 per ton by rail; $2.10 per ton by seaway) for other types of freight that could be transported more efficiently by water. In short, compared with the existing routes, a deep waterway would provide considerable cost savings for a wide range of commodities moving to a variety of ports. The *St. Lawrence Survey* estimated that a deep waterway could immediately carry 7 million tons of American traffic, and within a "reasonable period" American traffic alone on the seaway could be up to 10 million annual tons, with potential yearly savings on transportation costs of $36 million. The total cost was estimated at $429,474,515, and it was anticipated that the United States would be responsible for about two-thirds of that figure.

After securing re-election in November 1940, Roosevelt urged a resumption of treaty talks. By February, an agreement with Canada was all but complete, and the president had decided to proceed by executive agreement rather than a treaty, since the former needed only the approval of the majority of both houses of Congress, rather than the two-thirds Senate majority required for the latter.[80] Because the Supreme Court of Canada had not conclusively decided jurisdiction over power rights, the federal and provincial governments opted to negotiate the matter. The Ontario premier wanted the cost of the St. Lawrence joint works to be split 50-50, rather than the 70-30 formula contained in the 1932 Canada-Ontario accord. A new federal-provincial agreement was completed along those lines (and an interprovincial Quebec-Ontario agreement in 1943) that deemed Ontario the owner of the Canadian share of water power in the upper St. Lawrence.

On the same day, March 19, 1941, that the new federal-Ontario cost allocation was inked, Canada and the United States entered into the Great Lakes–St. Lawrence Basin Agreement. The agreement created the Great Lakes–St. Lawrence Basin Commission to oversee construction of a twenty-seven-foot waterway from the head of the Great Lakes to the Atlantic in conjunction with a combined single-stage hydroelectric dam in the International Rapids section. But, as its name suggested, this agreement aimed to comprehensively solve all issues in the basin, not just that of the St.

Lawrence. Canada was again given credit for the Welland Canal, and there were also stipulations governing water diversions and the maintenance of scenic quality at Niagara by the construction of remedial works, limits on the Chicago diversion, and parameters for the other diversions into the Great Lakes–St. Lawrence watershed. There were still issues to be resolved concerning Quebec and New York, but the Quebec government soon came to an agreement with Ottawa and also purchased the Beauharnois Canal and Power Works. New York State proved to be more of a problem, but by May 1941 PASNY and the US Army Corps of Engineers had entered into an accord.

In Washington, the executive agreement was introduced into the House Committee on Public Works at the start of June. During weeks of hearings and statements, supporters and opponents alike addressed the defence merits of the project. In the first week of August, the committee favourably reported the bill, but instead of the House voting separately on the St. Lawrence bill, it was lumped in with several other projects. Before a vote had taken place, the Japanese attack on Pearl Harbor in December 1941 brought the United States into the war. The omnibus bill to which the St. Lawrence agreement had been attached was indefinitely deferred, and it became clear that the agreement had virtually no hope of passage during the war. Consideration was given to authorizing the St. Lawrence project as a war measure through a simple exchange of notes with Canada. However, the War Department had suspended all Army Engineers projects that could not be completed before the end of 1943, and the Secretary of War rejected Roosevelt's scheme, since the advantages of the project would be three to five years away and it would consume valuable resources to construct.[81] This refusal meant that the funds for the St. Lawrence project were unavailable.

In the aftermath of the setback, the exhausted American proponents of St. Lawrence development divided between fighting for power or navigation. Several bills to approve the St. Lawrence agreement were subsequently introduced into Congress, but they went nowhere, as fighting the war was the first priority. In May 1944, US Secretary of State Cordell Hull requested that a new St. Lawrence bill proposal describe the project as a postwar, rather than a wartime, undertaking.[82] With the defeat of Germany and Japan on the horizon, proponents of the St. Lawrence development began to gear up for another protracted battle.

Nonetheless, other parts of the 1941 agreement were put into effect. The Long Lac diversion, which was intended to transport pulpwood and generate power locally as well as facilitate greater hydroelectric production at

Niagara Falls, had become partially operational in 1939. It was officially approved by the United States in November 1940 only because of the need for wartime power.[83] This diversion used an extensive network of dams and canals to reverse the flow of the Kenogami River south via Long Lac into the Aguasabon River, which debouches into Lake Superior. The enormous Ogoki diversion was completed in 1943 – it too employed large dams and canals to reverse the flow of a river, in this case the Ogoki River southward through Lake Nipigon into Lake Superior. Long Lac diverted an average of 1,500 cfs, whereas Ogoki averaged 4,000 cfs, making the latter the largest diversion into the Great Lakes watershed. The agreement increased the limits on the amount of water that Ontario could divert at Niagara Falls, to an amount roughly equivalent to the volume of water that the Ogoki-Long Lac diversions were putting into the Great Lakes system. These diversions were subsequently increased again during the war with exchanges of notes in October 1941 and May 1944, and a submerged weir to aid in water diversion was built above the falls.[84] As power shortages remained a chronic concern throughout the war, the two countries explored a range of other possibilities for coordinating resources and hydro development, including exporting power from Cornwall to Massena for aluminum production.[85] A new American lock at Sault Ste. Marie was also authorized, and the Canadian federal government gave Ontario and Quebec 999-year leases to develop the upper and lower parts, respectively, of the Ottawa River for hydroelectricity.

In anticipation of postwar action, the Canadian government had requested that its Advisory Committee on Reconstruction study the potential impact of the St. Lawrence project. The result, completed in 1943 by Norman D. Wilson and predicated on the combined single-stage engineering plans, was titled *The Rehabilitation of the St. Lawrence Communities*.[86] It recommended that the federal government, which held the main responsibility for rehabilitation under the existing agreement with Ontario, acquire all lands along the river less than 238 feet above sea level. However, the local and provincial governments should also be involved, and Wilson therefore suggested a dominion-provincial-municipal commission to supervise the rehabilitation.

The report postulated that over 14,500 acres of Canadian land would be lost and recommended that the towns of Iroquois and Morrisburg, at the western end of the inundated area, be relocated. Buildings should be moved wholesale where possible. As diking along the shoreline would not be cost-effective, the communities between Morrisburg and Cornwall would simply be lost under the water. It was likely, though, that some of the communities would relocate or reform to the north of their present locations, subject to various factors such as employment opportunities. The government could help facilitate this process, but it would not be responsible for actually relocating these communities. Some of the other noteworthy features of the report were proposals to combine all the cemeteries that would be flooded into one, and maintain the location of the Battle of Crysler's Farm memorial by raising it and placing it on an island. The contiguous chain of islands created by the raised water level should be turned into parkland, linked by causeway or bridge, and named Long Sault Park in honour of the eponymous rapids. The refusal of the American Congress to approve the 1941 agreement rendered this report effectively moot, but it would serve as the basis for future rehabilitation plans, with many of its elements retained outright or in modified form.

Conclusion

Proponents of a seaway and power project were disappointed that the 1941 executive agreement had not been immediately ratified, yet they could find some solace in the steps that had been taken toward the St. Lawrence dream in the preceding decades. In the immediate post–First World War period, government planners had begun to conceptualize a dual hydroelectric and deep navigation scheme that would radically reconfigure the St. Lawrence River. After the Wooten-Bowden and IJC recommendations in the early 1920s, the United States was the most consistent suitor in terms of an accord but was frequently met with Canadian ambivalence and obfuscation, particularly when King was prime minister.

The 1932 Great Lakes Waterway Treaty was the result of a change of government in Ottawa, but Canadian opinion, and the governments in both Ontario and Quebec, remained obstacles to a joint development of the river with the United States. Both countries had also flirted with serious consideration of a new canal system entirely within their national boundaries. Although the American Congress rejected both the 1932 and 1941 St. Lawrence accords, provincial political opposition in Canada had dissuaded Ottawa throughout most of the 1920s and 1930s from embracing American seaway overtures. This resistance lessened in the years leading up to the 1941 Great Lakes–St. Lawrence Basin Agreement, and with the Second World War creating a higher demand for the benefits of a St. Lawrence project, Canada permanently embraced the idea. The exigencies of America's entry into war, however, led Congress and the US government to shelve the project for the time being.

"Did the River hear the faint scratching of the pens? Was the sound of cameras perceptible beside the falls of Niagara, in the whirlpools, eddies, and rapids?" asked noted political writer and journalist Bruce Hutchison shortly after the signing of the 1941 St. Lawrence agreement. "No, the River only smiled that morning and went about its spring business of breaking the winter's ice, for it knew men would never tame it," Hutchison continued; "they might use it, grasp a little of its power for themselves, build cities at its sides, and launch their ships upon its current, but nothing could alter by more than a few yards the irresistible current of the St. Lawrence or curb its yearning towards the sea."[87]

2
Watershed Decisions

The Second World War transformed the context of St. Lawrence Seaway and Power Project discussions. Both Canada and the United States benefited by their physical distance from the destruction in Europe and Asia during the war. The United States emerged from the conflagration in 1945 as a superpower, although long-time leader Franklin D. Roosevelt had died in April of that year. His successor, Vice-President Harry S. Truman, proved to be a vigorous proponent of a joint St. Lawrence project. The Liberal government was given a fresh mandate by Canadians in June 1945; this would prove to be Mackenzie King's last term as prime minister, as he stepped down in 1948 and was replaced by Louis St. Laurent. With so many countries devastated, Canada too was a major state for a time, a middle power according to its own rhetoric, as it entered the reputed "golden age" of Canadian diplomacy. The Department of External Affairs had grown enormously during the Second World War, and a major goal of post-1945 Canadian foreign policy was to ensure that the Americans did not retreat into isolationism; that worry turned out to be unfounded, as the United States, brandishing nuclear weapons, set out to recast the world in its capitalist liberal-democratic image, create open markets, and contain Soviet expansion, aims that Canada was all too eager to support.

The Second World War had prompted Canadian-American defence and economic integration at an unprecedented level. Despite a reduction of this intertwinement on some fronts in the immediate postwar period, Canada was in the process of turning, out of necessity, from Great Britain to the United States as its main ally. Between 1946 and 1951, Canadian exports to the United States more than doubled, from $888 million to $2.3 billion, with the United States absorbing 38 percent of Canada's total exports in 1946 and 59 percent in 1951.[1] The war had resulted in a voracious consumption of natural resources worldwide and within North America, entrenching the mobilization of natural resources on an unprecedented scale and setting in motion patterns of extraction and development that a dawning Cold War would only accelerate. In Canada, the state-led centralization of control over hydroelectricity for wartime aluminum production imposed path dependencies on postwar hydro development characterized by federal intervention and control of hydroelectric resources.[2] For both Canada and the United States, stewardship and conservation meant managing natural resources so that they could be most efficiently exploited, manipulated, and harnessed. Water was commodified, valued for what it could become: power. And power, in turn, became commerce and armaments, prosperity and security, and symbols of state control and legitimacy.

Canada was experiencing an economic, social, and population boom, which manifested itself in a more palpable Canadian confidence and nationalism. Aided by this self-assurance, the St. Laurent years were an era of megaprojects, reflecting a belief in the state's ability to configure nature to its needs. In addition to the St. Lawrence development, these included the Trans-Canada Highway, the TransCanada pipeline, and radar defences stretching across the northern reaches of the country. Those that stood in the way, such as First Nations, were of little consequence. Much of the growth in industry and manufacturing depended on natural resources such as oil, uranium, iron ore, and hydroelectric power. Canada had them in abundance. Because Canada was rich, and wanted to be richer, the St. Lawrence project was seen as not only viable but also necessary for meeting the country's increased transportation, industrial, and power needs. These needs, as expressed in pipelines, highways, and waterways, left long and visible scars across the country. Few cared at the time.

After the war had ended, deep waterway proponents hoped that the discovery of iron ore deposits in the Ungava region of northeastern Canada, coupled with proposals to make a seaway self-liquidating through tolls, would make the 1941 Great Lakes–St. Lawrence executive agreement more attractive to US legislators. Since Congress continued to spurn bills to develop the St. Lawrence, chiefly because of specific interests opposed to a seaway and/or power project, the State of New York and the Province of Ontario emerged with a proposal that called for a hydroelectric development not formally connected to a deep waterway. As this chapter shows, frustrated by congressional delay in authorizing an American seaway role, the Canadian government began seriously contemplating the construction of an all-Canadian seaway. Although hampered by the Truman administration's stalling tactics and interference with the requisite Federal Power Commission licence, this contemplation turned to action, culminating in a 1952 application to the International Joint Commission for a New York-Ontario hydro project and, by extension, an all-Canadian seaway.

Tolls and Ores

Attempts to obtain congressional approval for the 1941 St. Lawrence agreement had recommenced during the latter war years – after initial failures immediately following America's December 1941 entry into the Second World War – and continued into the postwar period. In October 1945, President Truman used his Message to Congress to request that it pass St. Lawrence legislation.[3] Despite arguments that the war had shown the necessity of the St. Lawrence development, and that postwar conditions required it, Congress remained unconvinced. US opponents alleged that, on top of being uneconomical, vulnerable to aerial attack and sabotage, and open for only seven months of the year, the seaway could not accommodate many, if not most, types of large vessels. An American 1934 governmental estimate had indicated that approximately 60 percent of the world's ocean-going tonnage and 70 percent of its freight cargo tonnage would be able to use the proposed seaway.[4] However, in the postwar period, the size of vessels likely to ply a potential deep waterway had increased considerably, and the extent to which all Canadian or American officials really believed that there would be serious traffic from ocean-going vessels on a finished seaway is debatable.

Both countries were obviously preoccupied with the war recovery effort, but a breakthrough of sorts came in January 1947. N.R. Danielian, the author of the Department of Commerce's 1941 *St. Lawrence Survey*, had recently become vice-president of the pro-seaway National St. Lawrence Association. Along with the influential senator Arthur H. Vandenberg, Danielian proposed that objections to the St. Lawrence agreement based on cost could be met by making the project self-liquidating through tolls, an idea that had been all but abandoned in the 1930s. There were divisions of opinion in Ottawa about the tolls prospect, but the Canadian government recognized that such levies could help counter claims in the United States that a seaway would be unfairly subsidized in comparison to other methods of transportation. In March 1947, the King government reluctantly informed Washington that Canada would agree to the tolls principle, subject to further negotiations.[5]

In addition to tolls, there was a significant financial and national defence inducement that reframed the seaway debate in both countries: iron ore. Indeed, the recently discovered iron deposits in the Ungava region, straddling the Quebec-Labrador border, could be most efficiently transported to the steel mills of the Great Lakes region via the St. Lawrence. The magnitude and availability of the iron ore deposits were soon definitively shown and agreed on at the highest levels. Drilling in the Labrador range had revealed a gargantuan 150 to 250 million tons of proven ore; it was hoped that this estimate would rise by the end of that calendar year to as high as 300 million tons, about three-quarters the amount the United States had used in the Second World War.[6] Compared with other ore possibilities, such as those in South America, the Ungava ore would be more economical, accessible, and secure in case of international conflict. Advocates contended that transportation costs for most cargo, but especially ore, via the seaway would be approximately one-half of the cost of shipping it to the Midwest via Philadelphia or Baltimore.[7]

The strategic importance of minerals and metals underscored the importance both countries attached to the seaway in terms of continental defence and economic mobilization. After 1948, but particularly after the onset of the Korean War in 1950, American authorities sought out Canadian minerals and metals to fill the stockpile targets that had been set as a direct result of the Paley Commission, which was established in the United States to determine what natural resources were needed for national security.[8] Iron ore was one of the most important resources identified. In the decade following the end of the Second World War, American companies pumped $1.4 billion in new investment into Canadian mining and smelting.[9] The Canadian government, as well as the governments of Quebec and Ontario, also coveted the possibilities of developing the ore. To support ongoing studies on the St. Lawrence project, the Canadian cabinet established an interdepartmental committee in June 1947 consisting of top-level officials, and it began meeting frequently with representatives of the American government. They reached a general concurrence on estimates for grain and iron ore traffic, and both sides thought general annual traffic on the seaway would be in the neighbourhood of 30 million tons.

But tolls and iron ore were not enough, at least not for the time being. On February 27, 1948, the US Senate voted 57-30 to recommit the St. Lawrence legislation to its Committee on Foreign Relations for further study. Although the resolution remained alive on paper, this effectively ended the seaway's prospects for the immediate future. Seaway proponents on both side of the border were disappointed, as they had been optimistic about the resolution's chances in the Senate. There were numerous reasons for the defeat of the St. Lawrence resolution. Although the tolls provision had been introduced in order to gain the support of some fence-sitters, it was too vague to significantly sway the vote.[10] Moreover, the need for Quebec-Labrador iron ore remained contested and, because it was inconsistently presented, the national security argument failed to "strike a responsive congressional chord."[11]

The Power Priority Plan

An alternative plan, with far-reaching consequences, was about to be added to the mix. New York governor Thomas Dewey and Premier George Drew of Ontario were working on a procedure that would see the American state and the Canadian province develop only the hydroelectric power resources of the International Rapids section (IRS) of the St. Lawrence River.[12] Power and navigation had long been fused in the minds of planners from HEPCO, but during the interwar years, and then in the first years of the Second World War, HEPCO had raised the possibility of developing power alone. Under this scheme, if any navigation works were to be constructed, they would be the separate responsibility of the federal authorities.

There was little doubt that Ontario, and the country in general, needed access to power. Canada's industrial base and economy had grown tremendously during the war, led by Ontario, and growth continued apace after 1945. The central Canadian province was already experiencing power shortages, and if the capacity of the country's most important industrial area was to develop to its full potential, new sources of power were required. In 1945, southern Ontario had consumed almost 10 billion kilowatts, would consume 11.8 billion kilowatts in 1949, and estimates put the 1951 requirement at 15.6 billion kilowatts.[13] The waters of the St. Lawrence provided the most viable method of meeting this need, and of ensuring that Ontario would play the leading role among the provinces within the Canadian federation.

In the wake of the most recent congressional defeat of the combined project, Dewey and Drew indicated their intention to proceed with their "power priority plan." The Ontario–New York proposal required their respective central governments to submit the provincial/state plans to the International Joint Commission. However, adopting such a procedure would preclude the unratified 1941 St. Lawrence agreement, since it was based on the coupling of the power and navigation aspects. Because of conflicting evidence, the degree to which HEPCO fully supported the Ontario–New York push for immediate power development is disputable.[14] Nevertheless, an examination of HEPCO archival sources indicates that the commission had, in direct correspondence with PASNY, supported a power priority plan.[15] The Canadian government believed that support for the Ontario–New York scheme would close the door to a joint Canada-US development and, under the 1941 agreement, Canada was given credit for the significant amount that it had spent on works such as the Welland Canal. If the Province of Ontario and the State of New York were to develop power exclusively, the cost of power and navigation would be increased, the possibility of developing hydroelectricity at the Lachine section in conjunction with the Province of Quebec might be lost, and there was a good chance that a seaway would not ever be built. But arguments within the Canadian government also ran in the other direction. Accepting the proposed Ontario–New York power project might pressure the United States into approving a new waterway agreement. Alternatively, if the power project was undertaken, Canada could conceivably complete a canal on its side of the border according to its own timetable, just as it had done with the Welland Canal.[16] The King cabinet opted to wait until it actually received an application from Ontario before deciding on a further course of action, and informed the Americans that it continued to favour the joint project.[17]

A few weeks later, in mid-June 1948, Ontario submitted the power priority application to the Canadian government. Its New York counterpart soon made the required submission to the Federal Power Commission (FPC), which had to approve such hydroelectric development plans within the United States. The Canadian embassy apprised the Americans of Ottawa's desire to stall until Congress had again considered the 1941 agreement. However, the State Department was finding it hard to delay a decision indefinitely, since the Department of War, Department of Commerce and Trade, and the FPC all advocated the transmission of the PASNY application to the IJC. The Department of the Interior was more hesitant, in part because of its reluctance to grant a state the right to harvest the hydroelectric power. In August, the FPC announced that hearings were set to commence on Project 2000 (the PASNY application) in October.

Truman eventually took care of the issue, at least for the time being. The president held a press conference to declare that he would not approve a project that separated navigation and power; however, this did not necessarily mean that the American leader would refuse to allow the New York application to go forward. Although he did not elaborate on the reasons for his objections, Truman feared that the navigation aspect, which he considered more vital for the United States than the power development, might not happen if the two were separated.[18] This bought the Canadian officials some breathing space, and it appeared that no further word about the official US position would be immediately forthcoming, since Truman was locked in a close presidential campaign with, as it happened, New York governor Dewey.

Prime Minister King retired in November 1948. He had been in power for much of the previous three decades and had been largely responsible for Canada's position on St. Lawrence development since the First World War, with the exception of Bennett and the 1932 treaty. The replacement for Canada's longest-serving prime minister was the steady, unspectacular, fluently bilingual Louis St. Laurent, a highly competent chief executive officer for an age of economic growth and nation building. Canadian diplomat Charles Ritchie recorded in his diary that although he and the Quebec-born St. Laurent, who had been a lawyer before his conscription into the King government during the war, were of different minds and natures, he respected and admired the prime minister, who had "more a lawyer's mind than a politician's, and he is completely free from the vanity and grudges of political life."[19] As the narrative traditionally goes, with St. Laurent came the putative "golden age" of Canadian foreign policy during which a "liberal consensus" in Canadian society allowed the country to play a more activist and internationalist role based on the principle of functionalism. Canada was certainly less isolationist than it had been, yet functionalism was just a sophisticated term for Canada acting in its self-interest and capacity. Moreover, the extent to which the age was really golden, as well as the extent to which St. Laurent's international outlook represented a fundamental departure from his prime ministerial predecessor, have been exaggerated.

On December 3, President Truman, having recently retained the presidency, indicated that he would continue to pursue joint action rather than move ahead with the power priority plan.[20] This meant that the White House was not going to take any action on the New York application. The New York Power Authority applications to the FPC and the International Joint Commission would stay "on ice"; in other words, the applications

FIGURE 2.1
Robert Saunders.
© *Ontario Power Generation*

would be held in abeyance while the president sought congressional validation, and the FPC would not give a decision on the licence, though the proceedings could be quickly resurrected if congressional hearings proved inconclusive.[21]

With the immediate prospect of acquiring St. Lawrence hydroelectric power diminished, HEPCO asked if Niagara diversions that had been taking place on a temporary basis since the Second World War could be made permanent. Right before Christmas 1948, Canada and the United States exchanged notes endorsing a 4,000 cubic-feet-per-second (cfs) diversion of water from above the falls, and another 2,500 cfs diversion to the DeCew Falls hydro station during the non-navigation winter season. Niagara power offered a possible panacea for Ontario's needs, and HEPCO was about to build a massive new hydroelectric development beside its existing Beck station, which PASNY would soon duplicate immediately across the Niagara Gorge. Although Robert Saunders, who had recently become chairman of HEPCO (after serving as the mayor of Toronto), had not promised that pressure for an Ontario–New York power plan would be temporarily relaxed if a Niagara accord was quickly approved, it seemed that this might be the quid pro quo. Saunders went to Washington to lobby for the project. This form of provincial diplomacy came to involve an informal understanding with an American newspaper bureau official who advocated HEPCO's point of view in the American capital and provided information about opposition in the United States to the St. Lawrence development.[22] As a result, the Canada-US Niagara Diversion Treaty was signed in 1950, entrenching hydro diversions and prescribing remedial works to mask the effect of the decreased water going over the cataract.[23]

Given the relative lack of means for measuring popular opinion at the time, the most consistent and accurate gauges of it were the two national governments, particularly their foreign agencies, along with newspapers and the occasional poll. According to the US embassy in the Canadian capital, there was general support north of the border for a quick start on the St. Lawrence project. This was substantiated by the Canadian government's various efforts to assess public preferences, as well as Canadian newspapers. The *Globe and Mail* remarked that "Canadians are sick and tired of waiting for Congress,"[24] and a *Financial Post* questionnaire asked prominent Canadians, "Do you favour the St. Lawrence Seaway and Power Projects proceeding now?" Out of the twenty responses to the poll, only four were negative. These sentiments were not restricted to Ontario, as Montreal and Quebec newspapers increasingly supported developing the river, as did most western newspapers that reported or commented on the subject. However, newspapers in Atlantic Canada, such as the *Halifax Chronicle Herald* and the *Glace Bay Gazette,* were less well-disposed toward the project, and there was speculation that Ottawa might be bluffing about going it alone.[25] Nevertheless, all available evidence indicates that, as of early 1949, the majority of Canadians favoured promptly proceeding with a dual project but also appeared to support a power priority plan if necessary.

The growing popular desire for action did not fall on deaf ears. To bring some pressure to bear on the United States, in January 1949, Minister of National Defence Brooke Claxton stated in a public speech at Sault Ste. Marie that St. Lawrence arrangements should be "started as soon as possible and pressed to completion."[26] Claxton's statement signified an important change in the government's approach to the St. Lawrence Seaway, though one that was all but imperceptible to those outside cabinet. The Liberals still strongly desired the combined navigation and power proposal.

But should its prospects fade, they might favour the development of a power project alone and an all-Canadian seaway. In light of Truman's announcement, the Canadian cabinet affirmed that its approach should be based primarily on defence considerations, since "this was considered the language that Congress and American public opinion was best able to understand."[27] The Canadian prime minister used a February 1949 meeting in Washington with the president as an opportunity to apply pressure, and stated in the House of Commons that Canada would have to seriously consider a power-only route if the combined project could not be obtained. The dual-purpose project remained the goal of the Canadian government, St. Laurent said, but planning for the separate scheme was ready to "shift into high gear ... at the first intimation from Washington that hope for the combined scheme is waning."[28]

The persistent ability of Congress to hamstring a St. Lawrence project that had strong presidential support is striking.[29] By the middle of May, Ottawa had received discouraging appraisals of the situation on Capitol Hill, as well as further affirmations of the defence utility of a power and waterway development. At a May 17 meeting of the interdepartmental St. Lawrence committee, a memorandum titled "An All-Canadian St. Lawrence Waterway" was circulated, setting out in detail the case that could be made for a purely Canadian water route. The memorandum began by outlining the disadvantages of the Canada–US agreement of 1941. First, it would be subject to the whims of the US Congress, as had been the case for decades. Second, it would cost Canada substantially more than a Canadian canal and Ontario–New York hydro complex. Even if the United States constructed all the works in the IRS, that would not sufficiently compensate Canada for the other works in the wholly Canadian section, and "Canada would thus, in the end, have paid for the lion's share of a 'joint' waterway."[30]

A waterway built by Canada alone would initially extend only as far as Lake Erie, the memorandum continued, but would be made self-liquidating through tolls unilaterally controlled by Canada. Moreover, immediate action would be possible, since only the approval of the Canadian government was needed. It would cost $575 million to build this project. If Ontario and New York were to pay for the power and common works in the IRS, and taking into account other variables, a waterway completely on the northern side of the St. Lawrence River would cost Canada $210 million, as a low estimate, and $275 million as a high estimate. Proceeding alone would thus carry distinct advantages in almost every respect except initial cost.

Considerable debate among members of the St. Lawrence committee ensued, particularly the extent to which the threat of unilateral Canadian action would hasten Congress. Even though they decided to hold off for a month before relaying the idea to cabinet, the concept of an all-Canadian seaway was now considered a realistic option by the senior members of the Ottawa bureaucracy represented on the committee. A letter from St. Laurent to Truman was dispatched, asking for an expression of the president's view on the probability of "securing early action on the combined St. Lawrence project" and, failing that, his view on separating the power and waterway projects and proceeding with the Ontario–New York plan.[31] In conjunction with his campaign for the Canadian federal election in June 1949, St. Laurent made a series of public statements designed to reinforce the letter's message. Truman replied to St. Laurent's missive a week later, stating that he continued to favour joint action.

"To a Degree Which Might Influence Cooperation"

The motives of the Federal Power Commission came into sharper relief. The FPC announced it would resume consideration of PASNY's application. Yet, just before Christmas 1949, the FPC examiner recommended rejecting the application. Moreover, he advised that a dual-purpose development was more advantageous and that the federal government of the United States rather than the State of New York should exercise jurisdiction over any hydroelectric development on the St. Lawrence. The FPC was supposed to be a politically neutral body, but Canadian officials strongly suspected that the FPC's decision was influenced by a negative recommendation from the Truman administration.[32]

The Truman administration had been leading itself to believe that its legislative effort had a realistic chance at success, though it was equally interested in impeding a separate power development and Canadian waterway, at times misinforming Ottawa about the bill's actual prospects. The president told the secretary of the interior that he "never will agree to the development of the power project without the seaway."[33] Concerns about jurisdictional and governmental control of water resources complicated the American stance. Conservation of natural resources was a priority for the Truman administration, with conservation understood as utilizing natural resources productively so as to promote economic growth and national security.[34] As Truman adhered to a New Deal philosophy regarding natural resource extraction, his administration preferred that the federal government control power development.³⁵ The joint seaway-power scheme's division of costs potentially violated two major elements of the federal government's power policy: first, that power development should be part of the multipurpose and economic development of all natural resources in a given area, and second, that power developed by the federal government should be sold and distributed so as to maximize widespread use at the lowest possible rates, with preference given to public bodies and cooperatives.³⁶

But with Thomas Dewey as governor of New York, PASNY had shifted from the goal of producing and distributing power to a focus on the former, creating an ongoing ideological conflict – between the state and the federal government, and within the federal Department of the Interior – about public versus private development of hydroelectric power.³⁷ Even if New York was granted the right to harvest the power, the Truman administration wanted assurances that it would be shared with New England.³⁸

In January 1950, PASNY appealed the FPC's December decision, and the commission announced that it would undertake a review. Although FPC action remained an obstacle for the prospects for the Ontario–New York power priority scheme, Canada had gone further down the road toward support for the separate power works – and to some degree an all-Canadian waterway. It seemed all but guaranteed that the St. Lawrence dual project could not be approved until 1951 at the earliest, and that too appeared improbable. In early 1950, Guy Lindsay, a high-ranking official in the Canadian Department of Transport, suggested that an all-Canadian waterway might be the best option if the 1941 agreement did not pass Congress.³⁹ This was a weighty endorsement, as Lindsay was the Canadian government's leading engineering expert on the St. Lawrence, having worked on it for twenty years and served since 1949 as the chairman of the St. Lawrence interdepartmental committee. Both Canada and the United States had in the previous decades considered building a deep waterway completely within their respective territories, and the American threat to do so had even motivated Canada to enter into the 1932 St. Lawrence treaty.

When congressional hearings resumed, a number of witnesses, including previously opposed corporations, extolled the virtues of the Labrador (i.e., Ungava) iron ore deposits to the House committee, as did a report from the US National Security Resources Board released during the hearings. According to the board, the Labrador range would be indispensable for meeting America's future needs, given the dwindling Lake Superior supplies, particularly in the event of a national emergency. The St. Lawrence Seaway was deemed necessary to meet these demands and provide a defensible route. The report also emphasized the benefits of the power aspect of the project, concluding that "a prudent regard for national security requires that the power phase, as well as the transportation phase, of the St. Lawrence project be authorized now and that construction be initiated promptly."[40] The Truman administration echoed this by promising to present the St. Lawrence to Congress "dressed in uniform" – that is, to frame the project primarily as a military measure.

The Canadian government was eager to capitalize on the potential of the ore reserves. A report from the Department of Trade and Commerce disclosed that approximately 355 million gross tons had been proven by 1949 and concluded that the completion of a Great Lakes–St. Lawrence waterway and power project would be of great value to both the peacetime economy and national defence.[41] Another lengthy study by the Department of Commerce and Trade also strongly favoured the St. Lawrence development (see Table 2.1), and a report a year later argued for the necessity of the project on economic and defence grounds, either as a Canadian or joint endeavour.[42]

Canadian patience with American delay was wearing thin. The Canadian prime minister indicated this at a June commencement address at St. Lawrence University in Canton, New York.[43] A few weeks later, Ambassador Hume Wrong informed the US undersecretary of state that "failure to act on the St. Lawrence Agreement here by the end of 1951 would cause disappointment and complaint in Canada to a degree which might influence cooperation in other respects." Several Canadian cabinet ministers had also recently gone public with the view that they "might feel compelled to press for priority treatment of the power phase of the project." Secretary of State for External Affairs Lester Pearson underlined the fact that, although Canada still preferred to construct the St. Lawrence works in tandem with the United States, political pressure was building to the point that if the US Congress did not pass legislation during its current session, "such failure would be apt to have an adverse effect upon United States–Canadian relations in the broader sense."[44]

Over the course of the summer of 1950, the Canadian prime minister made a pivotal decision for the future of the St. Lawrence. According to Minister of Transport Lionel Chevrier, St. Laurent had at some point in August or early September told him that Canada should build the seaway alone. "I think," St. Laurent had asserted, "the Americans should be made

TABLE 2.1 Estimate of potential traffic on the canal systems of the St. Lawrence waterway (in thousands of short tons)

	St. Lawrence	Welland	Sault Ste. Marie
Downbound			
Wheat	6,000	7,000	10,000
Other grain	2,200	3,100	3,300
Flour and mill products	2,200	1,900	1,000
Iron ore	—	—	60,000
Iron and steel	1,586	1,000	126
Pulpwood	—	114	793
Soft coal	3,000	4,123	—
Coke	200	48	32
Petroleum and products	50	1,687	4,000
Autos and parts	790	740	—
Fertilizer	75	75	—
All other	2,000	1,500	1,000
Total down	18,101	21,287	80,251
Upbound			
Iron ore	20,000	19,000	345
Paper	850	980	—
Woodpulp	300	300	—
Pulpwood	865	690	—
Lumber	375	100	—
Hard coal	500	56	343
Soft coal	500	30	15,500
Petroleum and products	1,014	475	476
All other	2,000	1,500	2,000
Total up	26,404	23,131	18,664
Grand total	44,505	44,418	98,915

Source: LAC, RG 25, file 1268-D-40, pt. 14 (FP. 1), vol. 6345, Report: "The St. Lawrence Waterway and the Canadian Economy," Department of Trade and Commerce (Economic Research Division), Government of Canada, January 1951, 54.

aware of our determination to get the seaway built."[45] Although there appears to be no explicit directive to this effect, the records of the Canadian government, as well as its ensuing policies, clearly support Chevrier's recollection. At the same time that he announced this new approach to Chevrier, St. Laurent had charged the minister with preparing public opinion in both Canada and the United States for the possibility of Canada going it alone.

Chevrier, not coincidentally the Member of Parliament from the St. Lawrence city of Cornwall, opened his publicity campaign midway through September 1950. The 1941 agreement was the best solution, he claimed, but if Ottawa was convinced that no progress could be made on the combined scheme, Canada should forward the Ontario application to the IJC. Furthermore, the minister stated, "Canada should also explore the possibility of constructing a deep waterway on the Canadian side of the boundary."[46] This was widely reported in the press, but there was plenty of speculation by contemporaries, and subsequently by historians, that Canada was only attempting to bluff or pressure the United States into taking action. Since there were mixed messages emanating from Canada, the United States had reasonable grounds for suspecting that talk of an all-Canadian seaway was just posturing.

Important elements of the St. Lawrence committee, including Chevrier and C.D. Howe (the Canadian minister of trade and commerce, and one of the most powerful politicians in the country), backed a solely Canadian waterway. St. Laurent wanted to proceed pragmatically by preparing the groundwork for a Canadian route without closing the door on American involvement. According to St. Laurent's biographer, if the Americans reacted in time, "then a joint venture would still be possible; if not, they would have had ample warning, and the national pride of Canadians would have been stirred sufficiently to accept the total financial burden of the enterprise."[47]

The all-Canadian seaway proved to be extremely attractive to Canadians on several levels. In fact, for many Canadian citizens, the St. Lawrence symbolized something beyond the sum of its parts: a wholly national seaway epitomized the possibility of the new Canada. After the deprivations of the Great Depression, followed by the trying experiences of the Second World War, a seaway and power project offered a very tangible means of achieving both prosperity and concomitant national self-confidence. The idea of their own seaway resonated with Canadians for practical and symbolic reasons, as it would enable economic growth while providing a valuable addition to Canada's defence capabilities. It also represented a Canadian capacity to strike out independently of the United States. Moreover, the seaway project and its anticipated results were intertwined with a burgeoning faith in the ability of technology, science, and engineering to bring about the "progress" that captivated governments and countries throughout the world.

During the interwar period, the central Canadian provinces had resisted a St. Lawrence waterway because of federal-provincial disputes over water and hydro rights, as well as for partisan political reasons. Quebec Premier Maurice Duplessis, who had traditionally opposed developing the St. Lawrence, remained in office. Opinion in Quebec, which may not have been as opposed to the St. Lawrence project in previous decades as the province's political elites claimed, now seemed to be of the view that Montreal, and the province in general, would benefit from shipping the Ungava ore to the steel factories of the Great Lakes region.[48] The iron ore reserves had led the American Hollinger-Hanna Group, along with the aluminum and automobile industries in the United States, to end its opposition to the seaway. Numerous companies joined with Hollinger-Hanna to form the Iron Ore Company of Canada, and this conglomeration of US interests signed a development deal in 1951 with Duplessis, sufficiently minimizing the political and business opposition to the St. Lawrence project in Quebec.[49]

Ontario was desperate for the benefits of a St. Lawrence development. Much of the country's industrial and population growth was taking place in or near the Great Lakes–St. Lawrence basin, particularly in southern Ontario, and factories required electricity and new outlets to make and move their products. Central Canada, and Ontario especially, remained Canada's dominant manufacturing region. Automobile production, for example, became the largest secondary manufacturing industry in Ontario. In response, Canada invested in a range of transportation networks in the decade after the end of the Second World War, including significant investments in railroad expansion. Since the current means of transportation were operating at full capacity, rail interests in Canada did not actively try to block the growth of water transportation via the St. Lawrence.[50] By way of illustration, in 1947, Canadian railways carried 153 million tons of cargo, and Canadian canals (Sault Ste. Marie, Welland, St. Lawrence) moved 130 million tons.

The economies of Canada and the United States were soaring, though it is all too easy to retrospectively overlook the many worries that the growth and conditions might not last. The intensification of the Cold War led to a massive rise in defence spending by both countries. Helping to fuel the Canadian economy through this postwar growth period were American investments and branch plants. In 1950, 76 percent of the $4 billion of foreign investment that came into Canada was from the United States. In October 1951, the two North American countries signed the Statement of Principles for Economic Cooperation, the first of a number of industrial defence agreements that imported US capital and technology on a far larger scale than had the Hyde Park Declaration. Canada was by this time more comfortable in alliance with the United States than it had been in the past, but many Canadians were also apprehensive about the commitments and perceived loss of sovereignty such US expenditures implied or required. This bolstered a robust Canadian nationalism, some of which took the form of anti-Americanism. The Massey Commission, which was created to investigate the status of Canadian culture, and the subsequent Gordon Report on Canada's economic prospects, spoke to the concerns of at least some segments of Canadian society that resented American encroachment and interference in the Canadian economy and society, and that felt American influence needed to be mitigated. This growing apprehension was articulated in the enthusiasm for an all-Canadian seaway.[51]

National self-confidence, combined with economic growth, technological advancement, and rising expectations in terms of living standards manifested itself in an accelerated exploitation of natural resources in general. Private and public industries developed energy resources such as uranium, petroleum, natural gas, and atomic power. Total hydroelectric output in Canada rose 50 percent between 1948 and 1954, led by developments such as the Kemano-Kitimat project built to power Alcan's aluminum plant and the new Sir Adam Beck 2 power plant at Niagara. Both the American and Canadian states viewed the natural environment, and the St. Lawrence in particular, as a great storehouse to serve society's needs and advancement; it simply needed to be corrected, capitalized, and channelled. But each country also viewed the relationship between environment, technology, and nationalism in different ways. The Canadians had a sentimental attachment to the St. Lawrence, seeing it as a repository of national identity, an impulse not shared by the Americans. Whereas the evolving Canadian nation had in the past focused on the St. Lawrence River for its development, the United States had traditionally ignored the St. Lawrence.

"Injure Our Relations"

Washington had learned by the end of August 1950, at the latest, that Canada was seriously considering abandonment of the dual Canada-US project. Chevrier and Howe maintained pressure on the United States after the November congressional elections, as did speeches by Ontario Premier Leslie Frost and head officials from HEPCO, such as Robert

Saunders and R.L. Hearn. Both Queen's Park and HEPCO continually badgered the federal government, and public opinion in Ontario was decidedly in favour of Canada proceeding alone on a waterway, though a few newspapers speculated that a unilateral declaration to this effect was only a pressure tactic.[52] As Canada's most populous province, its industrial and manufacturing heartland, and home to the constituencies of many key members of the ruling party – including Howe, Chevrier, and Pearson – Ontario's voice carried a great deal of weight.

On December 19, 1950, the FPC rejected the PASNY licence on the grounds that navigation works should accompany a power project and that PASNY might not sufficiently share the resulting hydroelectric power with surrounding states. The FPC recommended instead a combined power/navigation project. The Truman administration's preferences plainly affected the FPC's licence rebuff, which, despite the commission's prescribed insularity from political considerations, was not surprising, since all commissioners were presidential appointees.[53] The impact of political influence is revealed by a range of evidence, including the Canadian record of a conversation between Ambassador Wrong and E. Robert de Luccia, chief of the FPC's Bureau of Power, in which the latter stated that the commission could reverse its December 19 judgment depending on the stance of the White House and Congress.[54] The waters were further muddied by the fact that the US government was in the midst of a review of national water policy via the Water Resources Policy Commission, which called for a number of projects similar to those created by the Tennessee Valley Authority.[55] The FPC rejection was followed by another defeat of the 1941 agreement in the House of Representatives. The all-Canadian project appeared to offer the quickest start on the St. Lawrence project.

Although the Korean War tends to overshadow the St. Lawrence controversy as the pre-eminent factor at the time in Canada-US relations, the latter was taking its toll on North American diplomacy. In 1950, Canada had joined in the Korean War under the auspices of a United Nations force, despite in reality being an American-led mission.[56] The ensuing Canadian "diplomacy of constraint" confirmed the patterns of Canadian-American collaboration and helped convince Ottawa that it had a special ability to shape American policy through "quiet diplomacy."[57] The problems created by the St. Lawrence issue and Korean War prompted Secretary of State for External Affairs Lester Pearson to initiate a "comprehensive internal review" of the bilateral relationship in March 1951. The Department of State initiated its own assessment of Canadian-American relations over the previous decade; it concluded that a "unique partnership" had developed, particularly in terms of Canada's strategic and economic importance to American national security.[58]

In a May 1951 speech titled "Canadian Foreign Policy in a Two-Power World," Pearson foresaw the end of relatively "easy and automatic political relations" with the United States. He indicated that the speech was primarily intended for home consumption and was a warning to the people of Canada that open differences with the United States might develop. This speech is widely cited as a reference to the Korean conflict, but Pearson divulged separately to US officials that the St. Lawrence was an equally major irritant in the two countries' bilateral relations.[59] The feeling was mutual in the US State Department: it had identified the St. Lawrence as the primary problem in the Canada-US relationship about a year and a half earlier.[60] Just days before Pearson's speech, a Canadian diplomat told an American State Department official that Ottawa was "very discouraged" because of the St. Lawrence situation, and the official reported that

> Canadian leaders are expressing the view that unless [the United States] is prepared to cooperate with Canada on this matter which is so important to them that there are avenues of leverage open to the Canadians which they should not hesitate to use which, of course, would result in the diminution of the cooperation which we have been receiving from Canada on various fronts ... a very unfortunate atmosphere on the matter is developing in Canada.[61]

Other warnings soon followed – for example, Pearson told the State Department that the reaction would be strong in Canada if it was felt that the US president "had become responsible for blocking the project."[62] The US Department of State understood that American obstructions could lead to a deterioration in North American relations, cautioning the White House that "intense resentment caused by [the United States obstructing a Canadian development] might well spread into other areas of our relationship with Canada and to some extent jeopardize the essential cooperation now existing between the two countries." If Ottawa was to gain the impression that the United States was "putting the Canadians in a box in which it appears to them that they cannot get the project either through Congressional action or through cooperation with the administration ... it would probably injure our relations with Canada more than any other single incident which has occurred during this century."[63] This might have been hyperbole, but it emphasized what was at stake, and the US undersecretary of state told the president that "the Canadian Government is firmly of the opinion that Canada has a *right* to build the seaway itself ... and that it would be unjust for us to stand any longer in their way."[64] However, the State Department's suggestions concerning Canada and the St. Lawrence file generally fell on deaf ears in the Truman White House, which was fickle in its dealings with Ottawa on the seaway and power project. To be fair, Truman was seeking to advance what he believed to be in the best interests of his administration and his country – as was the St. Laurent government – but his methods would prove to be somewhat disingenuous.

The "unfortunate atmosphere" to which the American official referred was indeed developing in Canada. The St. Laurent government felt strong pressure to pursue a wholly Canadian seaway in order to satisfy popular opinion, a sentiment reflected in the media. In May 1951, the *Globe and Mail* argued that "there is no doubt that Canada can handle the project alone ... Ottawa should lose no time in making such a decision known." The *Montreal Gazette* opined that "recent public declarations on the subject of the St. Lawrence Seaway project have emphasized anew how unwarranted and arbitrary is the persisting blockade of proposals for immediate undertaking of its much needed and quite feasible power phases alone," and the same city's *Daily Star* asserted that "the weight of opinion is in favour of going it alone."[65] Moreover, a survey of Canadian business leaders and industrialists indicated that they were of the same opinion, and a Citizens' Joint Action Committee was eventually formed in Cornwall to lobby for a wholly Canadian waterway.[66]

Over the ensuing summer, Canadian public opinion continued to rally around the idea of going it alone. The *Lethbridge Herald* reported that "the average Canadian is whole-heartedly in favour of the seaway."[67] The *Windsor Star* declared that "Canada is big enough, strong enough, and rich enough to do it alone and should do so," and the *Globe and Mail* contended that a failure to immediately undertake the St. Lawrence project "would be a betrayal of our national needs and our national future."[68] A number of other newspapers were more qualified in their embrace of the project because of concerns about Canada's ability to build unilaterally – as the editors of *Le Nouvelliste* wrote, "The only thing which may make Ottawa hesitate is the cost of the undertaking." But even those stressing careful consideration realized the potential: "The successful completion of the seaway, although it would strain Canadian resources severely, would be a

FIGURE 2.2   Citizens' Joint Action Committee. © *Ontario Power Generation*

tremendous fillip to our national pride and it would be an advertisement of our strength which no other country could ignore."[69] As the feasibility of solely Canadian construction became apparent, many of those who had previously expressed hesitation endorsed the undertaking. A minority of newspapers scattered throughout the country, particularly in the coastal provinces, conveyed opposition to the project, although much of this dissipated over time.[70] A Gallup poll indicated that a majority favoured a

Canadian solution: of the 62 percent of Canadians surveyed who knew what the seaway plan entailed, 55 percent were in favour of Canada going ahead alone, 14.5 percent were undecided, and 30.5 percent were opposed.[71] Over the next few years, the percentage in favour steadily climbed.

Buttressed by popular opinion, and aided by a tentative recommendation from the St. Lawrence interdepartmental committee, the Canadian cabinet spent several months considering the steps necessary for an all-Canadian route.[72] A preliminary Canada-Ontario agreement on the cost distribution of the hydroelectric aspect was reached by early September 1951. The combination of the Canada-Ontario agreement and the apparent lack of hope for the 1941 Canada-US arrangement led the St. Laurent government to finally take the plunge and decide in favour of proceeding on the basis of an all-Canadian project. Some cabinet members preferred a joint binational development, but the majority favoured a Canadian development.[73] St. Laurent went to Washington to ask Truman for his support in advancing a Canadian waterway and the Ontario–New York hydroelectric project. In his memoir of the seaway saga, Lionel Chevrier recalls that "it was amusing to us to see the almost shocked US reaction to our proposals after we had spent many months trying to warn them about our intentions."[74] St. Laurent opened by discussing Ontario's power needs, and offered that his government was ready to ask the Canadian Parliament for authority to begin construction of the seaway as a Canadian project. The prime minister stated that his government "would definitely prefer the approval of the 1941 Agreement" but pointed out that opposition in Canada to constructing a wholly national waterway had "practically vanished" and, to capitalize on Canadian popular sentiment, it would be useful "to get the Canadian people talking about a constructive project of the magnitude of the St. Lawrence development, which might help to prevent overconcentration on their troubles over prices and short supplies." The Canadian prime minister did offer that, if Canada built alone, the Americans could still help with "materials, manpower, and financing."[75]

Truman reaffirmed his strong preference for joint action on the seaway but agreed to support the Canadian plan if Congress was given one last chance to approve the wartime agreement: "My heart's in this [joint] project. I think it is vital. I want the 1941 agreement, but if we can't get that, I want to do the next best thing."[76] Dean Acheson, the US secretary of state, had revealed separately that Truman would not stand in the way of a Canadian waterway, with two qualifications: Canada would need to provide a commitment that it would build a deep waterway in addition to the power installations, and there was to be no discrimination exercised against US ships using a Canadian seaway. Despite this apparent co-operation, the White House's main motive for supporting the Canadian initiative was to spur Congress into taking action on St. Lawrence legislation. This support would prove transitory when American lawmakers failed to act accordingly.

The conservative press in Canada, which favoured an independent waterway, attacked St. Laurent for refusing to stand up to the United States – "the lure of Yankee dollars" had led the prime minister to "save dollars which nobody asked him to save."[77] According to historian Dale Thomson, "Chevrier's carefully planned campaign to stir up public support for the project was almost too successful, and threatened to boomerang."[78] After stoking the Canadian public's desire for an all-Canadian seaway, there was considerable political danger in abandoning it for a bilateral undertaking. Nevertheless, Truman had promised to support a Canadian seaway if an early beginning on the cooperative alternative proved impossible. Ottawa was optimistic. It appeared to be just a matter of time before a St. Lawrence project would become a reality. Moreover, on the way back from Washington, the prime minister stopped in Toronto and Quebec City for productive talks with Premier Frost and Premier Duplessis.

"This Agreement Anticipates an All-Canadian Seaway"

Canada immediately sought to capitalize on the momentum. In addition to taking the necessary internal legal steps toward the goal of an all-Canadian deep canal, over the next year Ottawa focused on convincing the Truman government to join in making submissions to the IJC for a separate New York–Ontario hydro development. After further discussions, and Canadian indications that they intended to carry on alone, the White House "intimated that they were ready to proceed on the basis of Pearson's statement that an early beginning of the joint project could *not* now be achieved and that the United States would cooperate with Canada in having the seaway built by Canada alone."[79] But the Canadian and US governments had different interpretations of what constituted an "early beginning." East Block officials believed that it meant immediately, whereas the US authorities wanted more time to evaluate the congressional prospects.

Constitutional impediments concerning the State of New York necessitated changes to the pending Canada-Ontario agreement, which was formally signed on December 3. Under this agreement, the Canadian share of the power produced from the International Rapids section of the St. Lawrence River would be managed by Ontario. The province would handle power works, with the central government responsible for any navigation facilities, and the terms of the agreement spelled out the various stipulations for the exchange and compensation of land between the federal and provincial governments.

The day after the conclusion of the Canada-Ontario agreement, two acts were introduced into the House of Commons – the first to ratify the just-inked accord with Ontario, the second to create a federal St. Lawrence Seaway Authority – with the expectation that American participation in the waterway would not be forthcoming. Within a few weeks, both pieces of legislation received the unanimous assent of Parliament, though they would not be enacted until 1954, demonstrating the wide-ranging support that had developed in support of Canada going it alone. In the apt words of author Carleton Mabee, previous opposition "had melted away, in anticipation of watching a newly adult Canada remake the continent without Uncle Sam's aid."[80] The St. Lawrence Seaway Authority Act incorporated a federal agency with the responsibility to build and operate, either wholly by Canada or in conjunction with an appropriate authority in the United States, a deep waterway between Montreal and Lake Erie. The seaway authority was given the right to establish and charge "fair and reasonable" tolls, designed to make the authority's expenditures self-liquidating.

The bills provided concrete and definite steps toward the long-coveted St. Lawrence scheme, but they also exhibited a national confidence in the country's ability to single-handedly carry out a task of such immensity. According to Lionel Chevrier, "Going ahead with the seaway was notice to the US that we were now a first-class power. Some of my colleagues felt it was high time the US was made aware of this."[81] These sentiments resonated with Canadians and their elected representatives on an almost unprecedented level: all private members who took part in the parliamentary debate expressed approval of the pending bills, and the leaders of the opposition parties proclaimed unanimous and enthusiastic support. One Member of Parliament from the Prairies hoped that Canada could "prevent the Americans getting their hands on the seaway at all. I would rather we sacrificed our chance at hydro power on the river to keep the seaway all-Canadian." Another wanted "to see the US being shoved out of the limelight." Outside Parliament, public opinion also clearly embraced the Liberal bills. A US State Department memorandum written in 1952 succinctly summarized the popular sentiment: "Canada's decision to build the St. Lawrence Seaway as an all-Canadian project has seized the imagination of Canadians. It is a symbol of their new-found strength."[82]

With the necessary legal mechanisms in place, the Canadian government set out to take the remaining required steps to obtain the hydroelectric and navigation development anticipated by the St. Lawrence legislation. A formal note was forwarded to the United States on January 11, 1952. It requested American cooperation in preparing concurrent applications to the International Joint Commission, with the caveat that "such a preparatory step would in no way prejudice the possibility of proceeding with the project on the basis of the 1941 Agreement in the event that Congress should approve that Agreement."[83] In reply, the US government confirmed that, if the 1941 agreement was not approved at an early date, it was prepared to cooperate in preparing concurrent applications to the IJC.

The White House, however, was anxious to forestall precipitous Canadian action until at least the end of March 1952, lest the Canadian plan "be regarded as a club over the head of Congress." There was still confusion about whether the FPC licence would need to precede an IJC application, and it remained unclear whether PASNY would have to make a new application to the FPC. Although this drawn-out process was undoubtedly a partial result of genuine legal difficulties and grey areas, the American government's procrastination was chiefly responsible for the delay. To President Truman, the exchange of notes "represented both a hope and an unpalatable possibility."[84] The hope was that the threat of Canadian action and the sanction of the IJC would lead enough recalcitrant members of Congress to approve the 1941 agreement, while the unpalatable possibility was that this could backfire if congressional support remained unattainable and the president needed to either approve the Canadian plans or take measures to impede them.

The Canadians too were attempting a balancing act. Canada intended to proceed alone, but since it needed the cooperation of the American federal government to surmount the remaining legal hurdles, the Liberal government had to at least appear willing to allow US involvement. Chevrier made it clear in a national television address that "we are not closing the door on United States participation in the seaway. That participation is still the logical, the desirable, choice. But it is results that count."[85] But Ottawa's calculated hedging on its commitment to proceed unilaterally gave Washington reasons to believe that the St. Laurent government genuinely preferred to construct in tandem. Indeed, in the American government and media there were references to Canada's position as a bluff, based on the presumption that Canada could not afford to build independently, or was threatening to do so mainly for domestic political purposes.[86] Further confusing the situation were several contradictory remarks made by Canadian officials.[87]

The most important objective from the Canadian point of view was submitting the power applications to the IJC. The Canadian ambassador petitioned the US secretary of state to "ask the President if he did not agree that prospects of favourable congressional action were remote and hence that full reliance should be placed now on the second alternative of a Canadian seaway."[88] After a week without a reply from Truman, apparently because the president's subordinates had not passed on the entreaty, Canada backed down from its request. When it was finally brought to the president, Truman offered that the Canadians had been very patient and cooperative and that the executive branch of his government should follow through on its undertakings.[89] He thought that the Canadian hope for authorizing action no later than May 1 was reasonable, stated that he was now prepared to designate PASNY as the American entity for hydroelectric development, and agreed to see Lester Pearson and Lionel Chevrier a week later.

At that meeting, Pearson indicated that the Canadian reference to the IJC on the hydro aspect would not include the Canadian waterway, though he was prepared to give the most "definite assurances" that his government would go ahead with the seaway. Revealingly, Pearson added that "it was only in the past couple of years that [the Canadian government] had realized what a great nation Canada had become and that before they hadn't dreamed of building the seaway alone."[90] Truman consented to the request for the simultaneous forwarding to the IJC of the Canadian and American power applications, though the president then delayed the application a few weeks further, avowing that this was the last postponement.[91] From Truman's perspective, IJC approval would remove the power aspect of the St. Lawrence project from congressional consideration, which could mute some of the domestic objections, and the United States had secured a guarantee from Canada that it would build a seaway in conjunction with power works if an American seaway role could not be obtained. However, securing American involvement in the seaway was ultimately why the president went ahead with the IJC applications, for he hoped that the realization that Canada was committed to proceeding alone would jolt Congress into action.

Officials from both countries met repeatedly to hash out an IJC application compromise, which was achieved by an exchange of notes on June 30, 1952. That same day, both countries submitted applications to the IJC.

However, largely because of the FPC's wishes, the United States declined to name its power-developing entity. Canada reluctantly gave in on a few issues that had been sticking points in the lead-up to the IJC application: how to handle the existing fourteen-foot-long Canadian canals that would be affected by the new waterway and hydro development, and a $15 million contribution toward the costs of dredging to be undertaken in the international section by the power-developing entities.[92] These would later prove to be thorns in the Canadian side, as would the decision that had been made a few days previous to refer concerns about Lake Ontario water levels to the International Joint Commission.

Conclusion

The protracted IJC applications seemingly represented a partial victory for the prospect of an all-Canadian seaway, but at the expense of a great deal of time, effort, and compromise. The drawn-out process stretched back to 1950, when the Canadian prime minister first authorized serious consideration of a solely Canadian route. The St. Lawrence dispute had become as significant and controversial an issue in the bilateral relationship as the Korean War. Truman had been officially apprised of the Canadian intention to pursue a separate course but failed to follow through on his pledged cooperation by adopting delaying tactics, such as refusing to designate an American authority to partner in the construction of the hydro works, and championing the 1941 agreement at the expense of the Canadian plan. This interference stemmed from several interrelated concerns for the Truman administration: reluctance to let a state entity develop the hydroelectricity, preference for the navigation aspect of the project ahead of the power development, fear that the waterway might be unduly delayed, and worries about the economic and defence repercussions for the United States if Canada alone controlled the seaway.

By 1952, the conception of a wholly Canadian initiative as simply the quickest available option for getting the St. Lawrence Seaway started had been superseded by the idea of the Canadian seaway as the best option, regardless of whether or not US involvement was possible. Thus, Canada had taken a watershed decision that it hoped would culminate in an all-Canadian waterway, along with a bilateral hydro development. Key cabinet ministers began to prepare popular opinion on both sides of the border for the possibility of Canada acting alone. The idea of an all-Canadian seaway tapped into various forms of Canadian nationalism, for the St.

Lawrence River held a pre-eminent perch in the Canadian pantheon, and the idea that the great stream could once again serve as the catalyst for nation building captured the national imagination. Canada proceeded on those matters within its control, including an agreement with Ontario for the construction of the power works in the International Rapids section and for the enactment of legislation to ratify this agreement and establish the St. Lawrence Seaway Authority. Eventually, the Canadian and American governments agreed to make joint submissions to the International Joint Commission for the power development. The Truman administration hoped that the threat of an all-Canadian waterway would motivate Congress to pass St. Lawrence legislation; the St. Laurent government hoped that an all-Canada seaway could recapture the empire of the St. Lawrence.

3
Caught between Two Fires

For Canada and the United States, the early 1950s were generally a period of amicable relations and rapid economic integration. Canadians looked to their southern neighbour for protection, leadership, and prosperity. The two countries were staunch allies; few neighbouring states enjoyed such cordial relations. The relationship had, however, hit a stretch of turbulent waters because of the ongoing St. Lawrence impasse. The seaway was inextricably bound up in the Liberal government's postwar nation-building goals. The project represented progress. It would provide access to hydro power while allowing for the movement of crucial economic and defence commodities such as wheat and ore. An all-Canadian seaway offered even more: it was seen as the means of fulfilling the age-old dream of the St. Lawrence.

In June 1952, Canada and the United States had submitted applications to the International Joint Commission for an Ontario–New York power project. Though that application was approved later in 1952, over the next two years Canada would diligently pursue an all-Canadian seaway but find its path repeatedly blocked by American obstacles, particularly those connected to the Federal Power Commission licence required by New York for the hydro development. The replacement of President Harry Truman by Dwight Eisenhower gave reason for hope, but in the end, a solely Canadian waterway was too much of an economic and security threat to the United States. Despite the widespread support across the country and within the government for a Canadian seaway, by the summer of 1954, Canada had reluctantly acquiesced in negotiations for a shared waterway; though the nation was desperate for the benefits of the St. Lawrence project, especially the hydroelectricity, the St. Laurent government realized the damage that would be done to bilateral relations if it proceeded alone. Nonetheless, during the ensuing talks that produced a joint St. Lawrence Seaway agreement, Canada managed to extract some important concessions.

Engineer as Hero

Various ideas concerning nationalism, environment, and technology underpinned the Canadian government's approach to St. Lawrence discussions. Revealingly, many of the officials actively taking part in shaping Canadian diplomacy were themselves involved in designing the physical shape of the seaway and power project. Ottawa's political and diplomatic approach to the seaway was based on the idea that developing the St. Lawrence for navigation and hydroelectricity was the most efficient, rational, and productive usage of the river basin. The waters would otherwise run wasted to the sea. Guy Lindsay and R.A.C. Henry, key figures on the St. Lawrence interdepartmental committee, were professional engineers in the Canadian Department of Transport. Two of the other leading officials in the internal Canadian debate about the future of the St. Lawrence waterway, C.D. Howe and General A.G.L. McNaughton, both had engineering backgrounds. Howe was at the helm of the Canadian megaprojects of the 1950s and personified the "engineer as hero" – the title of a chapter in a book commemorating Canada's war effort that was financed by government contractors and dedicated to Howe.[1] McNaughton, Howe, Lindsay, and Henry were all strong advocates of an all-Canadian seaway, and the phrase "bureaucrat as hero" – a uniquely Canadian construct that historian Alan MacEachern has applied to Ottawa's environmental management in a different context – seems equally apropos for this coterie of engineers.[2]

These technocrats conceived of the St. Lawrence as a resource to be harnessed and exploited to further Cold War security needs and large-scale industrial development. In this line of thought, mass societal displacement in the St. Lawrence Valley was a small price to pay for the production of hydroelectricity and the increased accessibility of iron ore deposits. Flooding out thousands of people was justified in the name of progress, as defined by the national government, and for the good of the nation, which seemed to be synonymous with those economic interests that would profit from cheap iron ore and electricity. The reorganization and resettlement of those affected by the power development would be for their own benefit. They would be placed in consolidated new towns – instead of scattered about in inefficient villages, hamlets, and farms – with modern living standards and services. Such plans imposed state-defined political, economic, and social values, and enabled the Canadian state to control how these communities fit into the emerging postwar order so that they could be more fully integrated into the dominant political and industrial capitalist structures.

There are a range of opinions on the extent to which Canadian views of the environment were formed in opposition to, or in alignment with, the United States, but the St. Lawrence case suggests that Canadian-led development of natural resources offered a means, real or perceived, of escaping American domination and exploitation. Hydro power in the twentieth century promised to deliver Canada from its "hewer of wood servitude to American industry and its bondage to American coal."[3] This was recognized in a 1952 memorandum by the US State Department, which stated that "there is a latent uneasiness [in Canada] about the extent to which US capital is joining in the recent development of Canada. The idea seems to be that Wall Street is muscling in on the Canadians' birthright and they will be left as 'hewers of wood.'" More specifically, there was "some criticism of the St. Lawrence Waterway as permitting Labrador iron ore to move to the States."[4]

The duality, or ambiguity, between water and the environment generally as a means of domination by the United States, or of escaping such domination, is also germane to the various views of the link or dialectic between nationalism and technology in modern Canadian history, discussed by prominent intellectuals such as Harold Innis, George Grant, Marshall McLuhan, and Ramsay Cook.[5] These scholars arrive at different interpretations of the interface of technology and nationalism, yet all agree that the two are inextricably entwined. The materialist view of nationalism has been described thus: "The requirements of industrial society introduce a measure of cultural homogeneity that is coincident with the unifying call of nationalist sentiment."[6] The notion carries important ramifications for the Canadian case. Taking the perspective that technology can be considered both a cause *and* an effect of nationalism, the argument that "technological nationalism has characterized the Canadian state's rhetoric concerning identity" is extremely persuasive and is applicable to St. Lawrence nationalism.[7]

Technological nationalism also corresponds to what historian R. Douglas Francis has identified as the "technological imperative" in Canada. Francis contends that in the early twentieth century, "technology had a negative identity when associated with the United States and American imperialism," since "technology was seen as instilling American values into Canadian society that were antithetical to traditional British Canadian morality. Technology was also seen as a source of power that had enabled the United States to dominate Canada and, through American imperialism, to control the entire world."[8] However, technology (and the modernity it represented) was, as Francis deems it, a "double-edged sword" because Canadian access to modern technology and, by extension, the environment, coupled with changed conceptions of technology in Canada over the first half of the twentieth century, held out the potential for the nation to evolve independently of the United States.

The evolution of the St. Lawrence development sustains Francis's contention that, by the post–Second World War era, there was a strong and unique link between technology, nationalism, and sovereignty in Canada, though these associations may well have been forming even before the twentieth century. The seaway and power works, as well as other contemporary transportation projects, had the potential to serve as nation-building parallels to the late-nineteenth-century transcontinental railways, or at least the resulting mythologies linked to the railways. Like these railways, which themselves had in many cases supplanted canal systems, the seaway could serve as a means of promoting and facilitating Canadian identity, national unity, progress, and prosperity while linking the country in an east-west orientation, in contrast to the north-south pull of the United States.[9] Moreover, a seaway carried on the Canadian state's penchant for coping with the country's challenging spatial reality – its vastness and environmental conditions – by heavily subsidizing or building large transportation networks that would most directly benefit private business and industry, even if they were framed as benefiting the nation as a whole. Although the unique Canadian intermingling of identity, environment, and technology certainly drew on the past, changing technological and industrial conditions in the post-1945 era were equally altering the way Canadians interacted with their natural environment, which in turn had important ramifications for Canadian conceptions of themselves and their nation.

McNaughton, through his roles on the International Joint Commission and Permanent Joint Board on Defence, served as a formal and informal interlocutor for the Canadian and American governments and agencies concerned with the St. Lawrence issue. The national security elements of the St. Lawrence project had brought it under the purview of the board, a bilateral mechanism for frank discussion of continental defence that had recommended a St. Lawrence dual project on a number of occasions since 1945. McNaughton was a strong nationalist who held a string of impressive and influential appointments, had extensive experience dealing with the Americans, and boasted a background as a hydroelectrical engineer. In fact, he had been selected for the chairmanship of the Canadian section of the IJC by St. Laurent, on the advice of Mackenzie King, primarily to help with the St. Lawrence impasse.[10] To the stronger Canadian nationalists, including the likes of Howe and McNaughton, US involvement in the St. Lawrence project and the negation of a solely Canadian route would yoke Canada more tightly into the American harness and further subservience to US foreign policy. An all-Canadian seaway promised the opposite.

"God, Himself, Couldn't Move the FPC"

President Truman had allowed the IJC applications to go forward only after there seemed to be no other recourse and at the urging of the State Department, which reminded him that failure to do so would jeopardize relations with Canada. However, by the end of 1952, the president saw a Canadian development as better than none at all. Truman was not running in the upcoming presidential election and realized that there would be no opportunity for Congress to authorize the joint agreement during his remaining time in office. Given his dedication to the seaway concept, perhaps Truman saw the Canadian scheme as the only remaining means of having a final St. Lawrence agreement attached to his legacy as president.

Nevertheless, the president's willingness to cooperate would prove to be short-lived. Within a week of the 1952 IJC applications, Prime Minister St. Laurent wanted to let the United States know that "we can no longer regard Canada as bound by the 1941 Agreement once we proceed with detailed engineering work on the all-Canadian alternative and that we are anxious to get on with this work as soon as possible."[11] However, he decided not to do so right away in light of an April 1952 "eleventh hour" appeal by Truman to Congress for passage of St. Lawrence legislation, and Truman's letter to congressional leaders calling the June 1952 application a "poor second best" option for the St. Lawrence undertaking. The prime minister was still anxious to make the Canadian position clear "without jeopardizing the whole project" and, accordingly, the Canadian ambassador gave the State Department a hint of what was in store by relaying that it was "becoming progressively more difficult for any reversion to the joint project."[12]

Others in the United States kept alive the hope of congressional authorization and thus a bilateral seaway. In mid-July, Dr. N.R. Danielian resumed his role as an American emissary, meeting with Canadian officials Hume Wrong, Lionel Chevrier, and Lester Pearson. Wrong, the Canadian ambassador to the United States from 1946 to 1953, told Danielian that the Canadian public had embraced the all-Canadian plan and that too much money, time, and personnel had already been committed for Canada to revert to a joint plan. There was no point in seeking congressional approval because the door was closing on an internationalized waterway.[13] Although Canadian officials remained willing in principle to consider a joint project, they would "resent" any delays or difficulties if the IJC was stalled in order to give Congress more time to take action.[14] Since there did not appear to be any scenario in which the United States could become involved without delays or difficulties, the Canadian officials were trying to hint that they had decided to proceed alone. Toward that end, government officials had avoided emphasizing, or mentioning entirely, the joint project in public utterances. Canadian diplomats couched blunt assertions in judicious language for fear that the Canadian approach would be frustrated if Ottawa did not at least appear willing to countenance American involvement. Internal Department of External Affairs memoranda were more straightforward: a joint project was now "impossible" and the seaway entirely on the Canadian side of the border was "the only way to proceed" to avoid delay.[15] Nevertheless, Danielian left Pearson with the "uneasy feeling" that the IJC hearings would be delayed in order to "give American supporters of the international scheme another opportunity."[16]

Throughout the autumn of 1952, the Canadians continued to receive conflicting indicators of the Truman cabinet's stance. On August 12, the PASNY application, which had been moved to the US court system after the 1948 licensing attempt, was remanded to the FPC. This had been done with the FPC's cooperation, but there were plenty of potential difficulties remaining, both from within and outside the power commission. In a conversation with Secretary of State Dean Acheson, President Truman disavowed his ability to influence the commission, stating that "God, Himself, couldn't move the FPC."[17] In truth, the White House would repeatedly demonstrate that it could move – or stall – the FPC when it was beneficial to do so. Danielian ominously warned Gordon Cox of the

Canadian Department of External Affairs that "certain major organizations ... are mobilizing to oppose Federal Power Commission action" and said that there was a "slim chance" New York would obtain a licence.[18] Danielian proceeded to name the various groups, organizations, and individuals collaborating to block the New York licence. This list included several officials from the Truman cabinet. Since the FPC process called for lengthy appeal periods, opponents of a licence, such as the Association of American Railroads and the National Coal Association, also had further means of delaying the licence even after the FPC had granted its approval.

The issues complicating the FPC matter affected the International Joint Commission. The chief engineer of the FPC, Roger McWhorter, was also the chairman of the American section of the IJC. McWhorter reportedly wished "to save the FPC the embarrassment of having to rule on this contentious item [the PASNY application] and therefore prefers to keep it in the IJC as long as possible."[19] In early September, McWhorter suggested adjourning for up to two months. The chairman of the US section of the IJC also made it clear that the St. Lawrence development would only be approved "over his dead body" unless the Gut Dam was removed.[20] This low barrier in the upper St. Lawrence River just downriver from Prescott and Ogdensburg was blamed by Lake Ontario shore dwellers for high water levels. Canada agreed to this, provided it would not prejudice navigation, power, and other interests downstream.

Although McWhorter privileged partisan interests, contrary to the role of an IJC commissioner, General McNaughton was doing the same as chairman of the Canadian section. McWhorter claimed to oppose approval of the dual Canadian-American applications on account of the cost distribution. Nevertheless, the other five members of the IJC had been convinced – it would seem in large part because of McNaughton, who was determined to force the applications through – that the IJC had no jurisdiction over the cost allocation, and they were willing to approve the application on the understanding that the US power-developing agency would be named later. Thus, on October 29, 1952, the IJC issued an order of approval on the joint hydro power applications by a 5-1 margin. Predictably, McWhorter was the lone dissenter, citing the allocations, lack of US input on future St. Lawrence tolls, and a general loss of US sovereignty if Canada proceeded alone. Nonetheless, IJC approval of the hydro works represented a major achievement in the history of the St. Lawrence development.

IJC applications had been predicated on the understanding that both governments would withdraw their applications if Congress approved the 1941 agreement; from the Canadian perspective, the 1941 agreement and the approach it represented (i.e., shared international waterway) was in turn invalidated, since the IJC had sanctioned the power applications. The Department of External Affairs identified several logistical, political, and legal reasons that made proceeding with the 1941 Roosevelt-King agreement undesirable and virtually impossible. For example, in its years of American legislative consideration, the agreement had undergone significant modifications and differed considerably from the original form to which Canada had agreed.[21] So long as there appeared to be two alternatives, the opponents of the development of the navigation and power resources of the St. Lawrence would endeavour to play one off against the other and thereby thwart the development itself. But since IJC approval had, in theory, made the 1941 agreement defunct, the approval had removed this divide-and-conquer possibility, as well as the FPC's grounds for denying New York a licence. The St. Laurent cabinet unanimously agreed that Canada should, before Truman's successor was known, signal its intent to abandon the 1941 agreement.

The cabinet verbally informed the State Department and the president on November 3, the day before the election, which it followed with official notification on election day. The note to Washington reviewed Truman's commitments to advance the Canadian waterway, communicating that Ottawa had concluded that "it would no longer be practicable to revert to the terms of the 1941 Agreement or to place that Agreement before Parliament for approval. The Canadian Government, therefore, considers that Agreement as having been superseded and does not intend to take any action to have it ratified."[22] Truman accepted the note, stating that the Canadians had proven themselves patient and were entitled to this approach.[23] At virtually the same time, Republican candidate Dwight D. Eisenhower was elected as the next president of the United States. Both State Department and Canadian officials initially believed that his election augured well for action on the FPC licence. But the day after the IJC announced its approval, the Federal Power Commission had suddenly and unexpectedly announced that the period for interveners in the PASNY licence application, which involved a lengthy process that included external consultations and an initial assessment by an examiner before the application went before the entire commission, had been extended another month, until December 1, 1952. This postponement, likely at the instigation of the US Department of the Interior, effectively delayed the hearings until later in December; the FPC then recessed until March 1953.

Why did the FPC avoid taking action? In addition to the various groups and individuals seeking to pressure the commission into denying New

York a licence for hydro power in the International Rapids section, the available evidence indicates that the FPC members, as presidential appointments, were also protecting themselves. Key members of the incoming Eisenhower cabinet had been canvassed and it turned out that many supported a joint project.[24] Eisenhower himself was undecided and noncommittal about his views toward the St. Lawrence project. He was not convinced of its security value or necessity, but his only statement since his nomination as the Republican candidate the previous July suggested that he did not think that the United States should forgo its interest in the waterway. The FPC commissioners thus had a strong motivation to delay until the president-elect's preferences were more clearly expressed.

The FPC postponement provided a "golden opportunity" for those in the United States committed to an American share in the deep waterway, which now included the leaders of the powerful US automotive industry. Canada's November 1952 withdrawal from the 1941 agreement did not preclude the possibility of a new binational agreement for a joint plan, and the US government did not seem to realize – or chose not to – that the Canadians had attempted to disavow *any* joint plan, not just those based on the 1941 agreement. Accordingly, in early 1953, congressional legislation was introduced calling for a new St. Lawrence accord between the United States and Canada, and congressional leaders requested that Canada "leave the door open."[25] Collectively, the Great Lakes–St. Lawrence Association and other similar interest groups threw their weight behind the new congressional attempt to effect American involvement in the seaway.

For their part, the Canadians were debating the repercussions of informing Eisenhower that "any further delay on the part of the United States in facilitating construction of the power project would have very damaging results for Canada-US relations," since Ottawa had already "taken irrevocable steps toward the construction of the seaway."[26] Nevertheless, Canada felt compelled to understate its seaway commitment in an attempt to avoid prolonging the FPC licence. There did not seem to be much alternative to this approach, and the drawback was that it could easily backfire.

A Louder Knock

Several events in the United States coalesced to further complicate the prospects for an FPC licence. In his remaining time as president, Truman supported the new congressional attempts at passing St. Lawrence legislation. Before having it scuttled by the Department of State, Truman had even started to formulate a letter to St. Laurent advocating a joint project.[27] Just before Christmas, the National Security Resources Board promoted the seaway as a defence requirement in its annual report, calling for the president to request Congress to appropriate funds for federal participation in both the seaway and power projects. Although US officials told their counterparts in Ottawa to disregard the National Security Resources Board report as irrelevant and the work of Democrats on their way out of office, important and influential groups in the United States were clearly pursuing the joint waterway with renewed vigour. Ottawa was cognizant of the American ability to circumscribe Canadian plans: "We feel we need not open the door yet to new negotiations for joint development of the sea-way ... [though if we] conclude later that we will need to open the door[,] we might wait for a louder knock."[28]

A considerably louder knock on the door was soon heard. The Canadian government learned on January 2, 1953, that President Truman, still in office until the end of the month, would use his budget message to exhort Congress to authorize American participation in the seaway and appropriate funds for the US share of the cost. Truman believed that "if the new Congress proposes practical arrangements for sharing the cost and construction of the seaway ... Canadians will, even at this late date, admit us to partnership in the seaway."[29] Canadian diplomats tentatively asked where the president had received such an impression, considering their recent indication that the 1941 agreement had been superseded. The American embassy replied that it had never been disabused of this notion by any high Canadian official.[30]

Canada was under no legal or formal responsibility to consider these new US proposals. But a failure to reply might give the wrong impression, and certain American interests would, in Lester Pearson's words, "find some way of delaying the present arrangements for power until we were prepared to revert to a joint scheme for the seaway as well."[31] Shortly thereafter, US Senator Alexander Wiley disclosed his own willingness to interfere with the FPC for the sake of American participation, throwing a thinly veiled warning at Canada: "I would propose delay in the Federal Power Commission licensing, if I felt that the Canadian Government would not allow joint participation in the seaway." But Wiley, who represented Wisconsin, was "so convinced of the fair-minded and friendly attitude of that splendid government toward the United States" that he felt that "even after such a licence is granted for the power project, our good neighbors to the north will definitely leave the door open for joint participation by the United States."[32]

The Canadian government rushed to determine a response that could be timed to coincide with Truman's budget speech, scheduled for January 9. They agreed on a conditional offer to the United States that maintained the precarious position of attempting to leave the door open just wide enough to placate the Americans, hopefully leading them to grant New York a partnership role with Ontario in developing the International Rapids section. When the US ambassador saw St. Laurent and expressed Washington's hope that the Canadian government would allow the possibility of future participation by the United States in the seaway project, the prime minister read out the text agreed on by cabinet:

> While the Canadian Government is, of course, prepared to discuss, in appropriate circumstances, joint participation in the seaway, the demand for power in the area to be served by the International Rapids power development is so urgent that the Canadian Government is most reluctant to engage in any discussion which might delay the progress of the plan now underway for the development of power in the International Rapids section of the St. Lawrence River at the earliest possible moment.
>
> Once an entity is designated and authorized to proceed with construction of the United States share of the power works, if the United States wished to put forward a specific proposal differing from that put forward by the Canadian Government for the construction of the seaway in the international section which proposal would not delay the development of power under arrangements agreed upon in the exchange of notes of June 30, 1952, and approved on October 29, 1952, by the International Joint Commission[,] the Canadian Government will be prepared to discuss such a proposal.

The note concluded with the warning that "the Canadian Government would naturally expect the discussion to be such as not to cause any serious delay in the completion of the whole seaway."[33]

US seaway proponents embraced the Canadian government's January 9 statement, but it became clear that they were "studiously ignoring" or misinterpreting the key points in Canada's press release – specifically, words such as *reasonable, specific, earliest,* and *delay* – that placed parameters on future US involvement in the potential waterway.[34] Moreover, the new legislative proposals that outlined an American seaway entity and authorized the financing, negotiation, and construction of a US canal and two locks in the upper St. Lawrence – named the Dondero and Wiley proposals after their congressional patrons and introduced into the House of Representatives and Senate, respectively, in January 1953 – were designed to win over opponents and fence-sitters by outlining a much smaller American financial commitment than had previous bills and bilateral agreements, removing the development of St. Lawrence hydro power from the aegis of the federal government, and omitting works outside the IRS. Previous St. Lawrence bills had provided for deepened navigation channels throughout the Great Lakes, but by restricting American participation to the IRS, and with New York absorbing the costs of hydroelectric production, Senator Wiley's bill brought the cost of American participation from over half a billion dollars to about $100 million. In addition to the Wiley and Dondero bills – which did not call for a treaty or agreement with Canada but were instead a unilateral declaration that the United States would share in the construction in the IRS – several other congressional resolutions were introduced and debated throughout January. These would eventually fall by the wayside in favour of the Wiley-Dondero resolutions, which when passed would form the act enabling the United States to build the Wiley-Dondero Canal and its two locks near Massena.

Eisenhower's opinion on the St. Lawrence project and the New York power licence remained an enigma. The famed Second World War general and NATO's first Supreme Allied Commander Europe took over the Oval Office on January 20, 1953, maintaining a noncommittal stance in the initial months of his presidency, failing to mention the seaway in his State of the Union address, and omitting it from his list of essential legislation for that session of Congress. But by the end of March, there was a consensus in the Eisenhower cabinet that it should take a stand in favour of the St. Lawrence Seaway because American economic and defence interests would suffer badly if the project was left to Canada alone.[35] However, the president wanted to wait until after the Senate Committee on Foreign Relations hearings slated for mid-April were finished before making a final determination of policy.

Eisenhower was influenced by both political and defence considerations, though the precise reasons for his disinclination to make a decision are a matter of conjecture. It may have been calculated indifference in order to avoid appearing beholden to special interests, and recent works have argued that, to avoid overhasty decisions, Eisenhower pursued a strategy of ambiguity and bluffing.[36] Certainly a pattern of that type is discernible within his handling of the St. Lawrence. Moreover, with congressional mid-term elections approaching, a seaway agreement was the type of accomplishment that could frame the Republicans as a dynamic choice in the minds of the electorate. Cost questions and the president's view on water policy also factored in, as Eisenhower wanted to limit federal expenditures on the St. Lawrence as much as possible and philosophically supported power construction by entities other than the federal government (e.g., PASNY), as his administration favoured partnerships between federal and state/local, and public and private, when it came to water developments.[37]

In terms of Canadian-American relations, the Eisenhower era brought with it no basic change, and the two countries continued to solve most bilateral problems quietly and informally. For instance, the two countries exchanged notes in 1953 to establish a Committee on Trade and Economic Affairs to consider matters affecting a harmonious economic relationship. In the recollection of one official who represented the United States in IJC deliberations with Canada, Eisenhower told him to "go fifty-one percent of the way ... When you're trading with those Canadians, be so fair that you could move on their side of the table and feel comfortable in your bargaining."[38] Economic and defence integration proceeded apace, and the White House and State Department carried on the approach of treating Canadian nationalism as an element that occasionally needed to be humoured, since Canada occupied a strategic Cold War position, both geographically and in terms of natural resources.

FPC hearings resumed in early February 1953. Yet, by the end of April, no ruling had been made. This excruciatingly slow pace was caused partly by the commission's rules of procedure and other items on the FPC's docket, but was chiefly attributable to Eisenhower's delaying the appointment of the vacant chairmanship of the FPC. The various and competing St. Lawrence resolutions introduced into Congress also motivated the FPC to prevaricate, and congressmen kept alive resolutions that muddied the waters because of the impact this had on the commission.

The Canadians were palpably upset with American temporizing. Minister of Transport Lionel Chevrier and Secretary of State for External Affairs Lester Pearson conveyed this in speeches at Port Arthur and in Parliament respectively. News of Chevrier's and Pearson's statements only seemed to further antagonize the FPC chairman.[39] On March 21, Pearson sent a letter to Dulles asking for an immediate start on the project, pointing out its potential impact on many of the critical materials supplying the defence industries of the two countries.[40] Pearson offered that the quick commencement of the hydro phase would not prejudice "whatever arrangements may be mutually agreed upon for the development of the deep waterway," calling for a seaway to be built by Canada or "under mutually agreeable arrangements by both our Governments."[41]

Canadian discontent with the situation then took the form of an explosive April 7 speech by C.D. Howe in New York. He excoriated the Americans: "The apparent unwillingness of your government to extend the small degree of cooperation required to enable Canada to proceed with this project puzzles us completely ... Canada's desire to further improve its outlet to the ocean can be, has been, and is being frustrated by lack of cooperation action by your Congress." The St. Lawrence waterway was, "and always has been, a Canadian seaway. Every important improvement has been built and paid for by Canada, from Lake Erie down ... Why then, should your country withhold its cooperation and thus delay completion of this vital Canadian transportation outlet?" In Howe's estimation, the recent American canal proposals would "only complicate the present situation," as "ownership by the United States of a short section of a very long seaway would not only add to the overall construction cost, but would complicate problems of maintenance and operation of the canal system." In conclusion, it was "obvious" to the Canadian minister "that continued ownership by one national authority of the entire seaway represents the most efficient procedure."[42]

Contrary to usual practice, Canadian officials had not forewarned the State Department, which was "flabbergasted" by such blunt language and interpreted it as a change in the position of the Canadian government.[43] In fact, there had been no change in Canadian policy, just a modification of tone. Howe was attempting to maintain the delicate position of leaving the door open just enough while moving ahead with the Canadian seaway and the hydro project. The minister had also reputedly heard that an FPC chairman friendly to the PASNY application was about to appointed, and as a result he felt he could speak freely without antagonizing the FPC.[44] There may also have been a measure of truth to the Department of State's speculation that, with a federal election upcoming in August 1953, Howe was trying to appeal to his northern Ontario constituents, who might "take a strong attitude regarding the US attempts to muscle in on the St. Lawrence Seaway."[45] Nonetheless, given Howe's status within the cabinet – and with St. Laurent out of the country on a forty-two-day world tour, he was also acting prime minister at the time – this speech must have been a calculated pronouncement.

The FPC remained at a 2-2 deadlock. There was one remaining vote on the five-person commission, that of the chairman, but his term had expired (though he was continuing to act as the chair on an interim basis). Hence, the person chosen by the Eisenhower cabinet as the next chairman would effectively cast the vote determining whether PASNY would be licenced. On April 22, 1953, the president finally appointed Jerome Kuykendall to the chairmanship of the FPC, a position he would not assume until May 15. The appointment augured well for the future of the PASNY licence, since Kuykendall was known to be favourably disposed toward it. The day after the appointment, Eisenhower stated publicly that his cabinet endorsed the idea of the St. Lawrence project for national security reasons and supported New York's participation with Ontario in creating the power works in the IRS. A National Security Council report had just given the project a clear endorsement, which, along with the support of the Joint Chiefs of Staff, joined the Permanent Joint Board on Defence's reaffirmation of its recommendation to develop the St. Lawrence.[46] Eisenhower was already impressed by the self-liquidation principle, which would reduce the cost of the project for the United States. The president requested that the report be forwarded to the current FPC chairman for "such action as he may deem appropriate," indicating the president's desire that the FPC grant the licence.[47] Eisenhower also suggested that, if Congress refused to allow American participation, his government would remove all obstacles to Canada constructing the seaway alone.

Dealing with Eisenhower

On May 7, St. Laurent and Pearson went to Washington to meet with Eisenhower. There was little direct deliberation about the St. Lawrence issue. The summit focused more on establishing a good personal rapport between the two leaders and dealing with trade and agricultural issues. As St. Laurent hoped, Eisenhower was ready to support the PASNY application, but the president also stressed his interest in making the St. Lawrence project a joint one and asked Canada again to wait longer. The prime minister emphasized that his government needed to get on with the job as soon as possible but would be willing to discuss American participation provided it did not result in further delay.[48] The day after the St. Laurent-Eisenhower meeting, the US cabinet subcommittee that had been formed to study the St. Lawrence issue reported in favour of American involvement in the seaway. The national defence argument sufficiently impressed the president, and the Eisenhower cabinet adopted US participation as policy, provided there was a Canadian commitment to complete the navigation works in its territory, and predicated on a self-liquidating project.[49]

A few days after the Eisenhower administration came out in support of an American seaway role, the examiner for the Federal Power Commission recommended that PASNY be granted a fifty-year licence. Although the entire FPC would still need to provide formal approval, this was now a near certainty. It was no coincidence that once the White House announced its support for the licence and American participation, the FPC quickly complied. However, both Ottawa and Washington knew that this licence would be appealed by various groups and interests, resulting in lengthy delays. As expected, in early June, four groups filed appeals against the FPC licence, and the Eisenhower administration remained unwilling to name New York officially as the developing entity until all the litigation was cleared away. The FPC continued its deliberations on the PASNY licence throughout June and into July.[50] Canadian officials were dismayed by the slow pace of the proceedings, but this time it was the statutory requirements of the Federal Power Act in the United States that made it necessary for the FPC to follow time-consuming procedures. On July 15, 1953, the FPC rejected the remaining appeal and, subject to certain conditions, approved the PASNY licence. Three days later, the US Senate decided to put off further consideration of the St. Lawrence legislation until January 1954. This meant that Congress would not have the chance to approve US involvement in the seaway for the remainder of 1953. Consequently, Canada had a potential half-year window of opportunity in which it could move ahead alone on the seaway, provided that the final technicalities surrounding the FPC licence could be dealt with quickly. This news was widely welcomed in the Canadian press, which played right into the hands of the Liberals, who were re-elected with a majority on August 10, 1953.[51] The widespread support for the all-Canadian waterway only added to the governing party's popularity.

But there were still routes for appealing the FPC's decision – the parties had sixty days to file a petition for review with the US Court of Appeals. Realizing they would need the cooperation of the American bureaucracy to hasten the appeals process, Canadian officials sought to bring pressure to bear by implying that continental defence might suffer.[52] Meeting separately with high-ranking American officials, C.D. Howe and Lester Pearson stressed the need for St. Lawrence power for Canadian industries, which in turn were vital to American defence concerns. Howe, the Canadian minister of trade and commerce (as well as defence production), made sure to point out to US attorney general Herbert Brownell that nearly one-half of Canada's production of defence goods came from the area that stood to gain from St. Lawrence electricity, and that these various goods and minerals were in turn indispensable to American defence production and undertakings such as northern radar lines.[53] Pearson, the Canadian secretary of state for external affairs, drew the attention of US Secretary of State John Foster Dulles to the potential ill-will that could develop in the Canadian populace as a result of additional delays, adding the threat that the Canadian public "would accuse the Government of a lack of diligence but [that] more importantly there would be a feeling developed against the United States which might be of such a nature as to make Canadian cooperation in fields such as continental defense more difficult."[54]

The National Security Council had said much the same thing in its April 1953 report. It argued that American defence could not be conducted independently of Canada, since the two "constitute a single defence unit," and warned that public opinion in Canada "increasingly supports the construction of an all-Canadian Seaway, partly as a symbol of developing Canadian nationalism." Furthermore, delay would serve only to harm bilateral relations that "in many fields – economic, political and military – have been close and harmonious and it is not in the US interest to damage them by delaying, in any way, a development of direct economic and political importance to the Canadian Government." Canada's view of the St. Lawrence as "the most important single facet in its relations with the United States" was confirmed by the State Department, which recommended prompt and favourable action by the FPC – otherwise the United States ran the risk of jeopardizing Canadian concurrence on key continental and strategic issues.

The overtures seemed to have an effect. Within a week of the Pearson-Dulles meeting, the US Department of Justice had adopted the Canadian suggestion and petitioned the Court of Appeals in the District of Columbia to hold hearings on December 15.[55] In early November, Eisenhower formally designated PASNY as the entity that would construct the American share of the power works and established the US section of the St. Lawrence River Joint Board of Engineers with a Canadian-American exchange of notes. This cooperation was a volte-face from what the Canadians had experienced in the past with the White House, though ascribing this assistance to anything other than national self-interest would be misplaced.[56] Even if the methods of expediting the legal process proved successful, the appeals would not be cleared away until some point in 1954.

By that time, Congress would have had another chance to approve administration-sponsored legislation for a joint seaway. Thus, expediting the process was also in the interests of the Eisenhower administration because it believed that the end result would be a hastening of a joint Canadian-American St. Lawrence project.

The Canadian window of opportunity elapsed. The Wiley bill quickly began gaining speed in the Senate in early 1954. The combination of the administration's support, iron ore prospects, the national security argument, and the lower cost to the United States (compared with the 1932 and 1941 agreements) was vital in acquiring the necessary congressional votes. National defence justifications came from important quarters, including the Joint Chiefs of Staff, which believed that "the concept of unilateral control by a foreign government, however cordial our relations may be, of an inland waterway touching the borders of the United States is inconceivable ... from a defence standpoint."[57] Yet, the chief reason was the knowledge that Canada was going to go it alone. Just as an all-Canadian seaway had seized the nationalist imagination in Canada, so too did the idea of the great river being controlled by the Canadians alone make many patriotic Americans balk. On January 20, 1954, after five previous Senate attempts since the 1930s to authorize an American seaway role, the Wiley bill was approved 51-33. The House of Representatives still needed to approve the companion Dondero measure, but there was considerable momentum leading in that direction, and passage of the Wiley bill was hailed as a breakthrough achievement by seaway supporters in the United States.

As had been the case on several occasions going back to early 1953, the United States asked the Canadian government for its input on the Wiley legislation.[58] Some of its principal parts sounded unacceptable from the Canadian perspective: it limited US seaway construction expenditures to $105 million, leaving Canada to swallow a much larger amount, and it put all the locks in the IRS on the American side. Ottawa refrained from expressing its reservations, however, so as to avoid implying a tacit acceptance of US involvement in general and of the Wiley-Dondero legislation specifically; the Canadian government did not want to commit itself "in advance by either rejection or acceptance of suggested forms of United States participation."[59] This was reinforced by a Canadian cabinet directive issued in January to its embassy in Washington instructing Canadian diplomats not to discuss anything connected to the Wiley bill with US officials.[60] Since the Wiley bill was the last chance for an American seaway role, the Canadians feared that they would be accused of acting in bad faith because their objections would be equivalent to a denunciation of American participation, which would be "irreconcilable with the much needed assistance of the United States Administration to end the pending litigation on the FPC's order."[61] The resulting silence led the US government to assume that Canadian sentiment was in favour of joint action and that the US legislation was in principle acceptable to the Canadian government.[62]

US Secretary of State Dulles, however, realized that "opinion in the Canadian Government is hotly divided on the issue of joint construction of the seaway, as against an all-Canadian seaway. C.D. Howe and Chevrier are leaders of the 'all-Canadian' sentiment and they are supported by a significant group in the Government, in Parliament, and in the Canadian Press." On March 15, Pearson gave a speech at the National Press Club in which he stated that an international waterway was preferable.[63] Yet, in a CBC interview, Howe said that a Canadian seaway was the best option.[64] Media stories about Canada's intent to push for an all-Canadian seaway were leaked or planted so that the Americans would hopefully get the message without the Canadian government having to be the messenger. A Canadian Press story quoted a cabinet source as saying that Canada was going to push for its own seaway regardless of what Congress did, though if the United States insisted, Canada might have to yield, since "we would rather do that than wreck the project."[65] The source sounded suspiciously like C.D. Howe or Lionel Chevrier, and the story was picked up by numerous media outlets. On the other side of the border, the noted columnist Walter Lippmann wrote that it would be better if Canada built the seaway alone, as it would help minimize the number of international complications. Lippmann argued that "there are strong reasons for believing that Canadians would not be disappointed, but on the contrary relieved, if in fact Congress decided to drop the joint project and to let Canada proceed alone with the seaway."[66] Given that Lippmann had discussed the article with the Canadian embassy during its drafting, it was safe to assume that it reflected the Canadian position.[67]

Pearson later characterized the Canadian government's situation as "caught between two fires."[68] Howe and Chevrier had important and strong allies in the interdepartmental seaway committee, including McNaughton and Henry. The Department of External Affairs, well-attuned to the need of maintaining good relations with the United States, expressed the least reluctance toward a joint project. The consensus that had earlier existed when a joint project seemed unlikely was dissolving, particularly in light of the assurances about an American role that had been previously extended to Washington. But disclosing this rift to the Americans in such obvious ways – through diplomatic correspondence and a prominent public statement by the minister of external affairs – was as much a tactical move to play down the Canadian desire to go it alone as it was the manifestation of an internal cabinet struggle. This approach was also aimed at cushioning the blow to the public if a bilateral project was the final result.

Those of different political stripes in Canada excoriated the federal government for considering US involvement. Canadian newspapers, particularly in Toronto, decried the prospect of the United States "buying in cheap" and renewed their clamour for an all-Canadian seaway.[69] The leader of the Labor-Progressive Party – as Canada's Communists styled themselves – wrote in the *Ottawa Citizen:* "The people of Canada are profoundly shocked to learn that the Government, without consulting Parliament, has entered into an under-the-table deal with the United States to scuttle the long-promised all-Canadian Seaway in return for short-term power concessions" and asked, "Is our historical Canadian river, the mighty St. Lawrence, to become a Yankee Canal? Are we to abandon the course of Canadian nationhood charted by the Fathers of Confederation? Is our country to be sold into bondage?"[70] Even though this could be dismissed as Marxist propaganda masquerading as Canadian nationalism, the central points of the critique, particularly that the St. Lawrence was a *Canadian* river, were shared across the political spectrum. For example, the conservative *Globe and Mail* lamented that the Canadian government had made a "well-calculated gamble, barter[ing] its chances for an all-Canadian seaway in return for a broad guarantee from the US Government to push approval of the power phase of the St. Lawrence project." According to the newspaper, this "deal" had "backfired to the extent that it is now virtually certain that Congress, opposed to any kind of seaway participation at that time, will now approve construction of the widely publicized canals and locks around the International Rapids."[71]

Keeping the Door Open

The door was about to be kicked wide open. The Court of Appeals for the District of Columbia Circuit denied the last petition for a rehearing of the PASNY case on February 19, 1954, though there was still a ninety-day period in which the appellants could petition the Supreme Court. After being tied up for two months in the House Rules Committee, on May 6,

1954, the Dondero bill was approved in the House of Representatives by a vote of 241-158. A week later, the Senate had voiced its approval and the president had affixed his signature. Combined with the Wiley bill, the US government had the authorization to participate in certain parts of the seaway.

At a meeting of the Canadian St. Lawrence interdepartmental committee a few days before the passage of the Dondero bill, McNaughton (now a part of the committee despite his role in the IJC) had argued that it "did not appear that there was a formula for joint construction of the Seaway which Canada could accept without prejudicing our national life," considering that the St. Lawrence had "traditionally" been developed by Canada.[72] Others concurred, but there was widespread agreement that the international repercussions of Canadian action along these lines would have to be studied.[73] Ambassador A.D.P. Heeney strongly disagreed with McNaughton and Henry and their allies, since the ambassador was under the impression that there had been an informal agreement within the government to consider any US proposal. According to Heeney, "Canada cannot refuse such an invitation to negotiate without exposing itself to charges of bad faith that would have far-reaching consequences for relations between the two countries."[74] Heeney then revealingly elaborated on the nationalist motivations for, and potential bilateral repercussions of, an all-Canadian seaway:

> I appreciate that such a decision will cause keen regret in many quarters in Canada. Ever since Champlain labelled it on one of his maps "La grande riviere du Canada," the St. Lawrence has been, and has been considered, an essentially Canadian river; and if recent events had turned out differently, there would have been wide satisfaction in the construction of an all-Canadian seaway ... But I do think that our reputation in the United States would be gravely tarnished if we refused to enter into negotiations with the United States looking towards the completion of a seaway in which some of the canals would be on the United States side of the river and some on the Canadian side. Our reputation would also suffer, I think, if we did not make an honest effort, in the course of such negotiations, to work out cooperative arrangements with the Americans that would be both workable and fair.[75]

The prime minister, who had kept relatively silent over the previous months, and a sufficient number of cabinet ministers shared Heeney's view about the need to hear out a US proposal. It was decided that, in the interests of staying in Washington's good graces and obtaining a start on the power project, the Canadian government was obliged to keep the door open to American involvement. However, in doing so, there was a reluctant admission that consenting to negotiations would likely result in the jettisoning of the all-Canadian alternative. Thus, the same day that the Dondero bill was passed in the House of Representatives, the Canadian prime minister reluctantly stated in the House of Commons that his government remained willing to discuss any specific proposal the United States wished to put forward, once an entity was designated and authorized to proceed with construction of the US share of the power works, and provided that the resulting discussions did not delay either the power or seaway project.

In the first week of June, the US Supreme Court denied the request to review the PASNY licence, which meant that the appeals had apparently ended. However, there was one complication: applicants could, within two weeks after rejection of the application, apply for a rehearing. But the court was rising for the summer the very day it denied the request; thus, there would still be a possibility of an application for rehearing within the two weeks after the court resumed in September. US authorities asserted that this was only a technicality. This it may have been, but the timing was calculated to maintain the threat of stopping the power project if Canada proved incompliant.

The State Department sent to Canada on June 7 a formal proposal to join in the St. Lawrence undertaking. However, the US proposal consisted simply of a short diplomatic note, with the Wiley bill attached, announcing the bill's passage and the creation of the St. Lawrence Seaway Development Corporation to direct US construction of its share of the St. Lawrence project.[76] It quickly became apparent in Ottawa that this did not constitute the requested "specific proposal." Nonetheless, the St. Lawrence interdepartmental committee was generally of the opinion that, though Canada could still decide to go it alone during negotiations, they were committed to at least entering into discussions, particularly with the Supreme Court rehearing threat dangling over their heads.

Interdepartmental committee discussions addressed an idea that had been circulating for several months: entering into negotiations with the United States in order to provoke a breakdown or ensure futile discussions so that Canada could then claim that it had no choice but to build alone. This approach was termed "productive disagreement."[77] A counter-argument was that there was no conceivable situation in which the terms of the Wiley-Dondero Act, which had been partially modelled on the

Canadian St. Lawrence Seaway Authority Act in order to provide compatibility, would be so unacceptable from Ottawa's point of view that Canada would be obliged to go ahead with an all-Canadian seaway. Moreover, this line of argumentation continued, even though Canada could break off talks if the results proved to be unacceptable, the government could not insincerely enter into these discussions with the aim of disrupting them. The cabinet believed that the American offer "is the best we can expect at this stage and that it offers a sufficient basis for going ahead."[78] Pearson therefore told the United States that Canada did not accept the note as a specific proposal but assumed that it indicated the US intention of preparing a specific proposal, and that Canada would be willing to consider this, provided it occasioned no delay.

Government-to-government negotiations ensued. The first round of talks was scheduled for July 5-6, 1954, with the purpose of preparing an agenda for an intergovernmental meeting. From the perspective of both countries, the real purpose of these Ottawa meetings was to feel each other out. Both sides were pleased with the general tone characterizing the discussions. Nevertheless, it quickly became apparent that the stated Canadian desires could not easily be met; the Americans realized the equity of the Canadians requests, but the Wiley-Dondero Act provided neither the authority nor the funds for compensating Canada. Department of External Affairs official Max Wershof advised that the Canadian government "should understand and accept the fact (unfair as it may be) that asking the United States to go back to Congress [to change the Wiley and Dondero bills] is like asking them to go to the moon."[79]

To the consternation of the Americans, the Canadians were considering the construction on the Canadian side of a future twenty-seven-foot canal in place of the existing fourteen-foot canals and were unwilling to guarantee that they would not, at a later date, duplicate the American navigation works. Canada hoped for a guarantee ensuring that it would be able to use the American canals on terms no more onerous or restrictive than was granted to American vessels, but for security, immigration, and customs reasons, the US contingent doubted that this could be achieved, since the required clauses might be unconstitutional. Nor could the United States give any assurances about favourable treatment of third-party shipping coming to or from Canadian ports, which was extremely important because of Canada's reliance on foreign shipping and commerce. In addition, a question that dated back several years was again raised in both countries, particularly the United States: Should the locks and canals be deeper than twenty-seven feet?

Although General McNaughton was not a part of the official Canadian delegation, he played an important role, meeting informally with American Deputy Secretary of Defense Robert Anderson.[80] McNaughton emphasized the difficulties, because of Canadian public opinion, that might ensue if the Americans stuck to too rigid an interpretation of the Wiley-Dondero measures. McNaughton warned Anderson that unless Canada was given the freedom and flexibility to build on its side, there would be strong repercussions, including the collapse of the joint negotiations.[81] Furthermore, he hinted that Canada would be more disposed to an agreement if the United States met the current Canadian demands of $15 million for common works and compensation for the loss of the fourteen-foot St. Lawrence canals. McNaughton's intervention evidently played a part in convincing the US government that Canada was not going to back down on certain points.

After the intergovernmental meetings, the St. Lawrence interdepartmental committee reviewed the advantages and disadvantages of both a Canadian and a joint waterway. There appeared to be many drawbacks to a bilateral waterway, including the recent assertion by Canadian engineers that construction would be cheaper on the Canadian side; in fact, both countries claimed that costs would be lower on their side of the river. From an engineering, economic, and navigation perspective, the committee deemed it better for Canada to go it alone, as indicated by a long list of the disadvantages of a joint seaway in the cabinet conclusions.[82] Moreover, Ottawa certainly had ample grounds for arguing that the Americans had not presented the requisite "specific" proposal, that the negotiations would delay the power project, and that Canada could by all rights proceed by itself.

The committee contended that there was really only one advantage to a joint project aside from having the United States absorb a portion of the cost: harmonious Canada-US relations. Going it alone would "cause very serious harm to our relations with the United States, particularly after the heavy effort that had been made to secure Congressional approval for what Americans considered as a cooperative venture."[83] This was the determining factor for the St. Laurent cabinet, which believed that it had the upper hand in terms of bargaining position, since the Wiley bill required assurances that Canada would build the connecting parts of the seaway outside the IRS, which might allow Canada to extract concessions during deliberations.

The next round of bilateral discussions was set for the second week of August, but in order to clarify some of the outstanding legal issues,

FIGURE 3.1   Groundbreaking ceremony in 1954. © *Stormont, Dundas, and Glengarry Historical Society, courtesy of Cornwall Community Museum*

particularly tolls and the rights of Canadian vessels in future American canals, Canadian and American officials met in Washington on July 23, 1954. The results were discouraging for both governments. Frustrated, the US delegation attempted to scare the Canadians by threatening that an all-Canadian seaway could not be built without the concurrence of the United States, pointedly remarking that Congress could at any time terminate the licence issued to PASNY.[84] Nevertheless, within a few weeks, ground had been broken on the joint power project.

This meeting prompted another round of discussions within the Canadian government about whether it was worth breaking off or stalling the negotiations and proceeding alone with a deep waterway. A consensus was beginning to form around the idea, apparently forwarded by Howe, of building one of the locks in the IRS in Canadian territory at Iroquois, and reserving the right to build twenty-seven-foot canals and locks on the Canadian side of Barnhart Island as indemnification, even though American officials had proven unreceptive to this plan.[85] Putting the lock at Iroquois Point could, the Liberals hoped, partially appease the Canadian public by taking a step toward an all-Canadian seaway, a consideration that exerted a strong influence on the St. Laurent government's approach to the negotiations. On July 28, 1954, the St. Laurent cabinet resolved that Canada should build at Iroquois first, which it hoped would result in the Americans accepting this as a *fait accompli*.[86]

"This Is What We Can Get Away with Politically"

After decades of discussion, the fate of the St. Lawrence was to be decided over the course of a few days of intergovernmental meetings in August 1954. The chairman of the Canadian delegation was Secretary of State for External Affairs Pearson, and he was joined by other ministers, including C.D. Howe and Chevrier, along with an impressive team of Canadian mandarins. The American contingent was led by Deputy Secretary of Defense Robert Anderson and made up of other high-ranking American officials. The object of these meetings was to find a mutually acceptable formula for modifying the existing international arrangements, the June 1952 Canada-US exchange of notes and IJC applications, which specified that Canada would build all navigation elements in the St. Lawrence. Aside from the overarching question of whether the seaway would be jointly built, the points of contention between the two federal governments were basically the same as in their July meetings: regulations and conditions of navigation for Canadian and third-party shipping in canals and locks to be constructed by the United States; the placement of all these locks in the International Rapids section on the American side; the continuation of fourteen-foot navigation on the Canadian side and compensation to

Canada for the destruction of its existing fourteen-foot facilities; and the $15 million payment that was to be made to the power entities for channel dredging beneficial to both power and navigation.

During the August 12 morning session, both countries reiterated their previous positions. Anderson indicated that the United States could perhaps, over the long term, repay Canada the $15 million dredging expenditure out of tolls. However, this was only a possibility, and in this scenario the tolls would need to be joint, and that ran counter to the Canadian preference for unilateral tolls. Turning to fourteen-foot navigation, Howe explained that the canals in the international section had been on the Canadian side for the last century, industry had been built up around them, and both the 1941 St. Lawrence agreement and the 1951 Canada-Ontario agreement provided for the maintenance of a fourteen-foot canal. As matters stood, Ontario would probably have to pay about $15 million to reimburse the federal government for the flooding of these older canals. Anderson replied that the United States was cognizant of this problem but could not go beyond its legislation. After further inconclusive talks, it was agreed to return to the matter at a later point. Pearson opened the discussion of navigation rights for Canadian vessels by foregrounding their importance, proffering a draft navigation treaty, and declaring that "if this were granted it would be infinitely easier to get Canadian public opinion fully to support this project."[87] The Americans appreciated Canadian apprehensions on this score but had serious doubts whether the United States could enter into "an inflexible treaty whereby the power of Congress to legislate would be circumscribed."[88] Even if Congress could be moved to approve such a treaty, it was unlikely to be constitutional. Canadian officials continued to stress their desire for a guarantee of Canadian, and third-party, navigation rights, but to no avail.

Thus, Pearson read out what the US side termed as the Canadian "haymaker": an *aide-mémoire* in which Canada informed the United States that it would modify the exchange of notes of June 30, 1952, by building all the navigation works from Lake Erie to Montreal, with the exception of the two locks in the vicinity of Barnhart Island – meaning that Canada would build the lock at Iroquois.[89] The *aide-mémoire* had been prepared with the expectation that the United States would be unable to meet the key Canadian demands. As Pearson explained, Canadian control of Iroquois would put the St. Laurent government in a better position to assuage Canadian public opinion, build an all-Canadian seaway in the future, and absorb the costs of channel enlargements and the replacement cost of fourteen-foot navigation. Howe was more blunt, stating, "We have come to the conclusion that this is what we can get away with politically."⁹⁰ American officials were taken aback by this proposal and initially declared that the *aide-mémoire* was phrased in language clearly directed at Canadian public opinion, believing it to be "unsuitable in an intergovernmental exchange."

Breaking for lunch, the American contingent huddled together to decide on its next move. It agreed that Canada had the right to build the lock in the vicinity of Iroquois, as did the United States. The Americans returned to the table in the afternoon with suggested modifications, and the two groups began to use the *aide-mémoire* as the basis for an agreement. When the United States expressed doubts and questioned the seriousness of Canada's intentions, Howe sharply retorted that his government intended to build immediately at Iroquois. A small group of American and Canadian members had also conferred to produce a draft consultation clause that was to substitute for a navigation rights treaty, and both sides found it generally agreeable during the afternoon discussions. Canada agreed to provide the assurances required by the Wiley bill – to build the remaining seaway works outside the IRS – as concurrently as possible with the power project, though the method for doing so would be worked out in the future.

As a result of the Canadian "haymaker," by the end of the first day, most of the substantive issues had been agreed on, and an agreement for a joint seaway had largely been hammered out. The United States would participate in the seaway but would not go beyond the Wiley-Dondero strictures, whereas Canada had indicated its resolve to build at Iroquois and take steps for a future Canadian deep-draught waterway. It had already been agreed that this new St. Lawrence accord would be cemented by an exchange of notes, and during the evening of August 12, the Canadian delegation gave the United States a note based on this new understanding. The talks continued the next day on the basis of the Canadian draft note. Progress was made in the afternoon of August 13 on reconciling the draft notes from both sides, but it was apparent that this would take more time to settle. Hence, the American contingent left three members in Ottawa to jointly sort out the specific language with their Canadian counterparts. This took several days, with the notes formally exchanged on August 17. In contrast to a treaty, an exchange of notes would not need the sanction of the Canadian Parliament or American Congress. In the notes, both sides recognized the rights of the other to duplicate facilities within its own territory in the IRS but agreed to consult each other in advance. Promises of consultation were also exchanged regarding any changes to

FIGURE 3.2 Seaway signing ceremony in 1954. Prime Minister Louis St. Laurent is in the centre. © *Stormont, Dundas, and Glengarry Historical Society, courtesy of Cornwall Community Museum*

third-party shipping within each country's respective navigation installations, and it was understood that both countries would endeavour to avoid placing unreasonable restrictions on the transit of passengers, shipping, or trade in the IRS. Since it was receiving no compensation for its existing canals, Canada reserved the right to install fourteen-foot navigation in the future but promised to notify the United States if tolls were instituted. The US reply did not agree to, but simply noted, Canada's intentions regarding Iroquois, fourteen-foot canals, and the potential for a future all-Canadian seaway.[91] The power produced at the Barnhart Island Power Dam would be split evenly, which McNaughton later called a "poor bargain," since Canada gave up the right to claim at Barnhart the extra water (and thus additional power production) resulting from the Ogoki and Long Lac diversions.[92]

On August 18, the two countries announced their agreement. The Canadian government had agreed to downplay its intentions regarding Iroquois and a future all-Canadian seaway in its diplomatic note, but a press release provided the St. Laurent government with a forum for domestic consumption in which it could be more forthright and elaborate on its intentions for Iroquois.[93] At a press conference, Pearson tried to put a positive slant on the joint waterway by boasting that Canada would be spending three times as much as the United States. Canada, as a good neighbour, had "agreed" to let the United States take part in the project and Ottawa had "wisely" held on to the long-term prospect of an all-Canadian seaway.[94] The tone of the press releases and conferences caused consternation on the American side.

Press criticism of the Canadian negotiating results were apparent even before the two governments had finalized their agreement with the exchange of notes, and the St. Laurent government was inundated with letters critical that Canada had allowed American involvement.[95] The *Globe and Mail* asked why Canada had backed down and entered into an agreement making Canadians "humble servants" or "grateful apprentices" who had, like Esau, "sold the nation's birthright for a mess of pottage."[96] Disappointment about the joint seaway was the prevailing mood, particularly in central Canada. At the same time, the public opinion backlash proved to be less severe than the Liberals had feared. The wound was salved by a mixture of excitement that the St. Lawrence Seaway had finally been approved, and cognizance that the reality of the Canadian-American relationship likely dictated that the project would be a cooperative one, that Canada had done very well in negotiations by retaining the possibility of a future national waterway and asserting its right to build the Iroquois works, and that the agreement was in many ways a reasonable outcome.[97] Over time, press and public opinion became less critical of the joint project and gave the federal government some credit for standing its ground in negotiations. For example, the editors of the *Ottawa Journal* asserted in October 1954 that the idea of an all-Canadian seaway in light of American opposition was "plain idiocy."[98]

Nevertheless, there was no doubting that the idea of an all-Canadian seaway had seized the public imagination. Department of External Affairs officials wrote several memoranda rationalizing the reasons for which Canada was embarking on a joint project.[99] For example, Canada had not tried to go ahead with a Canadian seaway because

> once the people and Congress of the United States resolved to participate in the seaway by building works in the international section, a Canadian design to "go our own way" and to build now the Barnhart works in Canada without any regard to the decision of the United States, however late in the

day that decision may have been taken, would have been a blow to good relations, and would therefore have done damage to Canada's broad national interests.[100]

This statement underscored the unique nature of the Canadian-American relationship as the reason for Canadian accommodation: if Canada had proceeded alone, it would have entailed repercussions for the country's broad national interests. In a speech in the United States two weeks after the August 1954 St. Lawrence accord had been settled, Pearson too made reference to the abandonment of an all-Canadian seaway as being in the national interest.[101] It is clear that the Canadian government viewed the national interest in this context as maintaining good relations with Washington, which were threatened by a Canadian attempt to build its own seaway in the face of American opposition. Another departmental memorandum, though couched in the language of a special bilateral relationship, contended that proceeding with the all-Canadian seaway would have been a serious affront to the United States, and "such an affront would surely have caused long-run damage to our national interest; we are intimately bound not only by tradition but by so many day-to-day and practical circumstances, to the destinies of our powerful and friendly neighbour."[102]

Conclusion

The relationship that the two countries shared in the early Cold War shaped the manner and method in which the St. Lawrence matter was settled, and in the wake of the joint agreement, there were numerous statements across the border about the long tradition of friendly and cooperative Canadian-American relations in which disputes were settled by consultation and agreement. In the oscillating tension between conflict and cooperation that characterized the transnational relationship, the St. Lawrence issue had taken conflict to its limits, threatening to spill its bounds into a full-blown crisis. The insulators that had been over the years built into the Canadian-American relationship through the many cross-border contacts at the personal and cultural level, the intertwining of continental defence, and the Canadian economic dependence on the United States helped prevent St. Lawrence negotiations from escalating out of control. To be sure, Canada had been willing to go it alone and would have done so if not for the insistence by successive American administrations that the United States should have an equal controlling interest in the seaway. In the end, the St. Laurent government had consented to American involvement chiefly because a refusal promised negative ramifications for the Canadian-American relationship.

This bilateral August 1954 agreement was much less comprehensive and coherent than its predecessor compacts – one Department of External Affairs official called it a hodgepodge – as it was restricted to the International Rapids section of the St. Lawrence River, did not cover navigation in the entire Great Lakes–St. Lawrence basin, and left several important questions unresolved, including which country would be responsible for major elements of the seaway and power project. These would have to be dealt with during the construction of the St. Lawrence project. The less binding form represented by the exchange of notes reflected the fact that Canada and the United States were really agreeing to construct two separate canal systems that were intended to work together. Each country would have a seaway authority build its portion of the deep waterway, financed through tolls, while Ontario and New York would jointly construct the power works. The placement of many engineering elements, such as locks, had also evolved and changed because of engineering, financial, or political concerns. No deep-draught vessel would be able to transit through the entire IRS without using at least one lock in each country, but the locks would nevertheless technically remain under separate national control. Canada would administer the Welland Canal and its eight locks, four locks in Quebec, and the one lock at Iroquois, whereas the United States would build two locks south of Barnhart Island. The Americans would therefore control only two of the seven locks in the St. Lawrence River, and only two out of the fifteen locks in the entire seaway. This was a substantial difference from the previous seaway agreement, as was the cost distribution. The 1954 agreement did not give Canada credit for the cost of the Welland Canal (over $130 million) as had the 1932 Great Lakes Waterway Treaty and the 1941 Great Lakes–St. Lawrence Basin Agreement. For the navigation works, Canada would pay $336.5 million and the United States $133.8 million, for a total cost of $470.3 million; including the cost of the power phase and other parts, the bill for the entire project was over $1 billion (all US dollars). The United States would have equal control of the seaway, despite contributing only about one-third of the cost and with only a portion of the whole seaway within American territory. The St. Laurent government had attempted to atone for this loss by insisting on building the Iroquois Lock on the Canadian side and by retaining the right to build an all-Canadian route at a future date. The river's transformation was at hand, and the two countries were about to grasp its power and alter its irresistible current.[103]

PART 2
Building

4
Fluid Relations

The St. Lawrence Seaway and Power Project was a complex and highly integrated navigation, power, and water control project. It turned the St. Lawrence Valley into a hybrid waterscape that blended the mechanical and the organic, with hydro dams and locks the means by which the river could be perfected and its water utilized to its full potential.[1] This transformation required a massive manipulation of the St. Lawrence River and its environs. In excess of 210 million cubic yards of earth and rock – more than was required for the construction of the Suez Canal – were moved through extensive digging, cutting, blasting, and drilling, using a range of specialized equipment and gigantic machines.[2] Approximately sixty-eight miles of channels and locks were built and others rerouted, and many miles of cofferdams and dikes were needed. It was one of the monumental engineering achievements of the twentieth century and, at a capacity of 1,880,000 kilowatts (2.2 million horsepower), the Robert Moses–Robert H. Saunders Power Dam could supply the energy needs of a city of about 1 million people. As a result of the various dams, the raised water level formed Lake St. Lawrence, which inundated thousands of acres of land on both sides of the border and required one of the largest rehabilitation projects in Canadian history: moving the towns, people, and infrastructure that would be under water. Dubbed the "greatest construction show on earth," the integrated St. Lawrence Seaway and Power Project was at the time the largest construction project in the world and stood for several decades as the largest transborder water control project ever undertaken.[3]

In the transitions from Depression to World War to Cold War, the image of the river was transformed from wild and untameable – canals went *around* the rapids in previous decades and centuries – to something that could be controlled. Rapids could be flooded out, the river could be turned into a lake, and water could be converted into electricity. It was the conceptual transformation of the St. Lawrence from a river to a seaway.[4] The project constructed in the 1950s was similar in many respects to that which was envisioned in the engineering plans of the early 1920s, but there were technical shifts and evolutions, such as from a dual-stage or a single-stage dam to a combined form, and the plans for remaking the submerged villages. The magnitude and scale of the post–First World War Wooten-Bowden Report had been decidedly ambitious, and the early 1940s plans – on which the completed version of the St. Lawrence Seaway and Power Project was based – were progressively indicative of a fully fledged high modernist mindset.

Since the 1920s, and particularly after the Second World War, the confidence that the "hydraulic bureaucracies" on both sides of the border had in their ability to control and manipulate the riverine environment only expanded. The fourth Welland Canal had been built over the span of almost twenty years (1913-31), with construction actually taking place in thirteen of those years. Construction plans for the St. Lawrence project in the 1930s and early 1940s anticipated taking up to a decade. In 1951, Canadian engineers saw the St. Lawrence as a seven-year project; by 1954, the projected length of the building phase had been whittled down to five years.[5] This contraction came at the insistence of Robert Moses, who was motivated by the need to pay back the revenue bonds that the Power Authority of the State of New York had sold (at an interest rate of 3.18 percent) to raise money for its portion of the work. Ultimately, power came online on July 1, 1958, and the whole project was completed by 1959. Most of the work, besides dredging and bridge building, was completed between 1955 and 1958.

Given its magnitude, the completion of the project on schedule was an amazing feat. The St. Lawrence project cost more than $1 billion: $470.3 million split between Canada ($336.5 million) and the United States ($133.8 million) for the navigation aspects, with Ontario and New York each spending $300 million on the hydro works.[6] Although both power agencies – the Hydro-Electric Power Commission of Ontario (HEPCO) and the Power Authority of the State of New York (PASNY) – managed to remain on budget for the duration, the costs of the two national seaway authorities rose. After the initial appropriation of $105 million, during construction the amount the United States was authorized to spend on the seaway was upped to $140 million, though in the end only $133.8 million was spent. The Canadian government had to go back and ask Parliament for more than the $300 million that had been initially granted. Some of the extra costs were attributable to mistakes, others to inflation, and part of it was the result of an accelerated construction schedule that included winter work.

The next three chapters cover the building of the St. Lawrence Seaway and Power Project. This chapter engages both the political and technical issues that had to be dealt with before construction could begin – such as land acquisition, dredging, and transportation networks – and then discusses the construction of the seaway up to 1958. Chapter 5 looks at the power project and flooding that created Lake St. Lawrence, with a focus on the Lost Villages and rehabilitation of the St. Lawrence Valley. Chapter 6 explores the engineering process behind the raising of the power pool and examines the political and construction steps needed to finish both the seaway and the power project after Inundation Day, and then looks at the long-term operation and legacy of the project, particularly the environmental impact.

Ripple Effects

Even though Canada and the United States had agreed on the broad contours of the St. Lawrence Seaway and Power Project in their 1954 agreement, it quickly became apparent that there would be a seemingly endless supply of modifications and additions that necessitated further negotiations at different governmental levels. Because of the bilateral and transboundary nature of the undertaking, multiple levels of government entities were involved in the project: the International Joint Commission (IJC), the two federal governments, provincial and state governments (Ontario, Quebec, New York), HEPCO and PASNY, several joint engineering and supervisory boards (St. Lawrence River Joint Board of Engineers, International St. Lawrence River Board of Control, International Lake Ontario Board of Engineers), the US Army Corps of Engineers, and the two federal bodies set up to administer the seaway, the Canadian St. Lawrence Seaway Authority (SLSA) and the American Saint Lawrence Seaway Development Corporation (SLSDC).

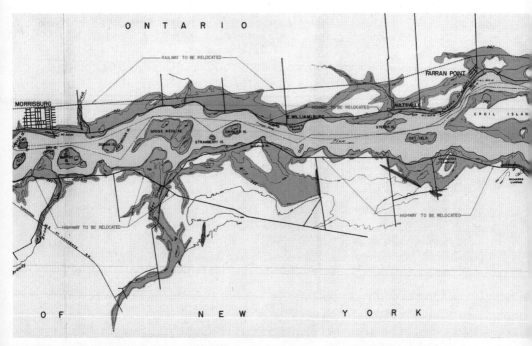

FIGURE 4.1   Portion of a HEPCO Blueprint of the International Rapids section.
© *Ontario Power Generation*

The SLSDC been officially formed in July 1954, as was the SLSA (even though it had been statutorily created in 1951). Since the Eisenhower administration was interested in reducing the federal government's role in such civil works projects, it was unclear whether the US Army Corps of Engineers would receive the designation as lead builder, even though it had built every federal lock over the previous century and had been at the forefront of the St. Lawrence project designs for decades.[7] The US Department of Defense was originally tasked with supervision of the SLSDC, though in June 1958 Eisenhower transferred responsibility for oversight of the corporation to the Department of Commerce, effective upon completion of the seaway.

The Great Lakes Division of the US Army Corps of Engineers (particularly the Buffalo District Office) was in September 1954 delegated some of the engineering responsibilities. These included aspects such as relocating highways and power lines, dredging in the Thousand Islands section, and planning the Long Sault Canal (soon renamed as the Wiley-Dondero Canal) and its Robinson Bay and Grasse River Locks (which were renamed

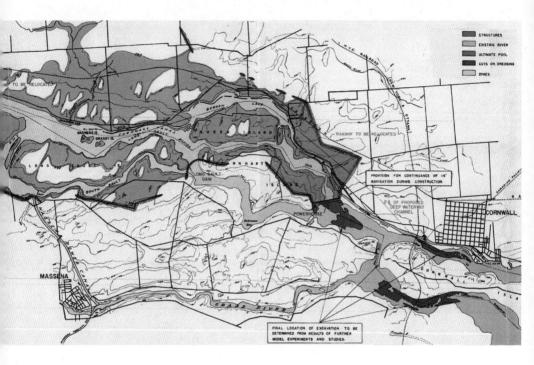

as the Dwight D. Eisenhower and Bertrand H. Snell Locks respectively before the project was completed). The corps was also represented on the binational St. Lawrence and Lake Ontario engineering boards. The organizational challenges of coordinating the navigation and hydro phases required the corps and PASNY to meet frequently during construction. Rivalries abounded among the different American organizations involved in the St. Lawrence project. One engineer contends that "the Power Authority was not staffed to deal with the great engineering issues ... They did not know what was going on. But they wanted to act like they had control over everything."[8]

In Canada, the St. Lawrence Seaway Authority incorporated the body previously responsible for the navigation aspects of the St. Lawrence project, the Special Projects Branch of the Department of Transport. HEPCO handled the Canadian work on the power project, and PASNY was its American counterpart.[9] Work on the power phase of the dual project started on August 4, 1954, with a ceremonial sod-breaking on August 10.[10] It took a few more months for the serious work to begin on the seaway.[11]

Much of the construction on both sides of the border was contracted out to private interests. PASNY had selected a private engineering firm (Uhl, Hall and Rich) and private contractors for its share of the power project rather than the Army Corps of Engineers, in part of because of the legal problem of a state entity hiring a federal organization such as the corps. Seven companies joined to form Iroquois Constructors, the main Canadian contractor for the power dam, and similar American conglomerates were established. Although HEPCO had assembled its own workforce for previous hydro projects on the Ottawa and Niagara Rivers, in this case it contracted out construction. To cope with the scale of the project, contractors and unions also created umbrella organizations to facilitate labour-management relations. HEPCO, for example, extended its Niagara project labour pact with the Allied Construction Council (which represented seventeen unions) to the St. Lawrence project, covering wages, working conditions, and safety regulations.[12] All construction work was overseen by the bilateral St. Lawrence River Joint Board of Engineers. After the construction was finished, the International St. Lawrence River Board of Control took over supervision of water levels and related issues.

Lionel Chevrier, who had been Cornwall's Member of Parliament (and minister of transport in the Liberal St. Laurent government) was appointed president of the SLSA. Both Chevrier and Lewis Castle, the newly appointed head of the American SLSDC, lacked experience in the national and international responsibilities that came with their roles as heads of the two seaway entities. Tensions developed between the two organizations. In the estimation of one Canadian official, Castle was "an intelligent person, extraordinarily naive and limited in international matters and believe[d] that being the presidential appointee consecrates the approach he ... always used as a Duluth banker."[13] Chevrier complained that the Americans were ignorant of the Canadian position and tried to run everything, especially in the early stages, citing an Eisenhower jest about Canada becoming an American state.[14] Conversely, many thought that Chevrier's was mainly a patronage appointment.

Finalizing plans for Montreal was one of the first orders of business. Chevrier approached Maurice Duplessis, the Union Nationale premier of Quebec, about developing hydroelectricity from the Lachine Rapids as part of the seaway construction. Since at least the 1920s, plans for the St. Lawrence project had included a power dam at Lachine.[15] But Duplessis claimed Quebec did not need Lachine power, and the province soon added its own massive hydroelectric complexes, first on the Manicouagan River and then in northern Quebec. For political reasons, the Quebec government also declined to play a role in the seaway, and the federal government handled all aspects of building the seaway in that province. Duplessis did eventually agree to help finance the highway and train tunnels running under the Beauharnois Locks.

Construction across an international boundary created some interesting problems. Government officials had long anticipated that binational construction in the St. Lawrence borderlands – in which a dredge, for example, crossed the border in the course of its work – could cause some legal difficulties in terms of bringing workers and equipment into the other country. The Canadian cabinet initially decided to remit all duties and taxes on materials and equipment and waive normal immigration requirements for the cofferdams needed to begin construction.[16] Ottawa initiated discussions about reciprocal general waivers on construction and, in response, the United States waived laws that prohibited foreign vessels from entering the United States. But, to the frustration of the Americans, Canada seemed reluctant to go beyond the cofferdam arrangements and wanted to proceed on a case-by-case basis. Ottawa's stance was largely predicated on the desire of Canadian construction firms to protect Canadian industry and labour from stronger US competition that could procure equipment at lower costs than their Canadian counterparts.

After lengthy consideration of whether the economic and labour implications of a reciprocal waiver served Canadians interests, the need to have HEPCO undertake dredging in American waters for cost-sharing purposes led the Canadian cabinet to agree to a general waiver in February 1955. But a series of complications, especially those regarding wet versus dry dredging and excavation, meant that no general agreement resulted. Rather, individual agreements to cover different aspects of construction trickled in over the following years of the project. These created a sort of patchwork delimited zone where certain aspects of the St. Lawrence project were deemed international and each country could bring in workers, supplies, and equipment free of custom duties and taxes.

The location of the Iroquois Control Dam immediately became an issue, as both hydraulic and foundation conditions were discovered to be less than optimal.[17] Based on model tests, the engineers proposed a shift in location of about two thousand yards downstream. Since this put the dam almost wholly in American territory, PASNY was given the task of constructing it alone. There was some concern that this would require reopening the 1952 IJC order of approval or would need congressional consent, but it was eventually agreed that the proposed change of location required only the approval of the St. Lawrence Joint Board of Engineers.[18] Nevertheless, changing the location had ripple effects, since PASNY's assumption of the entire cost and responsibility for the dam unbalanced the recent cost-sharing agreements that split evenly with HEPCO the bill for all power works.

Several entangled points of contention involving locks also gave rise to disputes. Various US interests, such as the armed forces, had been clamouring for larger locks throughout the St. Lawrence negotiations in the first half of the 1950s. In September 1954, the SLSDC invited representatives from different interests and agencies to express their views on lock size, and they preponderantly favoured increasing the lock size from the agreed-on dimensions (80-foot width, 800-foot length, and 30-foot depth) to a width of 100 feet, a length of 900 feet, and a depth of up to 35 feet. American defence and shipping interests, particularly the US Navy, were pushing for this increase, and since the SLSDC's expenses would decrease if Canada built the lock at Iroquois Point, the extra money saved by the United States could be spent on increasing lock size. Canada objected strenuously to larger locks because of the higher cost, estimated to be several hundred million dollars, and the complications that would stem from a need to change the Canadian St. Lawrence legislation to increase the dimensions of the Canadian locks. Enlarging the St. Lawrence locks also meant that the Welland Locks would have to be enlarged, at a significant cost, or they would cause a bottleneck. Even though Castle soon made a public statement to the effect that the United States would build to the same dimensions as those in the Welland Canal, other voices in Washington were not so willing to give up the bigger navigation works, and the United States would on several occasions use the lock size as a bargaining chip.

At the August 1954 negotiations, Canada had unilaterally declared that it would build navigation works at Iroquois despite the Wiley legislation calling for the Americans to do so. This dispute persisted through the fall of 1954, with Canada worried that the United States might still build a competing lock at Iroquois. Informal signals, however, from key American officials indicated that the SLSDC would "drag its feet" so that Canada could pull ahead in constructing its Iroquois navigation facilities, which would make it politically feasible to go before Congress and ask for the cancellation of the Point Rockaway Lock.[19] But soon after, the United States became more equivocal about the future of the Iroquois Lock, stressing that amending the Wiley bill to disavow the Point Rockway Lock risked giving eager internal seaway opponents another opportunity to modify or block American involvement. It appeared that American officials were using this as a way to extract a promise from the Canadians not to build parallel navigation works at Cornwall.[20] Canada grudgingly accepted American assurances, rather than an unequivocal promise, that they would seek congressional action to be relieved of their statutory obligation under the Wiley bill to construct an American canal opposite Iroquois.

Regarding the parallel navigation works, the Canadian St. Lawrence Seaway Authority had seized upon the requirement to provide navigation around the Moses-Saunders dam during construction as a way of building works that could later serve as a basis for an all-Canadian deep canal. This would be in competition with the American canal and locks on the south side of Barnhart Island. Washington pointed out that, during the August 1954 negotiations, Canada had committed to forgo twenty-seven-foot navigation facilities at Cornwall until increased traffic or interference with Canadian shipping warranted; but the nature of this commitment was purposefully vague and subject to interpretation, and the Canadian government felt it had the right to retain its freedom of decision. The dike that would be constructed along the Canadian side, starting at the Moses-Saunders dam, to hold back water for Lake St. Lawrence would need to be pierced so that the power entities could follow through on their obligation to maintain fourteen-foot navigation during construction. From the Canadian perspective, since the dike had to be breached anyway, it may as well be done in a way that would provide the groundwork for a future twenty-seven-foot lock. However, knowing the consternation that this would cause US officials, C.D. Howe downplayed, perhaps disingenuously, Canadian intentions during discussions with Deputy Secretary of Defense Anderson.[21] Howe indicated that Canada would not continue fourteen-foot navigation after the completion of the seaway and power project. Canada agreed that it would not build deep navigation works at Cornwall "under present conditions," a phrase that had been selected because it was ambiguous enough for each country to see what it wanted in the wording. Canada would not build a *through* system of navigation around the power dam, but reserved its right to acquire land and put in place some elements of an eventual all-Canadian waterway.

Ottawa's determination to retain the right to a future all-Canadian seaway created more construction difficulties than did any technological or natural obstacle. This was particularly obvious when it came to dredging in the north and south channels around Cornwall Island. Dredging would be required immediately below the Moses-Saunders Power Dam in order to lower the water level of the tailrace area (where water exited the dam after flowing through the turbines) for power production, and to slow down the current for navigation. Dredging in the international waters of the St. Lawrence-Great Lakes water system had historically been a point of bilateral contention, including the lead-up to the joint 1952 submissions to the IJC, and was further complicated by the need to distinguish between dredging for power purposes and dredging for navigation purposes. This remained an issue for the duration of the project's construction, since the directives for the location and form of this dredging in the existing bilateral agreements, as well as the extent to which the power or seaway entities were responsible, were vague.

To be sure, from Ottawa's perspective, it was more of a political than an engineering problem. The manner of the dredging would materially affect the flow of water levels and thus, according to the Boundary Waters Treaty, needed either IJC concurrence or a special binational agreement. Ottawa favoured a special agreement in case the IJC did not recommend that the dredging be done in a matter that would aid a future Canadian navigation channel (as well as access to Cornwall for deep-draught ships), which remained very much on the minds of the Canadian government and public. The Canadians figured that if tailrace water dredging had to be done anyway, why not do it in a way that would help create a future navigation channel on the north side of Cornwall Island? The American assessment that Canada seemed "to be throwing up a premature smoke screen" was not far from the truth, as other parts of the project changed levels in similar ways but did not elicit protest from Ottawa.[22] This Canadian plan engendered significant resistance from the SLSDC and the Eisenhower administration, as the perception that Canada was still going to complete an all-Canadian seaway would be a boon to seaway opponents in the US Congress, not to mention that it would create competing navigation works for the American locks in the International Rapids section (IRS). Protracted negotiations about the extent to which seaway or power entities were responsible for dredging, as well as the form and the cost of the dredging, continued for the rest of 1955.[23] Still devoted to the concept of a future all-Canadian seaway, and bitter about the 1954 deal it had to swallow for a joint seaway, the St. Laurent government was unwilling to back down, insisting that compensatory dredging take place in the north channel in a form that could later serve twenty-seven-foot navigation. The US government, with the major exception of PASNY, was firmly of the opinion that the power entities were responsible for excavations downstream of the dam, and the American seaway entity for the navigation dredging in the south Cornwall channel.[24]

Ottawa was willing to risk construction delay for the sake of retaining its position. Negotiations continued into 1956, threatening to hold up the power project. But realizing that both sides were deeply entrenched in their positions as much because of domestic political considerations as because of technical, legal, and economic issues, the two federal governments decided to agree to disagree. Canada outlined a straightforward modus operandi, which would constitute the "special agreement" required by the Boundary Waters Treaty. Canada maintained the position it had outlined in the August 1954 exchange of notes: it did not need US consent to duplicate navigation works, though it promised to "consult," and announced that it would undertake dredging in a form that would provide for a twenty-seven-foot channel in the north Cornwall channel. The Americans replied that Washington was "bound by events to take cognizance of the de facto situation which is created by the decision of Canada to proceed with deep-water dredging in the channel north of Cornwall Island."[25] Canada would do the dredging in its territory (which was estimated to be two-thirds of the dredging required downstream of the dam) and would not charge any of the cost to tolls. HEPCO and PASNY also consented to making a monetary contribution toward the cost of dredging.[26]

One of the most contentious issues involved in the joint construction of the St. Lawrence Seaway and Power Project had, after lengthy discussions, been disposed of. The Canadian press approved of Ottawa's decision to protect its right to an all-Canadian seaway; the American press did not pay much attention, though good will had been lost because of the nature of the dredging dispute and because of connected diplomatic leaks that had occurred in the previous months.[27] But there were still other controversial questions to be tackled, such as the location of a bridge for the New York Central Railroad line across Cornwall Island. The two countries were legally obligated to provide the railway with a replacement for the bridge displaced by the seaway and power project. Initial American plans called for a traffic and rail tunnel that would pass under the easternmost lock in the Wiley-Dondero Canal.[28] This route was designed to maximize traffic exposure to the grand new park system that PASNY chairman Robert Moses had planned for the American side of Lake St. Lawrence.

Moses had taken the chairman position in 1954 after an extensive career building parks, parkways, and bridges in and around New York City.[29] He was a quasi-dictator who treated the park – and the power authority – as his own personal fiefdom, but he also got things done, and usually got them done well. Indeed, the opportunity to create new parkland in the

Niagara and St. Lawrence regions drew him to the PASNY position. Unstable soil conditions mitigated against the original highway tunnel – the presence of marine clay was described as "a sort of oozing mass of earth" – and the SLSDC indicated to the SLSA that the highway connecting with the bridge at Pollys Gut (a narrow channel between Cornwall Island and the US mainland) would have to be built further west.[30] This was a problem for the Canadian government, however, as it would mean a much longer trip for the St. Regis (Akwesasne) First Nation. This group had just over two thousand members, divided between Cornwall-St. Regis-Chenail, an area spreading over the Ontario, Quebec, and New York borders. The Canadian members of this group, which occupied transnational space on Cornwall Island and adjoining lands on both sides of the river, would have to travel up to twenty-five miles in a roundabout fashion through US territory to access social and industrial services they required at Cornwall. This would negatively affect the community life of the area, as many members attended schools, picked up family allowance and pension cheques, and purchased groceries and supplies in Cornwall.

The Canadian government was, to be sure, most concerned about the increased cost of relocating facilities to St. Regis or of transporting the band members to Cornwall by bus or ferry.[31] But the Pollys Gut Bridge became moot when, in 1956, the New York Central Railroad was persuaded to abandon its line from New York to Ottawa. Without railway grade to worry about, a high-level traffic bridge could be built directly from Cornwall to the US mainland. This would be cheaper than the Pollys Gut alternative and cut the projected travel distance for the St. Regis band members by more than half. But it did lead to cost overruns for PASNY because of the planning and preparation that had already gone into the initial Pollys Gut Bridge scheme (a traffic tunnel was instead installed under the Eisenhower Lock), and the late start on the new bridge threatened schedules and deadlines. The bridge relocation infuriated Moses, who was also facing obstacles to PASNY's hydro development at Niagara Falls; he fired off his trademark acerbic letters, protesting that this new bridge would not allow traffic to wind through the expansive parkland system he was creating in conjunction with the St. Lawrence project.

"A Monumental Mess"

Land acquisition was a major challenge to St. Lawrence development. The power project required a mass relocation of people and infrastructure,

FIGURE 4.2   Robert Moses *(far left)*. © *New York Power Authority*

chiefly in the area to be inundated by Lake St. Lawrence (discussed in more detail in the next chapter). The seaway also required the acquisition of property, sometimes in conjunction with power works, at places such as the shore opposite Montreal, the Cornwall-Massena area, and Iroquois. Because of its inexperience, the SLSA used the lands division of the Department of Transport to acquire property.[32] PASNY secured the land necessary for the US share of the power project using the New York Department of Public Works, and out of that share gave the SLSDC the territory for the American seaway works.

Forced moving was relatively unexpected for the residents of the Laprairie Basin on the south shore opposite Montreal. Until the early 1950s, most engineering plans had called for the seaway channel at Montreal to run along the north shore, and earlier blueprints show, for example, a lock near Verdun and the navigation channel cutting through north shore neighbourhoods further to the west. Other federal plans dating to the late 1940s (and arrived at in conjunction with Quebec officials) featured various forms of a power dam across the entire river just downstream of the Victoria Bridge, with a canal and lock along the edge of territory belonging to the Kahnawake reserve.[33] Seaway officials decided that it would be easier and cheaper to build a navigation channel entirely along the south shore because of lower property values, less-developed infrastructure, and a lack of regard for the First Nations people that stood in its path.

Canadian owners of property taken for the seaway were offered current value plus 10 percent for inconvenience. By August 1955, 127 expropriations had taken place on the south shore near Montreal for the area needed for the Côte Ste-Catherine Lock, and more than seventy owners had received compensation.[34] In addition to arranging relocations, the SLSA constructed collector sewer systems and several modern water intakes. But the municipality of Saint-Lambert complained, and a formal investigation started in 1958 was the next year extended to include other south shore municipalities that were also dissatisfied, such as Longueuil and Saint-Hubert.[35] Complaints were about, among other things, the water and sewer systems, the need for a protection wall, and the depriving of "an agreeable view" of the river by fill (e.g., dirt and rock dumped to form canal embankments) that detracted from property values. The digging of the seaway channel also lowered the water table, drying up many wells in the area, so the SLSA provided water wagons.

The Canadian government needed to acquire property for seaway construction and locks at Iroquois Point (as well as neighbouring Matilda Township), which jutted into the St. Lawrence River and housed part of the old town of Iroquois, including its famous apple orchards, and had been the site of a fort several centuries previous. Seaway and power project expropriations overlapped at Iroquois, as a regulating dam was also being built from Iroquois Point across the river to Point Rockaway, and the town needed to be relocated as a result of the raised water level. In November 1954, the first federal expropriations of twenty-three families at Iroquois began. One landowner fought the expropriation of his orchard, and the court case involved the valuation of every single one of his thousands of trees.[36] The federal government also required land in Cornwall Township for the construction of a new bridge, and for the fourteen-foot navigation channel (which Canadian officials hoped would form the basis for an eventual deep-draught canal) built during the construction period, which incorporated parts of the existing Cornwall Canal. So as to prevent speculation, in April 1955, measures were taken to obtain fifty-two acres and approximately a hundred properties.[37]

Two Mohawk communities were affected by the St. Lawrence project: the transborder Akwesasne (St. Regis) reserve, and the Kahnawake (previously known as Caughnawaga, the anglicized version of "Kahnawake" used prior to 1982) reserve on the south shore opposite Montreal and astride the Lachine Rapids.[38] By virtue of its geographic location, Kahnawake dealt with only the Canadian government, whereas St. Regis had to deal with the national governments of both Canada and the United States, as well as those of Ontario and New York. A transborder bridge across Cornwall Island also necessitated land acquisition from the St. Regis (Akwesasne) First Nation. This complicated matters, since land in Canadian reserves was held in fee simple rather than outright. But the Canadian government decided against simply extinguishing Indian rights by an act of Parliament, which gave the First Nations some bargaining power.[39] Stateside, the reservation's land was held by treaty, and as a result there were complications in terms of state versus federal jurisdiction.

Chevrier, as president of the SLSA, met with representatives of Kahnawake on several occasions, promising the community new amenities. An independent appraiser was used and negotiations entered into with the inhabitants, and compensation advances were provided to help them obtain new homes. In mid-September 1955, the SLSA expropriated land where the seaway channel would run through the residential and historical section of the community, cutting it off from the river. Calls to move the navigation channel further out were rejected because of the cost. Further expropriations took place in February 1956. Kahnawake residents became "disillusioned by the callous disregard" shown by the Liberal government and by the forceful eviction of those who had not sold to the SLSA.[40] Members of the band council worked within the existing legal and constitutional structures, and applied for a court injunction against the SLSA's expropriation of land. They appealed to the Canadian government to honour past treaties, and even appealed to the British government, before taking their case to the United Nations.[41]

The Mohawk group sought to influence the media and public opinion, and enlisted McGill University professors to help their legal cause. They did find sympathy as the affair became headline news and the Conservative opposition in Parliament, including the soon-to-be prime minister, John Diefenbaker, criticized the Liberal handling of Kahnawake as "a monumental mess."[42] Louis Diabo, a Kahnawake resident, became an international celebrity by holding out, with the support of the Department of Citizenship and Immigration, for a better deal. In his memoir/history of the St. Lawrence undertaking, Chevrier writes that, when the SLSA first started surveying and expropriation proceedings at Kahnawake, there was general satisfaction. Then, however, the Mohawk residents, "perhaps with some prompting, took the position that here was a chance to make some money out of the seaway."[43] Chevrier's recollections were reflective of the

Canadian government's paternalistic misunderstanding of the Mohawk groups affected by the seaway, further symbolized when he smoked a peace pipe upside down.

A judge eventually denied Kahnawake's legal appeal for an injunction.[44] Frustrated, Kahnawake divided into factions, split between traditionalists and those who favoured a more confrontational approach.[45] There were changes to the band council, and more militant resistance came from those who refused to move for new bridge approaches. This delayed construction, so the SLSA seized the necessary properties. None of the varying strategies ultimately stopped the seaway from tearing through the community, and although the fight for better treatment met with some minor successes, the compensation was far from equitable or adequate. Residents felt pressured into cooperating, fearing the results if they did not willingly sell – as a result, about 80 percent of the cases were eventually settled without expropriation, and 177 families in total were bought out.[46] The community lost 1,262 acres (one-sixth of its land) to seaway expropriation, most of it along the waterfront. When finished, the Laprairie dike and seaway channel ran alongside the community's shore, effectively isolating it from the river.[47] This would be problematic for any community accustomed to river access, but it was particularly disruptive for a community that had for hundreds of years based its culture and way of life on access to the river. *Kahnawake* translates as "on the rapids," and the seaway robbed the community not only of territory but also of its meaning. This had an enormous environmental effect, radically reconfiguring the landscape and waterscape. Plants and wildlife disappeared, water and fish contained toxic levels of chemicals, and many cubic yards of spoil were dumped on the reserve and used for diking, creating "clay mountains."

Akwesasne lost little land (130 acres) compared with Kahnawake, but the St. Lawrence project still had significant repercussions for that community aside from the question of a bridge to Cornwall. They claimed Barnhart Island, the site of the Moses-Saunders Power Dam, as ancestral territory. This was not recognized, and in 1956 the Mohawks filed suit against New York State for compensation ($33.8 million) for loss of the riverbed and water power. In the next two years, the case rose to the state's highest court of appeals. Land seizures for bridge construction were major sources of grievance, as were the various borders, tolls, and customs issues resulting from the transborder placement of the reserve. The widening of the river channels, and New York's relocation of Highway 37, required the dislocation of several homes. Parts of the reserve were needed for sections of the Wiley-Dondero Canal and dredging below the power dam, and

FIGURE 4.3 Vessel in the completed seaway channel; Kahnawake is in the foreground, Montreal, the background. © *Library and Archives Canada / Credit: National Film Board / The St. Lawrence Seaway Authority fonds / e011068166*

protracted discussions revolving around an eighty-six-acre section of Raquette Point were finally concluded in January 1957. Some St. Regis Mohawks adopted more confrontational approaches as the St. Lawrence project neared completion. In August 1957, approximately two hundred Indians, including a number from Akwesasne, occupied land on Schoharie Creek near Fort Hunter in New York to protest their treatment resulting from St. Lawrence construction. But their demands were not met.

All this had a tremendous social and cultural cost and reeked of colonial assimilation. Compared with the Lost Villages – some of which were originally founded by United Empire Loyalists – the two reserves were treated as second-class citizens. The residents of Akwesasne and Kahnawake were granted fewer rights, but many showed less deference to authority than did the Lost Villagers. The seaway experience marked a major turning point in the history of both Mohawk communities. In Canada, it marked the end of their trust in the federal government and gave rise to a more radical nationalist movement that led to later Mohawk conflicts with the Canadian and American states.

Starting Construction

The evolution of the power dam plans is indicative of the ways that large-scale technological systems are compromises between the technology itself and political, economic, and environmental factors. Over the first half of the twentieth century, St. Lawrence power project planning had oscillated between a dual- and single-stage plan. Essentially, a dual-stage dam utilizes two hydro generation stations at separate locations; a single-stage dam features one larger combined installation. A single-stage project would be cheaper, better for navigation, and provide power at a slightly lower cost, but the Canadians had previously contended that a dual-stage development would allow for power to be more quickly produced (i.e., before the entire project was completed), and that it was better politically because it would result in substantially less flooding of Canadian land. The dual-stage design initially won out and was included in the 1932 St. Lawrence Treaty. However, in 1940, the Joint Board of Engineers decided that the combined single-stage project – a partial compromise between the single- and dual-stage plans – was the best solution from both engineering and economic perspectives. The dual-stage would result in the flooding of between five and six thousand acres in Canada, whereas the combined single-stage was expected to inundate between sixteen and seventeen thousand Canadian acres but be more cost-efficient, since it would yield greater navigation and hydro benefits. The combined single-stage project was incorporated into the 1941 agreement. Updated and revised versions of these plans would serve as the basis for the hydro project eventually constructed in the 1950s. The final form of the binational power dam concentrated in one location eighty-one of the ninety-two feet that the St. Lawrence fell between Lake Ontario and Cornwall.

At the seaway's eastern entrance, two locks, Saint-Lambert and Côte Ste-Catherine, were built into the new navigation channel on the south side of the St. Lawrence, with the former across from Montreal Harbour and the latter at the foot of the Lachine Rapids. The navigation channel, and the dike separating the channel from the river, continued through the Lachine section, and then a dredged channel went across Lake St. Louis.

FIGURE 4.4   PASNY aerial view of construction on the Wiley-Dondero Canal *(left)* and Moses-Saunders Power Dam *(centre)*. © *Stormont, Dundas, and Glengarry Historical Society,* courtesy of Cornwall Community Museum

Two locks ran beside the Beauharnois Power Dam, with a combined lift of eighty-four feet, and then a channel of sufficient depth through the Beauharnois Canal led to Lake St. Francis, where work consisted entirely of channel dredging. Then the seaway crossed into American territory, going south of Cornwall Island and into the ten-mile-long Wiley-Dondero Canal, built in place of formerly dry land, with two locks. As the seaway travelled west to the Iroquois Lock, the only Canadian lock in the IRS, more dredging and channel improvements were necessary, which involved removing or altering many islands. Similar improvements took place in the Thousands Islands stretch of the seaway channel, which ran mostly in American territory and was littered with shoals and rock. From there, vessels crossed Lake Ontario to an enlarged Welland Canal.

The St. Lawrence project required a major reordering not only of water-based transportation and mobility patterns but also of the interconnected rail and traffic infrastructure. Between Montreal and Cornwall there were seven bridges crossing the St. Lawrence that needed to have 120-foot clearance, with ten bridge projects in total on the Canadian side; achieving the requisite height required several engineering breakthroughs.[48] The SLSA was obliged only to build lift spans but realized that the traffic blockages that would result in Montreal would cause major problems. The whole southern part of the Jacques Cartier Bridge was raised eighty feet through an innovative jacking system, and in October 1957 a new span was inserted into the bridge with a minimum interruption to traffic.[49] This was the largest bridge-raising operation ever undertaken to that date, and the process attracted engineers from around the globe. The Victoria Bridge was given an additional approach, creating a Y-configuration, allowing traffic to be diverted to one of two vertical lift spans on each side of the Saint-Lambert Lock. A new, elevated span and approaches were added to the Mercier Bridge. Montreal's harbour received an expensive upgrade and facelift, including new grain elevators, wharves, piers, and sheds. Three bridges over the Beauharnois Canal were given movable spans, and a four-lane highway tunnel was built under the Beauharnois Locks. Initial schemes had featured a bridge running across the top of the Iroquois Dam, but these plans fell by the wayside.[50] Not too far away, a high-level international bridge was built from Johnstown, Ontario (beside Prescott), to Ogdensburg, New York. Having dispensed with the need for a Pollys Gut Bridge, and with traffic access to Moses's park on Barnhart Island provided by a traffic tunnel under Eisenhower Lock, the Canadian and American seaway entities decided in 1957 to buy out the Cornwall International Bridge Company and replace the dilapidated existing span with the new high-level Seaway International Bridge. This bridge had two spans from Cornwall to just east of Massena via Cornwall Island. The SLSA built the substructure of the south span while the SLSDC took care of the superstructure; Canada gave a span of the north bridge sufficient clearance to accommodate large vessels in a future all-Canadian seaway.

The power project required an assortment of machines, many larger than a house, including scrapers, power shovels, draglines, and earthmovers.[51] The necessary pieces of machinery also had to be moved to their respective sites, which itself often proved to be a major achievement. For example, in a well-publicized trip, a huge dragline called the Gentleman made the journey from Kentucky to the St. Lawrence by barge over the course of several months, which included two weeks stuck in spring mud.

FIGURE 4.5   Construction equipment at work on the St. Lawrence project. © *Dumas Seaway Photograph Collection, Mss. coll. 124, Special Collections, St. Lawrence University Libraries*

This mammoth piece of equipment, the largest used, weighed in at 650 tons and was capable of scooping 20 tons of earth at a time.

In addition to taking care of organizational issues, acquiring equipment, and expropriating land, the construction of extensive cofferdams was necessary before most of the actual work, which would generally be done in the dry, could commence. Cofferdams themselves were elaborate and challenging undertakings that involved large steel cells filled and covered with stone and soil. Forming the huge dikes on both sides of the river abutting the main power dam location were also among the first orders of business. The swift currents proved troublesome, sweeping away the fill for the cofferdam that was meant to still the famed Long Sault Rapids, which dropped the river about 30 feet, and divert water into a cut through Long Sault Island.[52] The cofferdam from the Canadian mainland to Barnhart Island stretched to 4,200 feet, reputed to be the longest in the world.[53] Large structures, such as the dams, were half built in the dry, and then water was allowed through the completed portion while the remainder was dewatered. When the Long Sault Rapids were silenced, sightseers flocked to the dry river bed to peruse the rock formations, finding sunken treasures such as cannonballs possibly lost during the War of 1812.[54]

FIGURE 4.6   A cofferdam under construction in the St. Lawrence River. © *Dumas Seaway Photograph Collection, Mss. coll. 124, Special Collections, St. Lawrence University Libraries*

Excavations created huge amounts of spoil, which needed to be put somewhere; this material itself contributed to noticeable changes to the landscape. The SLSDC estimated that a total of 27,250,000 cubic yards of excavated material would have to be deposited on the American shore alone, and some of this material would remain unstable and unusable.[55] It is hard to track all the spoil given the magnitude of the project and the different agencies involved, but the majority seems to have been dumped either onto the shoreline, the bottom of the river, land that was slated for flooding, or the many miles of dikes.[56] Engineering blueprints show that specific spoil areas included the raised areas beside the locks, to the immediate east of the town of Cardinal, on the south shore of the river opposite the Iroquois Dam, and various underwater disposal sites such as an area between Sparrowhawk Point and Toussaint Island. Marine clay spoil proved a nuisance in places such as Kahnawake and Iroquois because it was difficult to build on, though it could be landscaped and planted with trees. In the case of Iroquois, spoil from nearby excavations was dumped on the former town site, saving the abandoned area from inundation;

however, since this fill was marine clay, the deserted space along the riverfront was turned into parkland and an airport, giving the appearance that the town did not have to move at all. Iroquois and Matilda Township complained about the spoil piles, agreeing in 1963 to accept an improvement scheme for Iroquois Point as compensation.[57]

Ground was broken in April 1955 for the excavation of the ten-mile Long Sault Canal (Wiley-Dondero Canal) and locks.[58] Lock design was the most complex and demanding aspect of the job for the US Army Corps of Engineers.[59] The corps had been constantly revising its plans for these features based on soil and geological conditions, and model testing. When construction began, there were still many unknowns, and those who worked on the project recall the changes that were constantly being made.[60] A fault was found running under the proposed site of the Grasse River Lock, and there were unstable rock conditions where the Robinson Bay Lock was intended to sit; the location and alignment of the canal and locks were therefore altered. By the end of 1955, overburden excavation for each lock site was substantially more than half completed, and the excavation for the upper end of the Long Sault Canal was approximately half finished.[61]

Seaway contractors were generally working two ten-hour shifts, in daylight and at night under floodlights. Construction continued in the winter, especially on the Canadian side. During the winter of 1958, Ontario workers placed an average of 1,500 cubic yards (3,000 tons) of concrete daily in the power dam structure. The two powerhouses collectively formed the enormous power dam, which was the most monumental part of the whole project. Before the dam could be constructed, the bedrock underneath had hundreds of thousands of bags of grout pumped into it to form a solid foundation. The international border bisected the finished power dam in the middle, though it appeared as a seamless structure.[62]

Skilled and unskilled labour was needed: not only were engineers, carpenters, and machinists required but also personnel for the other types of jobs that came with such a large and complex undertaking, such as accounting and security. At its peak, HEPCO's workforce consisted of fifty-two hundred people. A total of twenty-two thousand workers from both countries were employed on the entire combined St. Lawrence project, though the highest recorded number of people working at one time was a little under twelve thousand. The demands were so large that more than half the workers came from outside the St. Lawrence Valley. Spin-off sectors – for example, service industries – also boomed. Waddington alone

FIGURE 4.7 Winter construction. © *Dumas Seaway Photograph Collection, Mss. coll. 124, Special Collections, St. Lawrence University Libraries*

FIGURE 4.8 Workers on the job at Moses-Saunders Power Dam. *Photograph by Eleanor L. (Sis) Dumas.* © *Dumas Seaway Photograph Collection, Mss. coll. 124, Special Collections, St. Lawrence University Libraries*

had seven bars during the construction years. Life on the job was hard and rough, with long shifts in hazardous areas often made worse by inclement weather. Although the entire project had a fairly good safety record relative to its scope, scale, and the expectations of the time period, scores of people were seriously injured, and accidents such as cables mangling or severing legs led some workers to call it "Cripple Creek." Forty-two people reportedly died as part of the project from accidents, drowning, or electrocution, though this was fewer than on other comparable projects.[63] Dissatisfaction with working conditions, changing specifications, and tight deadlines also led to labour disputes and work stoppages.[64] There were steel-industry and concrete-producers strikes in 1956 and 1957 respectively that threatened the supply of their products. Because of the problems caused by construction conditions, such as marine clay and glacial till, many contractors filed claims.[65]

Strikes, shortages, and adverse conditions, along with the controversy over an all-Canadian channel and dredging around Cornwall, had thrown off the construction schedule, and in summer 1957 the power entities were starting to panic, fearing that the work would not be completed in the time anticipated.[66] By the end of the year, it was still uncertain as to whether the hydroelectric project would be ready for the July 1, 1958, target date. Work continued up to the last minute, and the power pool was raised on schedule, with fourteen-foot shipping commencing in the new navigation works.

Construction proved to be immensely popular with the wider public. Numerous visitor lookout stations were built, and bus tours went to the generating stations.[67] Both power entities constructed permanent visitor centres for their respective powerhouses. Robert Moses had insisted that the initial aesthetic appeal of the power dam was not grandiose enough, and PASNY persuaded HEPCO to change the design to the "national monument" and projection of power that Moses envisioned. In addition to the impressive visitor's centre at the dam, he added a guest house for the trustees, and micro-managed elements such as the type of stone that would make up the exterior.[68] By the time the power dam became operational in summer 1958, over 1,800,000 people had visited the project.

Conclusion

The construction of the project involved a range of engineering, scientific, and technological advances. In 1960, the American Society of Civil

FIGURE 4.9   Moses-Saunders Power Dam under construction. © *Stormont, Dundas, and Glengarry Historical Society, courtesy of Cornwall Community Museum*

Engineers named it the most outstanding civil engineering achievement of the previous year, and in 2000 the American Public Works Association included the St. Lawrence project in its list of the ten most important public works projects of the previous century. Canadian engineers were at the forefront of applying the latest advances in soil mechanics for the design of dikes and embankments; jet piercing technology; the air-cushion method (developed by Canadian Industries Limited [CIL]) which enabled contractors to set off underwater blasts in close proximity to the walls of the Welland Canal; and new techniques for placing concrete, particularly in winter.[69]

Another of the major engineering advances was the extensive use of high-precision scale hydraulic models that replicated long stretches of the river in minute detail: the topography, the shoreline, the river channels and contours, and the turbulence and velocity of the currents in all areas of the river in its natural state at both high and low water levels.[70] The first computer used in Canada was employed to calculate backwater flows for an all-Canadian seaway in the early 1950s.[71] There were other technological and planning advancements to reduce and control ice formation. During the Lost Villages relocation, large-scale mechanical house-moving machines were employed in Canada for the first time, and the National Research Council used nine of the abandoned houses in Aultsville for controlled burns to test their resistance to fire. The results were used to revise the Canadian fire code and reputedly also led to the adoption of smoke rather than heat detectors.[72] Despite all these achievements, the major engineering advances of the St. Lawrence project lie more in the scale on which technologies and techniques were used. From engineering and administrative perspectives, the project was an organizational triumph, particularly the development of a project management system that was in many ways the forerunner of critical path methods. Indeed, it is remarkable that the entire project was finished on time given its scope, complexity, diversity, and size.

Nonetheless, there was a range of engineering and construction problems that often stemmed from faulty planning and lack of knowledge about specific and local conditions that could not be replicated on a model or predicted in an office. Many of the engineering estimates and schematics were wrong or needed to be revised. Moreover, these problems were exacerbated by the accelerated schedule for completion. Some of the chief problems were severe subsurface conditions that had not been properly identified or assessed ahead of time, such as difficult glacial till, sticky marine clay, and fault lines running under major works. The glacial till wore out drills far ahead of their expected lifetime, and the marine clay turned into a sticky mess, bogging down men and equipment when it got wet. Winter construction further complicated the situation. Heavier equipment was brought in, but many construction agencies still went bankrupt or had to significantly revise their estimates; many quit and filed claims against the construction authorities. As one worker later recalled, "A lot of it technically did not work out quite as nice[ly] as how they had planned it. Water and natural obstacles like soil and rocks can make the best-laid plans and scientific data useless."[73] Moreover, as Claire

Parham's oral history of the construction of the St. Lawrence project reveals, there was a good deal of trial and error, and shortcuts and rushed jobs were manifold.[74] The cement in the Eisenhower Lock, for example, soon had to be replaced and then constantly maintained. The river was being transformed, but not without a fight.

5
Lost Villages

Both the navigation and hydroelectric power elements of the St. Lawrence project were predicated on a dammed river. However, the people and built environment along the International Rapids section of the St. Lawrence River would have to be moved. This was necessary on both sides of the border, but the much smaller American population along the international stretch of the St. Lawrence meant that the inundation had far less impact in New York State. Between Iroquois and Cornwall, an area roughly equivalent in size to Manhattan was flooded. It would prove to be the largest rehabilitation project in Canadian history. Some 6,500 people were relocated from numerous Ontario communities that came to be collectively called the "Lost Villages."[1] Another fifteen hundred permanent residents east of Cornwall, mostly in Quebec, were moved because of seaway, rather than power project, construction.

The perceived ability to master nature implicit in reconfiguring the river extended to planning the model "modern" towns that absorbed the communities displaced by Lake St. Lawrence. The task of rehabilitating this area fell to Ontario and New York, which delegated most of the responsibilities to their respective power entities, HEPCO and PASNY. The method for relocation and rehabilitation changed and evolved throughout the 1950s, but in Ontario the final result was that all of the town of Iroquois and part of the town of Morrisburg were shifted, whereas the communities further east were completely submerged: the villages of Aultsville, Farran's Point, Dickinson's Landing, Wales, Moulinette, and Mille Roches, along with the hamlets of Woodlands, Santa Cruz, and Maple

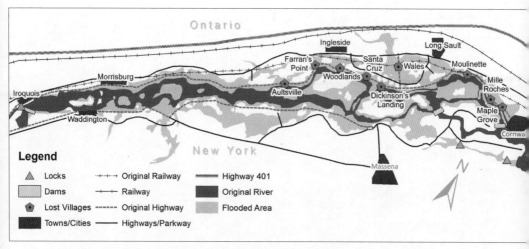

Figure 5.1 Map of Lake St. Lawrence and the Lost Villages. © *Cartography by the author, based on a map from the Lost Villages Historical Society*

Grove.² The denizens of this largely agricultural area (mills were the main non-agricultural form of employment) were placed in two newly created towns, Ingleside and Long Sault, as well as several subdivisions, such as Riverview Heights and Lakeview Heights. In addition, 225 farms, 18 cemeteries, 1,000 cottages, 37 miles of Highway 2 (the main east-west highway), and more than 37 miles of double-track CNR railway were moved. On the American side, approximately 18,000 acres were flooded, requiring the clearance of around 1,100 people, 225 farms, 500 cottages, and 12.5 miles of highway. No entire communities were relocated, in part because of dikes that protected Massena, though the waterfront area of Waddington was affected, and the town of Louisville lost about a third of its taxable land, including Louisville Landing, a historic port on the St. Lawrence.

The Lost Villages saga has been addressed in local histories, prose, music, and, especially after the recent fiftieth anniversaries of the openings of the seaway and the power project, academic study and fictional representation.³ Although it was a national – even international – story at the time, it has been largely forgotten outside the St. Lawrence Valley. Synthesizing the local and social histories with environmental, technological, and state-building approaches, and incorporating the previously unknown perspectives of the governments and utilities involved, particularly HEPCO and PASNY, allows for a full portrait of the impact of the St. Lawrence power project on the area around and under Lake St. Lawrence.

"One of the Greatest Experiments in Social Engineering"

Before the 1950s, the questions of whether the Canadian government or the Province of Ontario would be responsible for handling the Canadian land slated for inundation had not been definitively settled. Despite the 1943 Wilson Report's recommendation of a combined federal-provincial-municipal commission to oversee the rehabilitation, HEPCO was delegated the responsibility in 1951. Giving the "tough task" to HEPCO rather than to the provincial or federal government meant that neither would be saddled with the political repercussions. But it also seemed a conflict of interest.[4] Most of HEPCO's engineers were planning the actual power project, so its Property Branch was tasked with the reconstruction project, and it worked in conjunction with the provincial government, particularly the Community Planning Branch of the Department of Planning and Development.

HEPCO had experience to draw on. There were examples in Ontario's history of private power dams flooding out settlements, and other precedents in Canada where the government had expropriated land – for national parks, for instance.[5] In the early 1950s, HEPCO was in the midst of community rehabilitation projects, on a much smaller scale, at its Ottawa River power developments and the new Sir Adam Beck 2 power plant at Niagara. HEPCO consulted agencies with similar experience, such as the Tennessee Valley Authority (TVA) and the US Army Corps of Engineers. Interestingly, the TVA was the product of Franklin Roosevelt's administration, and his interest in state-controlled public power proceeded from his creation of PASNY while governor of New York to develop power from the St. Lawrence River. The TVA had flooded out five communities, though neither the authority nor the governments involved relocated the submerged residents – real estate developers did start two new towns nearby for those displaced – in part because, according to the chief engineer, "it has been our experience that in this day many of the inhabitants living in small communities are there by force of circumstance and have little desire to take part in the establishment of a new community if they can receive fair value for their real estate and move to a location they desire."[6] The Army Corps of Engineers reported that its preferred policy was to buy the individual properties at fair market value, let the owner assume responsibility for relocation, and leave planning responsibilities for new towns to the communities.[7] Ultimately, HEPCO and the province would assume a much greater degree of responsibility for the supplanted communities than had these US predecessors.

In May 1952, the Ontario government passed the St. Lawrence Development Act, which outlined the procedures for handling the area affected by the power project. At Ontario-HEPCO meetings leading up to the commencement of construction, it was apparent that HEPCO was mostly concerned with the technical and financial aspects of relocation, while the Community Planning Branch of the provincial government with what it termed the "human element": "The planning should not be purely technical but must have regard to the local social and historical factors and should reflect the local aspirations."[8] All sides agreed that "these relocations provided an excellent opportunity to demonstrate the benefits of wise community planning"; as matters progressed, it became clear that "wise" meant "modern." By reorganizing spatial and physical environments, planners aimed to "improve" the lives of area residents, as well as the residents themselves. HEPCO boasted that it would create parkland exceeding "in every way" the scenic value of the pre-seaway era.[9]

With the start of the project just months away, however, the affected communities were still in the dark about their future.[10] The people were understandably anxious for information, and eager to provide input and advice on what they wanted done. The popular chairman of HEPCO, Robert Saunders, promised the soon-to-be-displaced residents of "the Front" – as the area of Ontario bordering the IRS was known locally – that they would be treated well. He cited his power entity's record on frequency standardization and the rehabilitation connected to power development at Niagara, reassuring the public that every expropriated property owner had the right to be rehabilitated or compensated. Since 1945, Saunders declared, "Ontario Hydro [HEPCO] has completed 113,330 property transactions involving outright purchase or acquirements of rights-of-way. Out of all these dealing with property owners only 131 cases – or .118 percent – were referred to the official valuator."[11]

Underlying this approach was the idea that the people had to be relocated for the greater benefit of the nation in "one of the greatest experiments in social engineering in Canada's history."[12] The unquestioning attitude of the state toward the utility of the project was equally apparent across the border, although PASNY chairman Robert Moses was not as reassuring as his Ontario counterpart, showing in some respects the differences between the two countries in their approach to land acquisition. In a publication outlining the process in New York, Moses stated that a "sacrifice" had to be made "for the common good," while warning that PASNY would not "be cajoled, threatened, intimidated or pressured into modifying sound engineering plans to suit selfish private interests."[13] Moses considered anyone who did not willingly give over their land a speculator, and those who did not agree with all of his plans were part of the selfish private interests.

PASNY adopted a blanket expropriation plan, but without a bonus for inconvenience, with the goal of having title to all necessary land by the end of 1955.[14] The engineer's report identified four purposes for which land needed to be acquired: construction of major power facilities, construction of seaway facilities, areas required to deposit dredged material, and areas required for flooding purposes. The Federal Power Commission licence had stipulated that the American power entity would need the land up to a measurement of 249 feet above sea level; this was revised to 246 feet, and PASNY generally also took the 100 feet above this line, though it claimed that it did not "propose to prevent the reasonable use" of this area by the public once the project was complete.[15] The use of the 246-foot line (plus 100 feet above) meant that the amount of land taken was a little less than what had been envisioned in earlier plans based on 249 feet.

Easements were rejected in New York, just as in Ontario, as possessing the land outright (i.e., "fee title") was the only "practicable method." In New York State, homeowners were generally offered a market value price, and they could either take it or leave it. But PASNY and its agents had ways of exerting pressure, and residents were made to feel that the first offer was the best they could expect, for subsequent offers were likely to be lower. Agents were accused of first targeting the vulnerable or unaware by getting them to agree to deals that would deflate values in the area. Many people were naive or ignorant when it came to their rights and to the worth of their property. If property owners refused, they were – to a greater degree than in Canada – forced off their land and told that they could argue about compensation afterward. Internally, PASNY admitted that "in some cases" during the first year of property acquisition, the authority had been "unable to give much notice of intent to acquire as desirable and [has] had to move in to construct immediately after filing maps to acquire and before negotiations for purchase have been undertaken."[16] However, PASNY showed little public remorse and took measures to simplify and accelerate its land procurement program, which including hiring an outside appraisal company.

HEPCO considered the same blanket approach but found it too unwieldy for the much larger population it had to deal with. There were too many complications and variables, such as the difficulty in the pre-inundation stage of knowing the full reach of land that was required, the necessity of providing highway rights-of-way, the need to pay interest on

FIGURE 5.2   Moulinette. © *Lost Villages Historical Society*

expropriated land, and the unfortunate impact wide-ranging expropriation notices might have on the affected communities.[17] There were fears about land speculation, which generally proved unfounded, as growth in the area had been retarded over the years by the threat of the St. Lawrence project. For example, there were a number of vacancies in the towns along the Front; the largest employer in Mille Roches, the Provincial Paper Mill, was in the process of moving; and the Canada Cottons Mill in Cornwall ceased operations in 1959. On the other hand, despite having been exposed for decades to the idea that the St. Lawrence project was imminent, many residents stopped taking the possibility seriously because there had been so many false starts and failed treaties. For those who did not dismiss the possibility, the constant uncertainty was wearing.

The residents along the Front would surely have been chagrined if they knew that a great deal of Canadian territory, and some of the communities, could have been saved from inundation through the construction of dikes. These were used more extensively on the American side, and as late as

Figure 5.3   Mille Roches. © *Lost Villages Historical Society*

summer 1953 Canada and Ontario were still considering the installation of a larger system of dikes. However, HEPCO was against their use, except for the area immediately adjacent to the power dam. Although the archival record is scarce on the reasons, this line of thinking seemingly stemmed from cost concerns and worries that these earthen embankments would ruin the scenic appearance of, and access to, the new shoreline.[18]

Planning, and requests from the affected communities, accelerated when the power project was given the official go-ahead in mid-1954. HEPCO agents had already started approaching land owners. A July 1954 HEPCO report estimated that 2,035 transactions – that is, residences – would be required, at an average settlement of $10,000 to $12,000.[19] Representatives from the Front requested a meeting with the Ontario premier, Leslie Frost, citing rumours and complaints about HEPCO's approach, such as buying land it did not need so that it could later be sold at a profit, or offering ridiculously low prices for farms. The premier publicly assured them that residents would be treated fairly, stating that they would have their choice of cash or rehabilitation, and suggested that the government could change the 1952 St. Lawrence Development Act, the provincial legislation governing the process.

HEPCO had refrained from making a public statement on its envisioned policies and plans until it arrived at a formal agreement with PASNY on the power project. This took place in July 1954, and the two power entities set a shared $70 million cap on rehabilitation costs in the two countries, though expenses ended up totalling closer to $90 million.[20] After coming to this agreement, HEPCO immediately released to the public its extended vision for the area. The new or relocated communities would all initially have populations of around eleven hundred residents (aside from Morrisburg, which would be closer to two thousand) but would expand to populations ranging from three to twelve thousand, with long-term expansion ranging from five to thirty thousand. Professor Kent Barker from the University of Toronto helped with governmental planning, which envisioned the new towns abandoning the narrow-strip "main street" town design. Instead, there would be a clear separation of functions, as industry, stores, and municipal services would be concentrated and isolated from residential areas, and residents could expect a higher standard of living and better municipal services.

However, the lack of specific information, and the radically different town plans, served only to exacerbate local apprehensions. A HEPCO report later termed the spirit prevailing in the St. Lawrence Valley at this formative time as "militant." There were problems of miscommunication, and different or contradictory information from various arms of the government. Cross-cutting jurisdictions proved frustrating. Responsibility for certain aspects was sometimes divided between different branches of HEPCO, or between the power utility and the provincial government, or between provincial and federal authorities if seaway work was involved. HEPCO contributed to this situation by at times exploiting the lack of clarity. The reeve of Iroquois expressed confusion about these jurisdictional issues and declared HEPCO's plans for Iroquois as "wholly unacceptable to the people of the community."[21] A report was submitted to HEPCO and the Ontario Department of Planning and Development pointedly asking about several major issues that HEPCO had not elucidated, such as locations and compensation.[22] This had been drafted at a meeting of forty representatives from elected councils and planning boards of the communities that were to be adversely affected by the St. Lawrence project. The report concluded that, since the benefits of the St. Lawrence Seaway and Power Project were national in scope, and the project was to

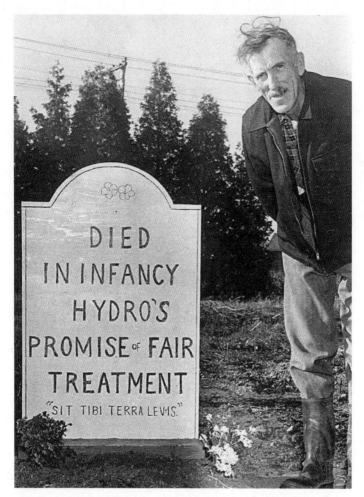

FIGURE 5.4 Manifestation of those unhappy with HEPCO's treatment of the Lost Villages and area. © *Stormont, Dundas, and Glengarry Historical Society, courtesy of Cornwall Community Museum*

be self-liquidating, the people of the area should not have to bear any of the expenses. It also declared that compensation should include replacement value plus extra for inconvenience and disturbance.

Numerous newspapers called on HEPCO to treat the people of the St. Lawrence Valley fairly. The *Port Arthur News-Chronicle* lamented that "the whole business appears to be one more illustration of the traditional great failure of the governmental machinery – the difficulty in working together, or letting the left hand know what the right hand is doing." Even though

Premier Frost wanted to be "decent and humanitarian" and although the HEPCO chairman might be sympathetic, the newspaper aptly predicted that "by the time it gets down to the lower levels and the actual business of negotiation the outlook is changed. Then minor officials are involved and they are out to do one thing: make an impression. And to them there is one way to make an impression: to save money."[23] On the other hand, the editors of the *Peterborough Examiner* "sincerely hoped that the Hydro [HEPCO] will not be too tender-hearted, for whenever the Hydro's heart is touched, it costs the Ontario taxpayers a lot of money."[24]

HEPCO was in the midst of internally debating how much of the cost would actually have to be borne by the residents and did not put much stock in the value of river views or history compared with new amenities. For example, in Iroquois, 206 of the 231 residential properties were worth less than $10,000 each, which meant that HEPCO would lose a combined total of almost a million dollars if it provided replacements in that value range; the commission was not willing to swallow this cost whole. Many people could not afford the new houses that HEPCO would construct in new Iroquois, or would need financial assistance. To HEPCO it was "immediately obvious that we cannot 'socialize' New Iroquois by giving new houses for old indiscriminately," since that course of action might result in people immediately selling their new house for a tidy profit. Nor could the townspeople expect improved services at the same tax rate, as a smaller population would likely be the initial result of the relocation.[25]

In historian Carleton Mabee's vivid description of the land acquisition process, HEPCO was both the pitcher and the umpire, and had virtually written the league rules as well.[26] HEPCO legal experts pored over the statutes that concerned expropriation and resettlement, chiefly the provincial St. Lawrence Development Act. Giving an award of an additional 10 percent as compensation had been established as court custom, as was defining "market value" as synonymous with value at the time of property seizure.[27] One provision within the act allowed a person to claim compensation if a relocated road took traffic away from his or her business, but HEPCO had itself successfully denied such claims in the past, with the assistance of the courts. Other means of watering down legislation beneficial to landowners were examined – for example, a memorandum recommended that, in any future attempts to amend the St. Lawrence Development Act, HEPCO's obligation to notify owners of their right to claim should be deleted. The author of a history of HEPCO writes that Robert Macaulay, a minister without portfolio in the Ontario government and the HEPCO second vice-chairman, was particularly instrumental in redrafting legislation to permit expropriations.[28]

"We Demand Replacement Value"

Throughout the late summer and early fall of 1954, HEPCO spent most of its time obtaining information that would allow it to appraise property values. It had, nevertheless, already started obtaining property in areas where construction would soon commence. These early efforts were not well received, and there were accusations about the acquisition agents and prices; in particular, indictments that they targeted the elderly and unaware were rampant. HEPCO realized that its approach was proving unpopular and its offers were perceived as unfair. As one newspaper put it, HEPCO had failed to consider "the psychological factor" and there were "serious faults in the first approach of Hydro [HEPCO] to the problem although the general policy was sound and fair."[29] HEPCO was surprised that some people were not attracted by new houses, modern towns, and higher living standards. Although the belief that the St. Lawrence project represented progress and technology was dominant at all levels, many highly valued their connection to the river and their way of life. This perspective found its way into the media. An *Ottawa Evening Citizen* article, for example, pointed to the people of the district who were "not interested in the uniformity and degree of regimentation which life in one of the new communities will necessarily entail." They would rather "have their old well in the back yard with its moss-covered bucket than modern waterworks … You can't put a price tag on history or buy memories at so much a yard, they'll tell you.[30]

Opinion differed in the various communities, but many residents were not that impressed with plans for the modern new houses, shopping centres, and general town designs. One respondent suggested that, since they were losing "intangibles" – such as life on the river – HEPCO should be generous with the "tangibles" – meaning compensation. Records of internal HEPCO discussions show that officials did sometimes recognize these "intangibles" but because it was so difficult to put a price tag on them, or because they were not accorded much value from the utility's perspective, they rarely bothered to try. Similar sentiments emerged from a mid-September survey of the St. Lawrence region by the provincial Department of Planning and Development.[31] The chief concern of the whole Lost Villages area was the basis for compensation, as most property owners wanted an equivalent structure to replace what they were losing. That was HEPCO's policy when it came to home appliances and frequency standardization.[32] Iroquois strung up a street banner proclaiming "We demand replacement value."

The Town of Iroquois would serve as a sort of guinea pig. The westernmost of the affected municipalities, Iroquois was slated as the first for removal. In his typical fashion, Saunders showed up in Iroquois in shirt sleeves one late August night to reassure the people that HEPCO was not wedded to its proposals and was responsive to the wishes of the Front. Several years earlier, Iroquois had hired Wells Coates, a Canadian-born British planning architect, who promoted his plans for a new Iroquois that would attract industry from Britain. Coates had exchanged correspondence with the Ontario Department of Planning staff, who even considered taking him on as a consultant.[33] Coates himself was wildly fanciful in his expectations for new Iroquois and did not have the town's support to the extent he claimed. He justly criticized HEPCO's plans for the community as too "ivory tower," pointing out that the western orientation of buildings was not suited to local conditions nor future industrial locations, noting the poor location of the highway and docking facilities. His proposal to completely move the town further east to take advantage of a natural harbour area was adopted by the municipal council. The plan had much to commend it, but the vision of an ultra-modern seventeen-to-twenty-storey building with a neon sign spelling "Iroquois" in large capital letters at the top, and a future population of between thirty and forty thousand, was not realistic.[34] Moreover, it required locating the town in the territory of neighbouring Matilda Township, which was resolutely opposed, whereas the HEPCO proposal was to locate new Iroquois within its existing limits, in pasture land a little less than a mile from its historic location.

The new location for Iroquois obviously had to be settled. At a meeting with HEPCO executives in early September, Coates continued to push for his alternative Iroquois scheme. Saunders presented himself as open to discussion but would not commit to Coates's plan, resulting in a war of words. The HEPCO chairman met many times with the council and residents of Iroquois and kept reiterating that HEPCO would "pay as much as we can justify – not as little as we can get away with." The townspeople made the counter-argument that there was no fair market value because the St. Lawrence project had depressed the economy. Others expressed dissatisfaction with HEPCO's proposed location, especially

FIGURE 5.5   Aerial view of the relocated town of Iroquois; the cleared area along the water and traversing the lock is the old town site. © *Stormont, Dundas, and Glengarry Historical Society, courtesy of Cornwall Community Museum*

moving the town off the river. Yet, the situation became even more confused as the community split into different factions. Saunders told Iroquois that it could choose its location but sought means of pressuring the town into accepting HEPCO's plan, highlighting Matilda Township's opposition to the Coates location and support for the HEPCO plan, and pointed out that the town would have to pay the extra expenses of building on marine clay at the Coates site.[35] It also became clear that the people were likely to follow Caldwell Linen Mills, the main local employer, to wherever it relocated. Recognizing an opportunity to get his way, Saunders began negotiations to entice Caldwell to move to HEPCO's proposed location. By November, a $1.3 million agreement was in place with the mill, and the Iroquois village council voted to accept HEPCO's site for new Iroquois.

In response to public feedback, HEPCO had begun exploring the idea of moving houses, an option that PASNY chairman Robert Moses had used in some of his previous New York City projects. In early October 1954, Saunders and other officials went to New York to see a Hartshorne house-moving machine in action. It cost over $100,000 to build, weighed 35 tons, and could move a house sixty-five-feet long by approximately thirty-six-feet wide, carrying a load of up to 150 tons at a speed of approximately six miles an hours. In an on-site estimate, a Hartshorne representative reckoned that the company could move 219 of the 279 houses in Iroquois.[36] The idea proved very attractive to HEPCO officials, who saw this as a means of effecting great savings in a way that would also be better received by the displaced people, particularly if they were given larger, landscaped lots.

Shortly after establishing the new Iroquois site, Saunders met with the councils of several other townships. In response to demands for replacement value, the chairman tried to convince them that this was not in their own best interest and urged that "it cannot be faced merely by saying they will obtain a new house for an old. That cannot be. In some cases it can be, where an old house is worth as much or more than a new one. It cannot be done as a general rule. We may as well face that."[37] He pointed out the situation that could result if a blanket new house policy were adopted: if there were two neighbours, one with a $3,000 house and the other with a $10,000 house and they both received new $10,000 houses, one made a $7,000 profit. The chairman kept stressing that the people would receive a "square deal" financially and that his organization was committed to doing "a real job": no person would suffer, but no one would make a profit. HEPCO would pay as much as it could justify, rather than as little as possible, and would be more generous than the letter of the law; but as trustees of the people of the province, it could not overpay or fall prey to speculators. The people could go to court over compensation if they so desired, but Saunders warned that past experience indicated that they would get only 75 to 80 percent of what HEPCO was offering. The utility offered to pay taxes on property it took over until the flooding in 1958. Saunders promised that those who were already on the water would still be able to have waterfront property, and promoted what HEPCO had done for the people at Iroquois.[38] The language of progress and economic growth was invoked to further assuage the fears of the residents; the people of the valley were reminded that their region would become "the greatest industrial area in the Dominion of Canada."[39]

Growing anxious about the situation, the Ontario government met with representatives of the St. Lawrence municipalities in December 1954. Stemming from this, the Ontario cabinet sought to "put our resources at the disposal of the municipalities so that the municipalities might, with their planners, in a sense be in a position that they could meet the Hydro [HEPCO] on even ground and come to resolve as to what should be done by a series of conferences."[40] In addition to opening information offices, effecting better consultation with the municipalities, and making planning experts available to the municipalities (at HEPCO's expense), the Ontario cabinet recommended that a new formula for compensation be established. A 15 percent, rather than 10 percent, award should be given for forceful taking. Homes should be moved where possible and the exterior improved at no cost to the owner, with separate provisions for renters and farmers. Amendments were made to the St. Lawrence Development Act to enact these cabinet recommendations and ensure "fair, just and equitable compensation." The Ontario Department of Planning and Development began working with the affected communities and their planning boards. The provincial government also established a St. Lawrence Board of Review, consisting of representatives of the government and the St. Lawrence communities, for matters relating to compensation and relocation.

However, after the spate of heated public consultations, Saunders indicated that HEPCO would henceforth meet only with the planning boards. From HEPCO's perspective, this may have been more effective than crowded, emotionally charged town hall meetings, but these boards and municipal councils tended to be made up of local business and economic elites who were not always representative of the wider interest, and the general public often had its input circumscribed as a result.[41] Moreover, this step did not look good in comparison to the moves the provincial government had just made to assuage local worries.

Saunders perished in a plane crash in January 1955, and his successors were not as adept at handling the commission's public relations. In February, various types of consultations took place with most of the communities. A survey of Farran's Point, for example, showed that of the 50 homeowners, 15 planned to sell, 16 would relocate, and 19 were undecided.[42] Similar proportions obtained in Dickinson's Landing and Wales, with around one-third undecided about what they would do. In total, 803 contacts were made, and of the 499 owners, 147 planned to sell, 231 would move their homes, 14 would sell and move away, and 107 were undecided.[43] The comments from Wales were fairly representative of the survey results, or at least representative of the HEPCO agents' impressions:

> The people of Wales did not seem to be greatly concerned over rehabilitation and for the most part volunteered information that would indicate

that they are confident that Ontario Hydro [HEPCO] will treat them fairly and carry out the rehabilitation program in a highly efficient manner. As in other towns they have heard rumours of the transactions where Hydro has already acquired property and the general opinion was that the prices paid were fair.[44]

This account did not paint an entirely accurate picture, as apprehension still remained, particularly regarding compensation and the approach of HEPCO's property officers. As one newspaper article contended, these agents were used to the hard bargaining and horse-trading that were common in other circumstances; but HEPCO had a responsibility to come in with a fair initial offer to residents who were unaccustomed to such dealings, under a great deal of stress, and unsure about their property value.[45] Iroquois members of council complained that "there was undue pressure by agents on property owners – no breakdown on how property values were arrived at – apparent wide discrepancies in offers made to property owners as between properties of similar value, and that in many instances the amount of 15 percent for forcible taking was not enough."[46] Some of these examples could certainly be ascribed to unscrupulous representatives: the unfortunately named director of property for HEPCO, Harry Hustler, complained internally that "one of the biggest troubles in taking on agents is in preventing them from asking deals for their own satisfaction."[47]

Preparations for the actual movement of Iroquois got underway as the snow melted in 1955. Some properties on Iroquois Point were needed by the federal government for the seaway lock there, but most were being moved back because of the power project. The relocation was carefully orchestrated to work with the complex organization and timelines required to coordinate the different parts of the St. Lawrence Seaway and Power Project. The number of houses that could be moved had been revised to 185, and further surveys were conducted to solicit the preferences of residents.[48] Those buildings that could not be moved would be destroyed. HEPCO had promised improved amenities for all homes, such as running water and sewers, and negotiations ensued with the town on these aspects. The layout of the new town remained an object of debate – not just the design of the streets and the different types of zoning, but where houses would be placed (e.g., with a good view, or beside former neighbours). Eventually, in many cases, HEPCO decided to organize the lots by the architectural style of home.[49]

FIGURE 5.6   House mover in action. © *Stormont, Dundas, and Glengarry Historical Society,* courtesy of Cornwall Community Museum

House Moving

House moving had been added to the previous options for homeowners, which included cash, rehabilitation, or a combination of the two. The first homes were moved in July 1955, and the house-moving process, using two machines and support equipment, proved to be a major draw for the national media as well as for locals and tourists.[50] Homeowners were notified of their probable move date a few weeks ahead of time and advised that they did not need to pack because of the stability and gentleness of the house-moving machines. People still enjoy telling about fine china placed on window sills or dishes on tables that were left undisturbed by the move. The movers lifted houses off their foundations by inserting metal beams into holes drilled into the house foundation; the building was then placed on a mobile house float equipped with ten-foot wheels and driven slowly to its new location. While a house was relocated and refurnished, its occupants could reside in stopover homes, which HEPCO provided, for a period of anywhere from a day to a few weeks. HEPCO would do minor improvements on relocated buildings of up to 15 percent of the valuation of the property. Later there were some moves involving unique structures, such as Christ Church at Moulinette or a complete bulk storage plant near Morrisburg.

At Morrisburg, the commercial section near the river and about one-third of the residential area had to be moved. Some houses were borderline in terms of necessity of removal, but after consideration was given to raising buildings that had basements near or below the expected water elevation, HEPCO decided to just buy out these residences.[51] A new shopping centre to replace the business and retail district was created at the north end of the town, off the river, as well as a new subdivision. HEPCO paid the cost of moving forty businesses, though without the 15 percent allowance. There was not enough room in the new towns and shopping plazas for the many duplicate businesses coming from the various Lost Villages, so some migrated to the new malls but changed their businesses or diversified. Others lost their stores or retired. Specialized enterprises replaced general stores, which had acted as important social congregating spaces. Many businesses had smaller retail space in their new locations, which HEPCO contended was compensated for by the more efficient layout and the improved features of the modern premises.

The form and location of the two new towns, as well as highways, railways, and railway stations, remained ongoing points of contention. Representatives of the Front, HEPCO, and the Department of Planning and Development debated many options over the course of many meetings, such as the location for Highway 2, which had been the main highway between Montreal and Toronto and served as the main street of all the Lost Villages, with the exception of Wales. The location and layout of New Towns 1 and 2 also still remained (they were named Ingleside and Long Sault respectively in January 1956).[52] The displaced communities on the eastern end of the rehabilitated zone, closest to Cornwall, would be amalgamated into New Town 2, whereas New Town 1 would be composed of the remaining communities running west to Morrisburg. The plans for New Town 2 were approved by members of the Cornwall and district planning board in early August 1955, largely without the input of the people moving there from Moulinette and Mille Roches, and two days later, after some minor changes to the street design, Osnabruck Town Council approved the plans for New Town 1. According to author Carleton Mabee, people moving to Long Sault had less influence than the occupants of Ingleside over the shape of their new community because it was the Cornwall Township council and planning board that made the decisions.[53]

Controlling the shoreline was an important concern. HEPCO determined that it needed to obtain all land up to a contour of 250 feet above

Lost Villages

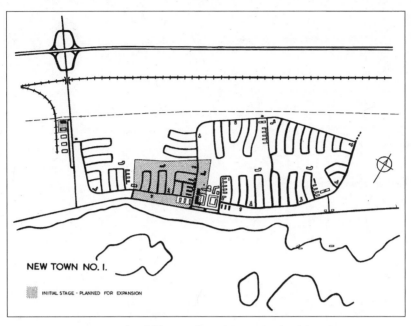

FIGURE 5.7   Long-term planned layout of New Town 1 (Ingleside). © *Ontario Power Generation*

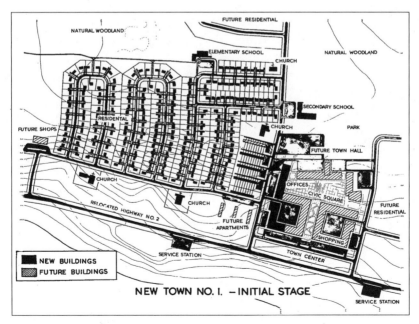

FIGURE 5.8   Initial layout of New Town 1 (Ingleside). © *Ontario Power Generation*

FIGURE 5.9   Long-term planned layout of New Town 2 (Long Sault). © *Ontario Power Generation*

sea level along the entire shoreline from Cornwall to Prescott, as it needed a buffer zone to accommodate changed water levels resulting from the operation of the power works. There were large swaths of ground between the new shoreline and the 250-foot mark, though no private property was allowed along the water, as it would open up HEPCO to liability claims resulting from erosion, seepage, or high water levels.[54] HEPCO also wanted to be relieved of the continuing obligation of management, control, leasing, and disposition of much of the land bordering the river. To deal with this situation but keep the shoreline under government control, the Ontario–St. Lawrence Development Commission was formed in 1955. It was chaired by George H. Challies, a local who had spent many years in the provincial legislature and had recently been a HEPCO commissioner.[55]

Over the course of the first year of the official rehabilitation process, HEPCO had responded to the desires of the Lost Villages in some respects, such as offering house moving (granted, this benefited HEPCO as well), and had moved toward greater recognition of the "psychological" factor, better communication, and more generous compensation. It had, however, resisted the loud public cry for replacement value or a standardized replacement program. The Frost government appeared as the more conciliatory half of the Ontario-HEPCO dyad, but this was because it had votes to worry about. Rivalry between Saunders and Frost, prior to the former's death, may also have factored into the evolution of this issue.[56] Nevertheless, both the power entity and the provincial government were fully subscribed to the idea that the state had a responsibility to bring the St. Lawrence Seaway and Power Project to fruition for the advancement of the province and the nation.

HEPCO's dedication to dealing with each property owner individually was both philosophical and pragmatic: it kept with the idea of the individual liberal property owner but equally served to keep home prices down and avoid the overpayment that would likely result from a blanket policy. In retrospect, competing discourses of liberalism, particularly regarding private property rights, are apparent.[57] The St. Lawrence Seaway and Power Project was certainly aligned to the quintessential liberal purposes of providing wealth-generating conditions, infrastructure, and social order. The appropriations and relocations were justifiable from the state's perspective, despite the apparent violation of private property rights, because they were based on a utilitarian calculus in which moving a few thousand people for a project that would tremendously benefit the entire Canadian economy was warranted. In resituating and reorganizing the Front communities, the Canadian and Ontario governments were, in their minds, prompting the villagers and surrounding farmers to participate more fully in the postwar industrial capitalist order. Conversely, the liberal discourse of the Lost Villagers, which called for replacement value and standardized prices while resisting individual transactions, attests to uniquely local interpretations of liberal, even collectivist, values.

Main street signs reading "We have to go – but watch us grow" replaced "We demand replacement value." This reflected the optimistic spin, forced or authentic, that people put on the situation in order to cope. Harry Hustler maintained, "I have not heard of any real criticism of our methods of acquiring property. The most criticism I heard about our methods was before we ever began to buy property." If indeed this was HEPCO's perception, it was because many people were afraid to voice their opinions, not only out of a general respect for authority but also because they feared there might be repercussions. For the same reasons, many were reluctant

FIGURE 5.10   Sign in Iroquois. © *Stormont, Dundas, and Glengarry Historical Society, courtesy of Cornwall Community Museum*

to appeal to the St. Lawrence Board of Review. Moreover, since the board was really an informal forum, its decisions were not binding, and there were complaints that it sided with HEPCO anyway. HEPCO was reasonably confident that, given its past experience and the experiences of other similar projects in and outside Canada, the board's decisions would be roughly in line with what the hydro commission offered.[58]

Voices of dissent could be heard across the river. Courts in the United States generally upheld the prices PASNY granted.[59] On the New York side, the complaints about land acquisition offers and techniques were noticeable enough that the New York governor brought them up with Moses, and President Eisenhower asked a member of the Federal Power Commission to inquire into the matter. Moses responded, in characteristic fashion, by stressing the urgency of the acquisitions and argued that "quite a few complaints which have emanated from the St. Lawrence area have been motivated by the most dubious political and personal considerations by owners and their representatives who are primarily interested in speculative increases in value."[60] Moses indicated that he would be glad, however, to investigate any specific complaints that had been brought up. In an exchange the following month with the PASNY chairman, US senator Robert C. McEwan, who represented a district bordering the St. Lawrence, wrote, "I can conservatively say that I have heard dozens of complaints over recent land acquisition policies and methods, and I can honestly say that I could not fairly label one of these complaints as a speculator."[61] The senator protested the need to take all riverside land, and criticized the acquisition agents' approach.[62] Moses retorted with a nasty letter, stating that McEwan's correspondence "consists almost entirely of garbled, rambling assertions unsupported by evidence."[63]

Some people in Ontario ventured to publicly oppose the process, taking their case to the Board of Review. One person brought with him HEPCO's offer for his property on Sheek Island, where there were a number of residences, farms, and cottages. The board concluded that HEPCO's offer of $22,080 was fair but added a special allowance of $1,560 representing a rental value and permitted him to remain on the property until the end of 1956.[64] In most cases over the following years, the Board of Review deemed HEPCO's offers fair. There were exceptions: for example, an April 1956 decision awarded one farmer a higher price for his farm in Matilda Township. However, HEPCO decided to reject this, alleging that the excessively high award of $26,000 would distort prices in the area, and took the case to the Ontario Municipal Board of Review.[65]

HEPCO's approach privileged those in certain classes who could afford to own their property, though one former resident of the Lost Villages speculates that lower-income groups may have materially benefited the most from the move.[66] A survey by the provincial government indicated that there were fewer than eighty people in the area on some form of public assistance, but also low-wage earners. These included residents of the "shack town" on the fringe of Mille Roches. The problem with the new housing in the relocated communities was that they were beyond the affordability of these low-income groups, particularly the elderly, and no cheap housing existed. HEPCO was legally required to provide assistance, so at new Iroquois it converted large houses into multi-unit dwellings, and in Morrisburg it built low-cost dwellings, which it subsidized according to income without the assistance of the federal or provincial governments.[67] Tenants who did not own property were offered an amount equal to 7.5 percent of the value of their premises, which averaged about $500.

Others were accommodated in empty homes that HEPCO had bought and moved to the communities, and those who accepted freed Ontario of continuing responsibility.

HEPCO had originally disclaimed responsibility for cottages in the area, since they were leased on federal land, but eventually relented and sought to relocate the summer cottages affected by the power project. Some of these cottages were moved to islands that would be created by flooding (i.e., heights of land along the old shoreline that remained above the new water level of Lake St. Lawrence).[68] Most of these were from Sheek Island, and a number were slid across the ice in the winter. By the end of 1956, eighteen cemeteries, containing 5,059 known graves, had been combined into St. Lawrence Valley Union Cemetery (which had separate sections for different denominations). A new cemetery was built for the hamlet of Maple Grove, and the Pioneer Memorial was placed by the province at Upper Canada Village to house the tombstones of early settlers.[69] On the American side, about a thousand headstones and remains had to be moved.

Both HEPCO and PASNY were going to be left with great swaths of land, though the leaders of the power utilities differed in the degree of enthusiasm with which they embraced the opportunity to create recreational opportunities. HEPCO eventually sold 5,100 acres (for approximately $659,000) along the riverfront to the Ontario–St. Lawrence Development Commission, and this parks commission was given authority to develop "surplus" land for recreational purposes, including a parkway linking nine of the eighteen new islands, along with beaches, camping facilities, and other parks. PASNY noted the impact that recreational facilities and tourist dollars had played in the economic development of post-construction TVA areas. Moses himself oversaw the planning of an extensive park system on Barnhart Island featuring beaches, campsites, and a marina, and these plans were communicated to the public via slick reports and pamphlets. PASNY also paid for other recreation areas such as new beaches for Massena and Waddington, allowed a leased boat dockage area near Massena, and created the Coles Creek marina.[70] But the recreational facilities were not developed to the level promised, and the expropriated land required for these parks were taken from the communities without compensation. Massena, for example, claimed that it would lose about $4 million in assessed valuation because of land it had lost.[71]

Unlike HEPCO, PASNY did little to assuage the fears of residents, and the power authority's land agents were perhaps even more despised than were their Ontario counterparts.[72] In St. Lawrence County, bitterness toward PASNY is still palpable and sustained by the power authority's ongoing legacy of empty promises. The Wilson Hill development was likely the only instance in New York where local opposition forced PASNY to substantially modify its plans. PASNY initially allowed quasi-waterfront property on twenty-year leases at Wilson Hill – the power authority technically retained control of the land directly abutting the water, as it did all along Lake St. Lawrence – first for those losing a house or cottage and unable to find a replacement through private sale, and then for the general public. An unsuccessful lawsuit was brought against this appropriation of land on Wilson Hill, though PASNY ended up paying approximately 24 percent more than the appraised value, largely to dispense with the case, and let the claimants buy back some of their unnecessarily expropriated land.[73] PASNY had originally anticipated that up to two hundred owners might relocate there but, by summer 1956, only forty-two had done so, as many were apparently dissuaded by the new water supply and sanitation works that had been installed and whose cost was being passed on to the new residents.[74] As a result, PASNY returned the lot deposits that had already been made, scaled back considerably the upgrades and services it planned for the area, and then reopened the area for cottages under new terms.[75]

Numerous houses and buildings were moved on the American side, though these relocations were arranged privately.[76] The Louisville Landing Community Church was the only place of worship affected. At Waddington, the only road lost to the rising water was River Street, though much of the land was reclaimed with spoil and a new waterfront recreation area was developed for the town. An old power dam stretching from the mainland to Ogden Island in front of Waddington that had hosted several mills also ended up under the water. The Ogden Mansion and several other homes on the eponymous island were lost to the project even though the land on which the mansion stood was not submerged.[77] A stretch of Waddington houses was allowed to remain on the new waterfront because it was cheaper than relocating them. These houses were jacked up – Russell Strait remembers that his family spent an entire winter with their house in this raised position on stilts, followed by a quagmire of mud in the spring – and then given foundations and fill to keep them above the new water level.[78] Some property owners at other places on the south shore of the international section of the St. Lawrence were granted rights-of-way to the water, but there were strict limitations on what could be done with this land and PASNY's permission was needed to build even just docks and boathouses. Many who lost waterfront property were promised that they would eventually have the option of reacquiring it, but a half century later there are very few instances of this actually taking place.

In Ontario, no private development, with a few exceptions, would be allowed directly on the river. Future use of the land would be controlled by restricting to whom and for what purpose the land could be sold. But much of the new riparian zone, including the area directly in front of the two new towns, would be flat, unsightly, muddy, and marshy scrub areas. This detracted from Ontario's plan to use the power project as an opportunity to "beautify" the area, and HEPCO looked into creating a more abrupt shoreline to lessen the effect. However, further study and cost later dissuaded them, and proposed excavations to deepen the water in front of the new towns were eventually dropped, with only minimal excavation work taking place in a few locations.[79] Ensuring that unsightly buildings and structures could not be built along the new highways or waterfront was a goal for both New York and Ontario, which aimed to ensure that, in the words of the PASNY chairman, "billboards, sheds, dog stands and other eyesores" did not spring up.[80] In October 1957, the relocated Highway 2 in Ontario was designated a controlled-access highway, which placed limits on building and property affronting the roadway.

Some locals were concerned about remembering the area's history. The Ontario Historical Society approached the Ontario–St. Lawrence Development Commission about preserving the archives, architectural details, and material culture of the soon-to-be-flooded zone. Parks and golf courses were created, thousands of trees were planted, and the Upper Canada Migratory Bird Sanctuary was formed in the hopes of attracting Canada geese, which it succeeded in doing.[81] However, most of these recreational areas had formerly been farmland, and those who had land seized but not flooded were justifiably upset.[82] Directly expropriating a piece of land for use as a park was not technically legal, but indirect means were found.

The movement to preserve the history of the region included archaeological digs on Sheek Island, a temporary museum in Morrisburg, and the creation of Crysler Park and a reconstituted replica pioneer village that was eventually named Upper Canada Village. Although the Ontario section of the IRS had deep historical roots dating back to the United Empire Loyalists, not to mention the First Nations, most of them had moved on and the area had become oriented toward milling, textiles, and agriculture. Nonetheless, the perception that this was an area founded by Loyalists ran deep, as did connections to the War of 1812. This made all the more profound the faith in progress and technology; it was worth erasing key parts of Canadian history, literally flooding the site of the Battle of Crysler's Farm from the War of 1812. The memorial there was relocated to a hill on the new shore beside Upper Canada Village. Historically and architecturally significant buildings were selected and moved to the site of this new living history museum (staffed by residents of the Front), though only fifteen of the original forty buildings were from the flooded municipalities. The creation of this historical park was the product of high modernist thinking that simplified space and time, directly contrasting the past with the modern future, and provided an intriguing insight into the juxtaposition of history and progress.

"Once They Accept the Inevitability of Moving"

The extent to which people were satisfied with HEPCO still varied widely. A July 1956 article in the *Globe and Mail* reported that various problems persisted, and some detected a tightening of HEPCO's purse strings. On the other hand, according to one newspaper article, most residents were satisfied "once they accept the inevitability of moving," for "we've had so much from Hydro [HEPCO] we've got to expect to give a little."[83] The reeve of Iroquois believed that the community "lost its big chance to grow industrially when it was weaned away by Hydro from the site the village council had chosen two miles to the east." "But don't get me or any of the rest of the people here wrong," he qualified; "we think Hydro has done well by us. There are many things we might have changed as a result of experience, but by and large we're happy."[84] The reeve of Osnabruck Township felt that "the terms 'discussion' and 'consultation' as far as we are concerned were meaningless" when it came to dialogue with HEPCO.[85] Complaints kept arising about HEPCO's process of informing people about when or how things were going to happen, as well as about the ongoing compensation issue.

HEPCO thought differently. An extensive January 1957 HEPCO report to the chairman (now James Duncan) on land acquisition argued that the commission had been generous – "though not overly so" – in terms of compensation and had erred on the side of benevolence when it came to the schools and churches it was installing in the new towns.[86] It cited as justification the eighteen cases that had been brought before the St. Lawrence Board of Review where the compensation had not been "importantly improved" – a reference to the slight increase in awards the board tended to grant – and one case where the claimant's demands were

FIGURE 5.11 New shopping centre at Long Sault. *Photograph by Eleanor L. (Sis) Dumas.* © *Dumas Seaway Photograph Collection, Mss. coll. 124, Special Collections, St. Lawrence University Libraries*

ludicrous. The author of the report was not sure that the creation of the two new towns was justified, labelling them a response to pressure to "assure the perpetuation of a community life to restore the hamlets." The decision to move homes was identified as the main reason they were able to subdue the "replacement value" demands.

By December 1957, the last of the houses had been moved, and the modern, new four-block shopping mall at Morrisburg had its official opening. A HEPCO news release boasted that "the Morrisburg Centre is attractively constructed of buff brick with the individual store fronts in aluminum and glass; canopies make it possible to visit the entire shopping district without being exposed to rain."[87] A new retail centre was opened in Iroquois, and shopping plazas for Long Sault and Ingleside were scheduled to be completed shortly thereafter.

The projected target date for inundation was July 1, 1958, but whether the seaway and power project entities were going to have all the necessary work finished for the raising of the power pool was uncertain. HEPCO worked feverishly to prepare the area that would be flooded, and to construct its share of the hydroelectric works and control dams. The vast majority of people had been relocated, but before the inundation, anything in the water's path that could create a hazard to navigation or power production had to be removed or razed, particularly as the water cover would be shallow in many places. For the resited cemeteries, relatives were given the option of transplanting headstones, disinterring the entire grave (about two thousand chose this option), or letting it remain in place; the many burial sites that remained were usually covered with stone to prevent erosion or movement under the water.

As Inundation Day approached, the reality of the situation began to hit home. The former town sites looked like war zones, shorn of the stately trees that had graced so many roads, denuded of houses and buildings. When the day came, people congregated at designated locations to witness an event that captivated many; inhabitants of the area certainly counted among this crowd, but some could not bear to watch, and others were indifferent. When the cofferdam was blasted open, the water anticlimactically trickled into the dried out and silenced run of the legendary Long Sault Rapids, which had been subdued by cofferdams since the early phase of the project. The *Cornwall Daily Standard Freeholder* called the blast "an unspectacular whisper."[88] It took over three days to fill the reservoir that the river had become. Over those few days, residents could take a last stroll down their old street or pick out landmarks, if they could recognize them on the decimated landscape, as the water slowly crept up, agonizingly swallowing what already seemed a past life. Even wildlife responded erratically to the creeping water level as it washed out burrows and havens.[89]

The inundation marked the end of the major phase of rehabilitation, though HEPCO still had to finish providing the community infrastructure and landscaping it had promised, as well as final clean-up. The new communities were unsightly and dirty, with ugly scars, mud flats, and frog ponds dotting the terrain. In July 1959, government officials touring the area found the scene somewhat depressing.[90] Complaints flooded in from the relocated communities about problems such as the deficiency of hastily constructed sewer and water lines. These grievances continued well into the 1960s, as did compensation appeals.

In the end, about thirty-eight thousand acres of land were purchased, with more than half ending up submerged. HEPCO reported that it moved 531 houses as well as auxiliary buildings (garages, barns, sheds) and built 350 new homes.[91] HEPCO constructed fourteen new churches and ten parsonages to replace the twenty-four that lay in the path of the flooding, and four new shopping centres with a total of ninety stores. HEPCO provided replacement buildings for town halls, recreational facilities, and municipal services. Miles of paved sidewalk and roads were installed in the new town sites, as well as sewage and water infrastructure (lines, pumping stations, treatment plants) and new "backlot" distribution and power lines. Over 4,350 transactions had taken place, and more than 99 percent of property owners had accepted the commission's offers.[92] Twenty cases had gone to the St. Lawrence Board of Review, and another was determined by the Ontario Municipal Board of Review. An internal HEPCO assessment concluded that "it is evident that the policies of the Commission in this extensive land acquisition programme were fair and reasonable and so acceptable to the property owners."[93]

Finding the Lost Villages

The people of the sunken communities were asked to sacrifice for the greater good. Indeed, there was a societal tendency to accept that this sacrifice was the price to be paid, and it is striking that the Lost Villagers bought into this mindset for the most part. Many people were not upset so much with the premise of the project but that *they* were moving. Others thought there should be greater compensation, or at least better treatment, because of the nature of their sacrifice. HEPCO engineers called the residents' attitude "cautious skepticism."[94] In hindsight, there was a societal deference to government and a willingness to believe its grand promises, and a pervasive belief that the St. Lawrence project would usher in a grand new era of prosperity, which was particularly appealing for an economically depressed area. In the rhetoric of the time, the villagers would benefit directly and materially. Many were simply naive when it came to the worth of their property value or their ability to oppose the state. Of course, there had always been those, both inside and outside government, who did not believe the more extravagant claims of the seaway's benefits. But there was a pervading view that it was all inevitable and so no use fighting. Such attitudes were reflected in North American society, including the academic, media, and government elites. For example, the 1956 Round Table on Man and Industry, a product of University of Toronto sociologists, affirmed the dominant "progress" view of the time in discussing the St. Lawrence Valley relocation, including the future prosperity of the region. It pointed to sociological and societal changes brought on by continent-wide modernization, and those that resulted more specifically from the influx of people for the St. Lawrence project and the consolidation of smaller communities into larger ones.[95]

It was a disorienting experience for those who lived through the relocation.[96] For many, the best way to describe it might be traumatic, though not everyone experienced it the same way or in negative terms. In hindsight, it was a situation characterized by ambiguity and contradiction, and the range of responses from St. Lawrence Valley residents resists easy generalizations.[97] There was a short-term boom during the construction period, as many people saw their business and economic prospects increase, rented out rooms to workers, or acquired higher-paying employment. The promised levels of long-term prosperity, however, did not develop, and as this became apparent, those living in the area grew disenchanted. Those who experienced the rehabilitation first-hand tend to agree that they did not realize what they were getting into until it was too late; they did not comprehend the change and havoc that the dislocation would wreak until it had begun to happen. For some, the reality of the situation was not apparent until they saw the first buildings leaving on gigantic machines, watched great trees being felled, or witnessed the water inching over the remains of their former houses.

The population grew tremendously with the influx of workers – by more than 30 percent in most of the riverside communities.[98] Both power authorities left labourers to find their own lodging. The only accommodation HEPCO built was rental houses for high-ranking officials in the Riverdale area of Cornwall, though it erected an employee hospital and provided limited financial support for municipal services. PASNY built 133 houses near Massena, which were rented to engineers and contractors at subsidized rates.[99] Although the economic benefits of the project certainly did not meet expectations, it could be claimed that the stagnant economic growth of the St. Lawrence Valley had been at least partially reversed. HEPCO could also boast that it had rebuilt the dangerous Highway 2 and made possible the future extension of a modern freeway, Highway 401, from Toronto to the Ontario-Quebec border, to absorb the through traffic. It had given modern towns to those displaced, paid the salaries of extra police officers for the area, and made contributions to municipal councils because of the additional work the project required of them. HEPCO did some small things that people appreciated, such as taking care of perennials and shrubs over the winter, but such gestures were often outweighed by ancillary negative effects: rents doubled, traffic increased, accidents rose, and crime rates climbed.[100]

The consensus was that material living conditions were better in the new towns, yet some complained of the sterility, a decline in community spirit, and dislocation from the river they had known.[101] Joy Parr focuses on Iroquois in her insightful study of Canadian megaprojects. She suggests that the locals were not adequately consulted and that there was an "unnatural homogeneity" about new Iroquois that undermined a sense of belonging and contributed to anomie. The villagers' political loss of control was exacerbated by the loss of "physical reference points for the selves they had been" and the lack of "benchmarks for the spatial practices of daily life, for the habits through which residents had embodied the place."[102] The embodied dislocation – the sights, sounds, and smells – from the river was painful and disorienting, particularly for those who had known the river and its environment intimately.[103]

Iroquois was unique among the new municipalities in that it was transplanted directly beside its former location. Moreover, it was the first community to be moved, and many of the flaws were worked out in the process of dealing with Iroquois. Thus, in several key respects, the Iroquois transplantation was not representative of the other communities, which, aside from Morrisburg, were completely new creations. Combining disparate municipalities into a new community, as was the case in Long Sault and Ingleside, caused problems and affected self-identity. Rosemary Rutley's collection of oral histories and recollections highlights the ways in which the new communities changed how people socialized.[104] The sense of community decreased not only because the community itself had changed but also because of is spatial set-up: the design of the new towns and houses resulted in people keeping more to themselves. On the other hand, shared experiences at the hands of HEPCO helped create some sense of solidarity.

Instead of the previous hamlets spread along the waterfront in long and narrow grids, the new municipalities were based on planning principles utilizing curved streets and crescents, with the major services and amenities grouped strategically together in centralized plazas, and schools, churches, and parks placed to facilitate easier and safer access. For example, children would have to cross fewer streets to get to class. As a planner for the provincial government during the project pointed out, HEPCO's plans were based on modern conceptions of suburban development and traffic circulation; conversely, a rural planning approach, with light urbanization as necessary and houses oriented toward the water in ribbon-strip development, would have been more accommodating to the realities of the area and the desires of the local population.[105] People did not like the new style

FIGURE 5.12 The town of Long Sault soon after inundation. © *Stormont, Dundas, and Glengarry Historical Society,* courtesy of Cornwall Community Museum

of houses, and the treeless landscape seemed alien.[106] Many became confused and lost because of the new street patterns. Others remember vividly how dirty, muddy, and dusty everything was for several years.

In Claire Parham's oral history of the construction of the St. Lawrence Seaway and Power Project, one worker suggests that the majority "don't have much to complain about. HEPCO rebuilt the roads, the churches, and the schools, and gave us new shopping centres. Many residents of the Lost Villages got new homes that were cheaper to maintain and more valuable than their old ones."[107] Another person who lived through it argues that "a lot of area residents complain that they were not fairly compensated for their property, but that is just crybaby stuff."[108] According to a former villager, "Most of the people were satisfied with how they were treated ... they got indoor plumbing. Hydro [HEPCO] put in a bathroom in their house."[109] Many people cherished the new homes with basements, the modern sewers and water treatment plants, the removal of highway traffic from downtown, and the better-equipped schools and amenities.[110] It was certainly possible that the way of life in the Lost Villages was already dying off and that the St. Lawrence project simply accelerated a modernization process that would have happened regardless whereby smaller hamlets and settlements became consolidated into larger towns, with some becoming bedroom communities. This seems to have been the case, for example, east of Cornwall.

Bonnie Clark, who grew up in Aultsville, recalls that no one knew the market value of his or her property, and HEPCO took advantage of this. Referring to HEPCO agents, she says, "It was not well done at all ... Those people were hated."[111] Lyall Manson, another Lost Villager, recounts a story, qualifying that it may have well only been a rumour, where the house of one fellow who refused to move was burned down while he was away.[112] Joan McEwan, who had taught school in Moulinette and then Ingleside, remembers that one could get a better price by holding out – but he or she might also get expropriated.[113] For David Hill, whose grandfather owned a barbershop along the Front, though HEPCO did a lot of good things and has since been community-conscious, there is still a lingering bitterness.[114] HEPCO purchased many houses, moved them to Long Sault or Ingleside, and then sold them. These repurchased homes were given the same treatment as the others, but Joan McEwan remembers that HEPCO "tried to do things pretty cheaply" and took a lot of shortcuts, citing problems like cumbersome furnaces and basements that were not deep enough to be very functional.

The flooding exacted a cost that cannot be measured in dollars and new houses. Jim Brownell, who was a child at the time and went on to become a Member of the Provincial Parliament for the area, remembers the destruction of the churches as being particularly sad. Other people list natural features they hated to lose, such as the stately trees or the omnipresent sound of the Long Sault Rapids. Above all, the flooding represented the loss of a way of life. For those who wanted that change, it was welcome; for those who did not, it was a negative upheaval. There was no monolithic viewpoint, as people were spread all across the spectrum. Just as in contemporary society, people view progress differently. Some are excited by increased opportunities; others worry about what is being lost. Some want their communities to mushroom into cities; others want to maintain the village atmosphere. In terms of value judgments, there is no clear right or wrong.

In contrast to those of a more advanced age, the younger cohort tended to have the perspective that better material standards of living made the move worthwhile.[115] As a result, there was a generational gap in how people of the Front experienced and handled the St. Lawrence Seaway and Power Project. Those who were young at the time of the flooding remember it as a period of excitement, but they also recall the stress on their parents and the extremely difficult experiences of their grandparents. To be sure, the whole process was most difficult on the elderly, and they took it the hardest. Most Lost Villagers can recount more than one senior who died shortly after moving to the new towns. They essentially died of broken hearts.[116] Vale Brownell and others who were children in the 1950s remember that adults in the 1960s did not want to talk about what transpired between 1954 and 1959, for it was a taboo subject.[117] Such attitudes may well have been coping mechanisms, and this communally disorienting experience was collectively repressed for some time. As the youth of the 1950s came of age, they began to realize what they had lost and were more open to dealing with the past. People increasingly lamented the drowned communities with local and popular histories, prose, and folk music.[118] The formation in 1977 of the Lost Villages Historical Society speaks to a desire to commemorate the communities lost and use their memory to forge a shared identity, and the society's creation of a living history museum at Ault Park featuring old buildings from the villages seems an intentional effort to contest the form and type of historical memory emphasized – or ignored – at Upper Canada Village.[119] But commemoration also speaks to the impact of nostalgia for a lost past, with its potential for altering perceptions. Sentimentality can skew historical memory.

Retaining their original houses was a source of comfort and familiarity for many, though it seems possible that having a house moved did not allow for a psychological break or sense of closure. There were reference points and signifiers indicating that people were at "home"; yet when they looked or went outside, their surroundings had changed. The house was the same, but it was simultaneously different, for the location and context had been altered. This was perhaps even more jarring for those in Iroquois: they still lived in Iroquois, rather than in a different community with a different name and very different location, but it was not the Iroquois they had known. Further east along the Front, the sight of the old highway wandering into the water was and remains surreal, as does the visibility – and walkability – of sidewalks and house foundations far out in the new "river" at times of low water levels and high water clarity. Farran's Point and Aultsville, for example, are barely submerged when the river level is low, leading former residents to question the necessity of the relocation. HEPCO ended up burning many of the houses that could not be moved, and the incinerations were hard on those who witnessed their residence going up in flames.

Conceptually constricted by the logic of the project and the beliefs of the time, HEPCO believed that it was being fair. After all, previous comparable North American projects had not provided new communities and amenities for those displaced but left them to fend on their own. Relative to the legal parameters it was required to work within, HEPCO was partially responsive. It initiated consultation practices, even if it did not always really listen. The chairman could claim, as he did at a public address in 1958, that his organization "fully understand[s] the upheavals, the discomforts, the heartbreaks which the building of this project has meant to you all," and sympathize with the loss of intangibles: "It is easy to talk about progress and the common good but it is hard to translate these sentiments into one's feelings concerning the old home in which one grew up, the nook in the garden where one played as a child, and the old familiar churchyard."[120] Some HEPCO officials, such as Harry Hustler, showed empathy for those forced to relocate, and as individuals expressed some solidarity with them.[121]

Nevertheless, many problems, misunderstandings, and unmet expectations stemmed from the fact that the various sides were often talking past each other. HEPCO staff had an engineering mentality that saw the greater standard of living, modern amenities, and technological advancement as sufficient compensation for what was being lost. As H.V. Nelles has demonstrated, HEPCO had a history of self-preservation, siege mentality, and lack of accountability that combined to perpetuate its institutional self-image as the representative of the public interest.[122] HEPCO's attempts to be responsive were often self-interested or reluctant half-measures: the 15 percent bonus was a decision taken by the provincial government, not HEPCO, and house moving benefited HEPCO's bottom line as much as it pleased the people. The commission did undertake a wide range of consultations (meeting, surveys, town hall forums, and so on), though the prime motivations for this were a desire for control, a veneer of consent, and knowledge of numbers; a fascination with amounts and sizes was apparent in the statistics-laden documents that were regularly released to the public. A survey indicated, without a trace of irony, that the door-to-door calls by HEPCO were "one of the most important steps" to the average family.[123] The HEPCO chairman did go out of his way to help a particular elderly woman, proposing to accommodate her in a hotel at the utility's expense. But this offer was forthcoming because the chairman was impressed that she accepted moving "without a complaint ... the unavoidable march of progress shows a spirit and stoicism which one can only witness with respect and humility."[124] In order words, she bought into the state's logic of the project.

Many residents were scared of challenging the government or its power utility, which contributed to HEPCO's perception that people were satisfied. There were, however, certainly isolated incidents of grassroots opposition, such as refusals to sell, appeals launched, demands for changes, and signs such as "Hydro Unfair." There seems to have been a general sense that the appeal route tended to favour HEPCO, and looking at the appeal records, this does appear to be the case. There is certainly evidence that HEPCO representatives targeted the elderly and vulnerable in order to set a precedent of low prices – though this seems to have been the actions of unscrupulous representatives rather than directives from HEPCO executives.

HEPCO was more experienced than its counterpart, the Power Authority of the State of New York, in this type of endeavour. In fact, the St. Lawrence power development and rehabilitation was PASNY's first major project, despite its having been in existence for a quarter century. The power authority generally did not help people relocate but left it to them to arrange moving, as did the seaway authorities. HEPCO had helped displaced farmers find new land, which PASNY did not. Yet, PASNY did reserve some of the St. Lawrence power for the surrounding area (as did Quebec for Beauharnois power), in contrast to HEPCO. As a result, New York attracted more industry to its St. Lawrence region, though industry in general benefited in Ontario, since HEPCO added St. Lawrence power to its province-wide electrical grid. PASNY's licence from the Federal Power Commission required it to distribute power to municipalities and cooperatives, and a "reasonable portion" to neighbouring states, which it satisfied by sending 12 percent of the American power to Vermont. A private power-producing firm, Niagara Mohawk, challenged PASNY's valuation of its property, eventually arriving at a $2 million settlement for the 4,500 acres swallowed by the St. Lawrence project.[125] Moses controversially made a deal with Alcoa in which the company contracted for about one-quarter of the power from the Moses powerhouse, and Reynolds Metals and a General Motors plant relocated to sites near Massena and signed power supply contracts, giving these three industries over half of the US power from the St. Lawrence development.[126] Nevertheless, there was not really a decrease in electricity costs on either side of the St.

Lawrence, except in Quebec, though HEPCO's electricity rates had historically been much lower than that of the United States. Visions for the economic and industrial expansion of eastern Ontario were predicated on attracting new industry to the region and when, by 1956, nothing was on the horizon, the Ontario Department of Planning and Development set up a committee to aid in recruiting industrial developments. Eventually, three new companies settled in the area. But Morrisburg, for example, had itself lost three industries because of the town's flooding-induced dislocation.

In New York, the majority did not vociferously oppose the land acquisition and rehabilitation. According to one of the few studies conducted, there was more dissatisfaction with the expropriation process in the United States than in Canada.[127] The extant public resistance in New York did seem more militant, and there appear to have been more incidents of physical resistance to land expropriations. This opposition tended to take more individualistic forms, instead of being channelled through community groups and organizations, and the opposition discourse tended toward rhetoric about Communism or socialism versus private property, liberty, and freedom.[128] Interestingly, it was HEPCO that took a more individualist approach to land acquisition – as compared with PASNY's blanket expropriation – which complicates somewhat the portrait of Canadians as the more communal society, as well as notions of the connections between liberty and property. But the different approaches were as much the result of pragmatic logistical and economic factors, as well as PASNY chairman Moses's steamrolling nature. In the decades following the St. Lawrence project, Canadian expropriation laws were standardized, with "market value" becoming the basis of compensation.[129]

The St. Lawrence was less intertwined with American and New York history and identity. Conversely, the river was the lifeline of central Canada's historic development and the link between its original major settlements, not to mention a further site of nation building vis-à-vis the United States through its role in the War of 1812. Such sentiments were not shared in the United States, even in the sections of upstate New York near the St. Lawrence, called "the North Country" by its residents. This area had historically looked toward and focused on the eastern American seaboard, and thus the Hudson and Albany Rivers, rather than the St. Lawrence; to put it a different way, the St. Lawrence River was Canada's front door, but America's back door.[130] However, a resident of Waddington recalls that the Canadian side of the river was called "over home," a moniker that speaks to the sense of a shared transnational St. Lawrence Valley identity prior to the seaway.[131] The seaway partially undermined this cultural affinity, in part by making it more difficult to freely cross the water border, but it did contribute to greater economic and industrial transborder St. Lawrence associations in which the North Country looked increasingly toward and across the St. Lawrence.

HEPCO did not follow through on its promise to provide waterfront property to those who previously held it. There was much more parkland than had been the case before 1954. Much of the former riverbank had been fronted by the old canals and government owned, though this rarely stopped villagers from using it for recreation purposes. These canals were mostly drowned by the power project, which effectively doubled the length of shoreline. Virtually the entire waterfront on the Ontario side of the International Rapids section became government-controlled parkland or was owned by HEPCO, which out of concern for water levels did not allow building along the water's edge, aside from two islands and a few other isolated pockets, including to the immediate east of the marina at Long Sault.[132] The two new towns were built back from the river and severed from the water by Highway 2.

In the minds of HEPCO and the Government of Ontario, the waterfront had been democratized and made accessible to all. In theory, everyone did have greater access to the water. But much of the riverfront was scrubland or mud flats. On average, the shallow areas along the shore were no more than five feet deep to an average offshore distance of about a third of a mile. As would be expected, there were no natural beaches on the new shoreline, since they had formerly been farmers' fields. A HEPCO memorandum revealed that "park-land is really left-over land; HEPCO has no obligation to beautify these left-over pieces."[133] Nonetheless, the park and camping areas created alongside the Long Sault Parkway were initially quite popular with tourists, and in subsequent decades it was common on weekends to see cars backed up to Long Sault or Ingleside waiting to enter the parkway. The public as well as displaced residents had to pay fees to use the Long Sault Parkway, which did not sit well with those who had their land taken.[134]

Conclusion

In retrospect, it is all too easy to criticize the governments involved for sacrificing the St. Lawrence communities on the altar of progress. They could not see beyond the momentum and logic that called for the completion of the St. Lawrence Seaway and Power Project, particularly since this perspective had general societal support. It is important not to overlook the fact that the various governments and authorities did believe they were actually improving the St. Lawrence Valley and its communities and, in doing so, providing a great benefit to the nation.

Nonetheless, one can interpret the St. Lawrence project as a socially and ecologically imperialist undertaking that followed the dictates of industry, big business, and modern capitalism. The Ontario government had a vested interest in developing industrial and economic systems that would justify and provide a return on investment for the resulting hydroelectricity. Since it was making a major investment in the St. Lawrence Seaway and Power Project, the Canadian state (both the federal and provincial governments) was interested in remaking society, as evidenced by the relocation of the Lost Villages, into more centralized and efficient nodes on the production and transportation network, and creating the conditions in which people would utilize the resulting electricity and transportation networks. Resettlement was a key strategy in high modernist projects, for it allowed planners to take communities scattered along the riverfront and organize them in more rational and efficient ways – consolidating a number of small villages and hamlets into larger municipalities and reorienting them away from the river and toward the new freeway. Government planners redesigned the towns with increased mobility in mind or, at least, mobility as they envisioned it.[135] The original plans that HEPCO offered the displaced communities underwent some revision in response to local desires; nonetheless, they still reflected the high modernist ethos that animated governmental and expert aims.

Ironically, Robert Moses, in an invited talk for the Canadian Club at Ottawa's Château Laurier while the St. Lawrence project was under construction, claimed that "we have approached the St. Lawrence task, especially in the International Rapids section, with genuine humility." He ventured closer to the truth in his following remarks: "At times it still looks like arrogance on our part, almost approaching the classical tragic Greek idea of 'hubris,' or insolence. For we are pitting against the rush of a mighty stream ... the vaulting ambitions of two democracies."[136]

6
Flowing Forward

On Inundation Day, twenty-five thousand people, one-fifth of them invited guests, gathered to witness the creation of Lake St. Lawrence. The flooding was initiated at 8:01 a.m. when thirty tons of explosives ripped through cofferdam A-1, a 600-foot artificial dam between Sheek and Barnhart Islands. Water started to flow into the dewatered stretches of the river, creating a 100-square-mile headpond approximately thirty-four miles long. Within a few days, several of the power turbines in the Robert Moses–Robert H. Saunders Power Dam had commenced operation. The last of the sixteen generating units in the joint dam, split evenly between Canada and the United States along with the $600 million cost for the entire power project, became operational in late 1959. This shared hydroelectric development was a 3,300-foot-long straight gravity dam from Barnhart Island to the Canadian shore near Cornwall with an installed capacity of 1,640,000 kilowatts from thirty-two generators (sixteen for each country) but eventually capable of 1,880,000 kilowatts.[1] The Beauharnois hydroelectric structure, downstream in Quebec, had an additional 1,700,000 million kilowatt capacity. The Long Sault Dam, roughly three miles upstream from the Moses-Saunders powerhouses and reaching from Barnhart Island to the American mainland, was a curved-axis spillway structure, with a maximum height of 145 feet and a length of 2,250 feet. At a height of 67 feet and a width of 2,540 feet, the straight-line Iroquois Control Dam, a buttressed gravity structure running from Point Rockaway to Iroquois Point, was slightly longer but not as high.

FIGURE 6.1 Opening ceremony at Montreal in 1959, with Queen Elizabeth II in attendance. © *Library and Archives Canada / Credit: Capital Press Service / The St. Lawrence Seaway Authority fonds / e011068165*

The initiation of power development marked a major accomplishment, but there were still fundamental aspects of the St. Lawrence Seaway and Power Project that needed to be finished. The navigation element, the seaway, was not yet complete. Behind the scenes, the engineers had been figuring out how best to control the water they were putting to work. Other transborder issues, such as tolls, also needed to be negotiated. These, as well as the initial operation of the entire project and its environmental impact, are the main subjects making up this chapter, which covers from Inundation Day onward.

"As Nearly as May Be"

The shape of the new river/reservoir was determined by a small coterie of experts from both Canada and the United States. This process was largely taken for granted, both by the public and by the governments involved, and the final result made it all appear a foregone conclusion. But beneath the surface, establishing the water levels was an uncertain and imprecise process, determined as much by personal and political motivations as it was by scientific expertise. Charting the evolution of the high-level engineering of the St. Lawrence waters, which is the focus of this section, shows the experts and planners to be products of their cultural and professional context, and exposes some of the contradictions in their logic of progress.

In addition to political and economic issues that had the potential to hold up construction, actual work on the interdependent St. Lawrence Seaway and Power Project had been unable to progress until the engineers had established a "river profile" and developed a "method of regulation" for the river and Lake Ontario. A "river profile" meant the side schematic view of a river's course from start to finish – the damming of the St. Lawrence changed the profile from a series of incremental slopes to just a few large drops, more akin to stairs. The "method of regulation" referred to the levels between which the water would be maintained by dams and control works so as to meet prescribed goals (e.g., hydroelectric production). The explicit goal was to maintain the water levels at an average that equated to "natural levels" but also to improve on nature by removing the extremes of high and low flows in order to create a predictable and orderly river. "Natural" was defined as that which had existed in the nineteenth century before the first man-made alterations to water levels – that is, before Canada installed the Gut Dam in the St. Lawrence River between Galop and Adams Islands in the early twentieth century.[2]

Yet, establishing exactly what constituted a "state of nature" was problematic from the outset. Not only did representatives of the two countries disagree on the historic impact of the Gut Dam but it was also difficult to find information regarding the natural levels to use as a baseline. The Joint Board of Engineers had set the elevation of 248.1 feet above sea level as the high water level in 1926, but there was concern that this measurement was unreliable because of the geological phenomenon of earth tilt, as well as a 1944 earthquake centred between Cornwall and Massena. Indeed, engineering and scientific studies indicated that natural factors must have played a much larger role in the recent rise in Lake Ontario water levels than had the man-made factors (i.e., diversions into the Great Lakes basin), though this assessment was partially motivated by the desire to escape liability for the damage done to the property of lakeshore property owners.[3]

Figure 6.2   HEPCO model of the St. Lawrence Seaway and Power Project. *Courtesy of Ontario Power Generation*

Nonetheless, since a minimum and maximum range of levels needed to be determined, 248 feet was taken as the extreme elevation. The various governmental and construction entities had to know the final expected levels before proceeding with digging channels and locks. Starting construction before the water levels had been determined risked costly mistakes; for example, the dredging costs for the power entities would rise several million dollars for each foot the water levels were lowered.[4] The 1952 order of approval by the International Joint Commission (IJC) had also provided that any concerned interests (e.g., shorefront property) would be given adequate legal protection and indemnity in their respective country. Largely in response to the complaints of the Lake Ontario Land Owners and Beach Protection Association, which represented shore owners, the IJC gave the Lake Ontario levels issue its own docket in addition to that of the St. Lawrence project. The Lake Ontario Joint Board of Engineers was formed in 1953, and the Lake Ontario issue became intertwined with the St. Lawrence discussions, as any decision about levels on the river would affect the lake. The corollary of restricting Lake Ontario water levels was the various downstream impacts; for example, lowering water levels by a foot meant the annual loss of 225 million kilowatt hours of power development at the Barnhart dam.

As the binational negotiations for a joint versus solely Canadian seaway reached their zenith in 1954, the IJC engineers were already busy utilizing models to simulate historical water levels on the St. Lawrence River and Lake Ontario. In addition to HEPCO's nine power project models, the SLSA built its own models for the seaway. In December 1956, the power utility added a new hydraulic lab that included models of the Montreal Harbour and Lachine Rapids, and the National Research Council in Ottawa created a model of a lock and another of the International Rapids section. The United States meanwhile had hydraulic model studies underway in Vicksburg, Mississippi, and Minneapolis, Minnesota.

It became apparent that there had been errors in the calculation of the interim Method of Regulation 5, as it would barely lower the maximum levels on Lake Ontario.[5] R.A.C. Henry, one of the Canadian experts on the engineering aspects of the St. Lawrence, commented on the process whereby the engineering representatives of the two countries had arrived at the previous "238-242" range of levels for the Barnhart dam: "In light of the evidence which is available on the subject it appears reasonably certain that the 238-242 range was actually a compromise between two conflicting views and was not based upon any positive and well-defined line of reasoning."[6] Yet, Henry and his colleagues were not immune from similar errors. Between 1954 and 1959, there were many engineering miscalculations, assumptions, compromises, and partisan preferences. Shortly after Henry's observation, in an internal Canadian meeting, HEPCO general manager Otto Holden stated outright that the engineers did not know what the natural conditions were. General McNaughton, who reputedly dominated the IJC and was known as a tough Canadian nationalist, emphasized that "the balance of conditions on Lake Ontario is so delicate that he could not feel assurance that the engineers could in fact keep the levels" within the specified range.[7] As a result, they strove to attain levels "as nearly as may be." However, in public they gave an impression of preciseness and confidence.

To be fair, the planners were in many ways products of their training and societal ideals, and were subject to dominant national and transnational ideas that promoted the collaboration of industrial capital and the state in order to maximize the development of natural resources in the name of economic and social progress. They believed they were wisely maximizing natural resources. There was great societal and occupational pressure on the "experts" to provide answers and do so in a confident manner: in

FIGURE 6.3 Aerial view looking north of the Moses-Saunders Power Dam under construction, the area behind the dam dried out by cofferdams. (The infrastructure for the fourteen-foot navigation and future Canadian seaway lock in the dike can be seen in the upper centre.) © *Ontario Power Generation*

FIGURE 6.4 (Facing page, top) View of the power dam under construction. © *Dumas Seaway Photograph Collection, Mss. coll. 124, Special Collections, St. Lawrence University Libraries*

FIGURE 6.5 (Facing page, bottom) View from the Canadian side of the power dam under construction. © *Dumas Seaway Photograph Collection, Mss. coll. 124, Special Collections, St. Lawrence University Libraries*

addition to hundreds of millions of dollars, many jobs and related economic factors, national and organizational pride, and the role of technology and expertise in capitalist/democratic and Communist Cold War tensions, their personal and professional stature was at stake. Whereas scientists studying pollution issues in the Great Lakes–St. Lawrence basin a decade later could publicly admit to uncertainty, open disclosure of doubt was unthinkable for the St. Lawrence engineers.[8]

All plans and specifications had to be approved by the St. Lawrence Joint Board of Engineers, but problems soon appeared because the power entities and construction agencies failed to submit their plans, and the board of engineers also became embroiled in the Cornwall channels dredging controversy. The Canadian and American sections of the International Lake Ontario Board of Engineers disagreed about the maximum level of Lake Ontario, squabbling over fractions of an inch. The Americans seemed to be largely motivated by political concerns emanating from the protests of Lake Ontario beach owners; the Canadian position was largely predicated on protecting Montreal interests, for any lowering of Lake Ontario levels would tend to raise water levels in the western Quebec section of the St. Lawrence.[9] The main future users of the St. Lawrence Seaway and Power Project – power production, navigation, shoreline property, and downstream interests – wanted different minimum and maximum water levels or varying ranges of stages (i.e., differences between high and low levels) and pleasing everyone seemed impossible. At the numerous public hearings that were conducted on the lake levels, many people came to voice their concern about the impact of higher water levels, such as shoreline erosion. However, the transcripts show that property owners were worried about their own property value, rather than ecological impact.

Nevertheless, in March 1955, the IJC told the Canadian and American governments that it was possible to regulate the St. Lawrence and Lake Ontario in such a way as to balance the various demands. A revised method of regulation was arrived at, labelled 12-A-9, but there were problems with that as well: for example, tests showed that under its parameters the seaway would constrict the channel at Montreal. There was discussion about increasing the upper limit marginally from 248.0 to, for instance, 248.3, but such precise goals appear in retrospect somewhat strange given the uncertainty about the evidence and tests used – engineers were trying to ascertain the historic conditions on which they based their arguments at the same time as they were making their arguments. The guiding principle of "as nearly as may be" continued to prevail. In July 1956, the IJC issued a supplementary order directing that Lake Ontario levels be maintained between 244 and 248, again adding the "as nearly as may be" rider. Yet, soon after, Method 12-A-9 was replaced by another method, 1958-A. The precise technical differences between these methods are not important here – rather, it is the frequency of changes and the decision-making manner that are noteworthy because they betray how messy and reactive the process of regulating the river levels actually was.

Attention also frequently turned to the ability of control works, such as the Iroquois Dam, to manage water flows in such a way so as to manipulate winter ice formation and prevent blockages. Here too the propensity of the engineers to act as if they could master their subject despite the imprecision of their knowledge (and their awareness of this impreciseness) was apparent. Intriguing was the way in which they labelled anything beyond their control or understanding as an "act of God," suggesting that if it was unknowable by their scientific techniques, it was beyond comprehension.[10] It was not so much the case that the governments and planners involved could not comprehend the complexity of their task, but rather that they chose to ignore or mentally exclude the uncontrollable aspects of the St. Lawrence environment in order to persist in their belief that they had perfect conceptual understanding. Put another way, they effectively bracketed those factors and contingencies for which they could not account or control.

Part of the problem stemmed from the faith that the engineers placed in their models. The planning authorities, including the IJC, were enamored with these models and believed them to be indispensable for determining the future fluvial geomorphology; thus, they were central to the engineering recommendations. Models were, however, often found to be wrong. Sometimes this was the result of incorrect knowledge, such as faulty gauge data, about the river on which they were based.[11] Because of the scale of the models, a slight error would be distorted out of proportion when applied to actual excavations or structures in the river. Attempts to simulate the turbidity of the river by increasing the model roughness factor also caused distortions.[12] In a further example, which exposes the rivalries that affected the various national sections of the different engineering boards, as well as the problem created by all the spoil from excavation, HEPCO complained that the US Army Corps of Engineers model of the American seaway installations in the International Rapids section did not "sufficiently exploit the river and the terrain and that the disposal areas [had] been unwisely shown. In at least one case, disposal could prove a hazard to navigation ... [the contractors] had simply decided somewhat crudely, to bulldoze their way in a straight line through the area regardless of its natural features."[13]

Even after years of experience with the St. Lawrence models, the engineers were still encountering significant problems in early 1958: "The model, although not near final verification, already showed inconsistencies in the prototype data" and as a result "it was again necessary to suspend continuous operation of the model due to incomplete, indefinite and unconfirmed prototype data."[14] Indicating the impact of improper extrapolation from models to actual river conditions, in March "it was discovered that, due to an oversight in establishing ... conditions, an area comprising approximately 1½ blocks was left in the lower channel."[15] Such incidents show that the planners and engineers were quite flexible and adaptive when they encountered changed situations or errors, but they also indicate the flaws in the conceptual approaches employed by the hydraulic engineers.

St. Lawrence planners had spent decades studying and analyzing the river and were attuned to many of the local environmental characteristics. While abstracting sections of the river into macro-scale models, the planners and builders of the St. Lawrence project relied on extensive studies of the specific conditions at particular spots in the St. Lawrence.[16] All told, 475 engineers collectively worked on the project, and many explored the river and its environs in great detail, relying heavily on specific place information produced by water gauges, soil and rock samples, soundings, test drills, and elevation measurements.[17] However, this knowledge, generated by "experts" and their technologies and methodologies, was the only useful and acceptable type of knowledge; local and therefore "unscientific" knowledge was ignored, for the planners were after a specific type of information. They had no interest in the embodied, experiential knowledge of those who lived in the river valley.[18] Given the size of the area that was affected, the St. Lawrence engineers were unable to know in detail every square foot of river, and in a number of cases the subsurface conditions took them by surprise. Nonetheless, the engineers believed that when they had sufficient knowledge of the situation on the ground – the types of rock, the composition of soils, the water flows and velocities – they could control the entire river and ecosystem.

Despite construction and engineering problems, strikes, supply issues, and legal and political delays that threatened the target date for the raising of the power pool, on July 1, 1958, the St. Lawrence River Joint Board of Engineers felt confident enough to authorize that it proceed. The filling proceeded very closely with the anticipated schedule, reaching the prescribed forebay level of 236 feet early on the morning of July 4 – it reached 238 a few days later, and then 240.5 in December 1959.[19] An elaborate gauging system tracked water levels and flows. Method of Regulation 1958-A remained as the provisional working model, though engineers were clear that it would likely need to be adjusted, and it was superseded by 1958-C at the beginning of 1962, which in turn was replaced the following

October by 1958-D, which too was scheduled to be replaced.[20] During the first year after the power pool had been raised, there were problems with ice forming "hanging dams" that reduced power output, and issues with low water levels downriver from the Moses-Saunders dam in the early 1960s.[21] It also quickly became apparent that the water level (and thus the power head) was higher on the American side of the Moses-Saunders dam than on the Canadian, and as a result the power entities agreed to equally share the output of power, rather than the inflow of water.[22]

PASNY and HEPCO concluded an operating agreement for the power works, and meetings between the two entities continued well into the 1960s to sort out the division of expenses and take care of remaining aspects of the project. Legal difficulties concerning redress continued, as it seemed that the legal structures in Canada provided no means of compensating those injured by changing water levels. Concerns about liability also led to debates about whether the IJC (through the St. Lawrence Board of Control) or the power entities would have responsibility for controlling the gates at the Iroquois Dam. The Board of Control was eventually given responsibility for the gates. But the method of regulation was not always satisfactory, as there were significant problems with low water levels in the river in the 1960s, and then high water levels in the 1970s.[23] These were attributed to natural supplies of greater variance than had occurred in the 100-year period upon which the engineers had based the various methods of regulation.[24] Nevertheless, in the longer term, compared with pre-project conditions, the St. Lawrence and Lake Ontario water levels were more predictable and controllable, and the range of water levels was compressed (i.e., extreme highs were lower, extreme lows higher). Despite the flaws and mistakes, in the end, the remaking of the river largely functioned as planned.

Taking a Toll

Old problems that had not been solved, and others that had just been put off, now had to be dealt with. The various Great Lakes basin diversions that had bedeviled the St. Lawrence negotiations in previous decades reappeared. Congressional proposals for an increased diversion of 1,000 cubic feet per second (cfs) at Chicago through the Illinois diversion were resurrected. The IJC and US Army Corps of Engineers both did studies on the effect of this 1,000 cfs diversion, finding that it would not have a substantial impact on St. Lawrence navigation and only a minor effect on power production.[25] Canada had several other reasons for opposition (such as the precedent that would be set) and objected with two notes in 1956. The US Senate did pass legislation that same year, but because of Canadian representations it was vetoed by President Eisenhower, and the Chicago diversion issue continues to this day.

Partially in response to the Chicago proposal, Ottawa brought up the issue of the Ogoki–Long Lac diversions. Canada had since the 1940s enjoyed the right to use the extra water flowing into the Great Lakes from these diversions; yet they had not been mentioned in the 1954 Canadian-American seaway agreement, and the Canadian government was making no concerted effort to insist on its rights to the water, though neither had it forfeited the water. Tentative feelers indicated that the United States wanted all St. Lawrence water to be split evenly at the power dam. HEPCO had remained silent on the issue, apparently because of worries that it would upset the apportioning agreement on the power costs that had been arrived at with New York. But the Ogoki–Long Lac issue ultimately faded away without conclusive clarification. The creation of Lake St. Lawrence also flooded out the control works that had enabled Alcoa's diversion via the Massena Power Canal. The previous diversion was discontinued, although an updated intake, which itself was a small dam, for industrial and domestic needs in Massena was installed and allowed to divert a certain amount of water.

Historically, the United States had undertaken most of the work to improve the connecting channels in the Great Lakes above Lake Erie. The 1941 agreement had provided for the deepening of these channels to a depth of twenty-seven feet by the United States. With a seaway agreement in hand, a US Army Corps of Engineers report recommended the deepening of the channels to twenty-seven feet, at an estimated expense of $109 million, in order to take advantage of the seaway.[26] Eisenhower was reluctant to approve these additional expenditures for fear that it would give truth to the charge of seaway opponents that the International Rapids section was only the first in a long line of expenses. Legislation was nonetheless approved in 1956. Canadian consent was required for some of this work, such as in the Detroit River and Lake St. Clair, and these improvements were done by the early 1960s. In connection with the construction of the seaway and these Great Lakes connecting links, major harbour and infrastructure improvements abounded in various cities, though many remained unable to take advantage of the larger vessels the seaway allowed.[27]

A binational committee to determine tolls was entrenched in the 1954 St. Lawrence agreement. Members of the US and Canadian toll committees were appointed in March 1955 to initiate preliminary studies. They investigated single commodity tolls and alternatives such as a system of tolls graded for different types of traffic. The United States initially wanted to be in charge of collecting all the tolls, but this role was given to Canada, as the seaway started and ended in that country. Talks progressed slowly in the following few years, both countries distracted by the construction of the project. The two sides rejected the toll structures used by the Suez and Panama Canals, and held conferences with interested parties and future seaway users to gather their input. The Canadian government preferred no tolls, a position that also found favour with US interest groups that had historically been pro-seaway. Seaway opponents predictably argued for higher toll charges. The Corps of Engineers also wanted to avoid tolls, whereas the American Saint Lawrence Seaway Development Corporation (SLSDC) defended medium- to higher-range rates because of its mandate to make the project self-liquidating. After several years of negotiations, in 1958, a tentative agreement was reached on toll rates, which was officially approved in March 1959, just before the seaway was officially opened. Paralleling the costs of the larger project, the division of tolls was initially set at 71 percent to Canada and 29 percent to the United States, with all toll proceeds from the Welland Canal accruing to Canada (see Table 6.1).[28] Toll rates were initially forty-two cents per ton for bulk cargo to travel the entire seaway (with lesser rates for shorter trips) and ninety-five cents for general cargo. This was a compromise between the requests of the high and low rate camps. Tolls were due for re-examination in 1964, but because of bilateral disagreements no agreement to change the rates was struck until 1978.

Table 6.1 Tolls on the St. Lawrence Seaway under 1959 agreement

Charge	Mtl-LO	LO-LE	Complete
Per gross ton of vessel	0.04	0.02	0.06
Per ton of general cargo	0.90	0.05	0.95
Per ton of bulk cargo	0.40	0.02	0.06
Pleasure craft	14.00	16.00	30.00
Other vessels	28.00	2.00	60.00

Note: Mtl-LO = Montreal–Lake Ontario; LO-LE = Lake Ontario–Lake Erie; Complete = complete length of seaway.

Opening of the 8th Sea

The celebration of the opening of the seaway in 1959 was an even larger celebration than had been the 1958 commencement of the power project. A binational working group to plan the joint international opening ceremony was formed in 1958, and architects were hired to assist in the planning and stage managing of every aspect of the ceremony. Both countries realized the opportunity this presented in terms of nation building within the larger context of the Cold War. As an American official wrote, this was a chance to "cement the unity of the Canadian and American peoples"; the "opening of the '8th sea,' through the genius of United States and Canadian engineers, is an answer to sputnik in the realm of psychological warfare. Properly publicized it will reassure freedom loving people everywhere."[29]

Navigation on the seaway started on April 26, 1959. Thousands turned out at Montreal over the weekend to witness the passage of the first ships, led by the *d'Iberville*, a six-thousand-ton Canadian government icebreaker that carried a large party of parliamentarians, media personnel, and Canadian and American government officials. The *Simcoe* of Canada Steamship Lines was the first commercial ship through the locks. The waterway got off to a rough start, however, as delays from inclement weather, unprepared facilities (e.g., the navigation channels had not been deepened to the full twenty-seven feet throughout), and unfamiliarity with the new waterway system caused accidents and delays. Lineups to use the Welland Locks sometimes stretched to fifty ships, and further dredging and deepening was needed in the Welland Canal. In August 1959, the Canadian St. Lawrence Seaway Authority (SLSA) announced that improvements to increase the capacity of the Welland, such as tie-up walls to reduce lock passage time, would be undertaken.[30] Canada suspended tolls on the Welland Canal in 1962, and canal congestion led to a decision in 1963 to twin the remaining locks (the flight locks – numbers 4, 5, and 6 – were already twinned). This plan was later abandoned, as were even more ambitious plans for a completely new canal route immediately to the east.[31] In 1965 Canadian officials opted to construct a bypass that took the navigation channel out of the City of Welland, and it was finished by 1973.

The ceremonial opening took place on June 26, 1959. The navigation works had not been completely finished – the SLSDC estimated that it had accomplished 94 percent of its portion of the seaway by the end of 1959 – and both countries still needed to complete 1 million cubic yards of clean-up dredging.[32] The raising of the power pool in 1958 had been

FIGURE 6.6   View from a Canadian Steamship Lines bulk carrier, looking north from Welland Canal Lock 8, 2011. © *Daniel Macfarlane*

open to the wider public, yet the people from the flooded areas were largely excluded from the 1959 official opening celebrations, which were reserved for invited guests and dignitaries. There were all types of pomp and speeches, and a packed itinerary to please the needs of the various political entities involved. Organizers had realized that the ceremonies could be timed to coincide with the visit of Queen Elizabeth II. The royal couple and Prime Minister Diefenbaker met President Eisenhower at the Saint-Hubert airport, and they travelled together to the Saint-Lambert Lock. During a half-hour ceremony, the Queen and president were each given a book bearing the names of almost sixty thousand people who had been involved with the St. Lawrence project.[33] Both gave speeches affirming the grandeur of the project, which went "to the heart of the continent"[34] – literally as a waterway and figuratively as a national and bilateral issue – and served as "a magnificent symbol to the entire world of the achievements possible to democratic nations peacefully working together for the common good."[35] Afterward, they boarded the royal yacht *Britannia* and entered the first lock, accompanied by a crescendo of fireworks, bells, sirens, and

FIGURE 6.7 Opening ceremony at the Saint-Lambert Lock in 1959. © *Library and Archives Canada / St. Lawrence Seaway Authority fonds*

gun salutes, embarking on a five-hour escorted cruise to the Lower Beauharnois Lock, and then further west along the seaway.

The next day the party proceeded to the Eisenhower Lock, at which point they disembarked to travel to the PASNY reception centre at the Moses-Saunders Power Dam, as well as to a ceremony on the dam itself. The Queen unveiled the stone marker at the international border bisecting the powerhouses. The marker had been the subject of much contemplation, including by a study group composed of prominent members of the University of Toronto to consider the wording of the inscription: "This stone bears witness to the common purpose of two nations whose frontiers are the frontiers of friendship, whose ways are the ways of freedom, and whose works are the works of peace." Lunch followed for the Queen in Cornwall, then a driving tour with short stops at Long Sault, Ingleside, Morrisburg, and Iroquois, where she reboarded the royal yacht and proceeded to Brockville.

A detailed history of the seaway after completion would require its own full study, but a broad sketch of its operational history can be provided

Figure 6.8 Opening ceremonies at the Moses-Saunders Power Dam. © *Dumas Seaway Photograph Collection, Mss. coll. 124, Special Collections, St. Lawrence University Libraries*

here. Looking in particular at the estimates from the 1920s to the 1950s for traffic, trade, and tolls, and then at the actual usage of the seaway after 1959, expectations were wrong in numerous respects (granted, many of the more optimistic prognostications and statistics came from public or commercial interests of suspect objectivity). The Canadian Department of Trade and Commerce, for example, predicted in 1951 that 44,500,000 tons of shipping would annually move on the seaway. That same year, the US Secretary of Commerce had testified before Congress that seaway traffic would be between 57 and 84 million tons annually, with toll revenue totalling between $36 and $49 million per year. The SLSDC estimated the waterway would initially carry 36.5 million tons, and 52 million tons by 1963. Based on any of these projections, the seaway would eventually be self-liquidating.

In its first year, the seaway carried 20,100,000 tons of cargo, compared with the 8,300,000 tons carried on the old St. Lawrence canals in 1958. This was, however, well below expectations, and the SLSA lost over $6.6 million in its first season.[36] In the 1960 navigation season, there was actually

FIGURE 6.9   Snell Lock. © *Dumas Seaway Photograph Collection, Mss. coll. 124, Special Collections, St. Lawrence University Libraries*

FIGURE 6.10   Inside of a seaway lock. © *Dumas Seaway Photograph Collection, Mss. coll. 124, Special Collections, St. Lawrence University Libraries*

FIGURE 6.11   Eisenhower Lock, traffic tunnel running underneath it. © *Dumas Seaway Photograph Collection, Mss. coll. 124, Special Collections, St. Lawrence University Libraries*

a slight decrease in cargo tonnage through the Montreal to Lake Ontario section, though there was an increase in traffic on the Welland Canal, and the St. Lawrence Seaway Authority suffered a net loss of almost $9.5 million.[37] Traffic and tonnage did increase significantly in 1961 and 1962, ushering in a more prosperous phase and a pattern of fewer but larger ships.[38] Nonetheless, tolls were still falling one-third short of expectations and were able to meet only half of the annual operating and maintenance costs.[39] Despite estimates that over $100 million had been spent to improve Great Lakes harbours and ports, many were unable to adequately handle the new water traffic. Over the first decades, the seaway fared better but still certainly short of the lofty expectations.[40] With the exception of 1964, over 54 million tons were carried each year in the 1960s. The best years were in the late 1970s, when cargo volumes reached 70 million tons.

Those who predicted in the 1950s that the waterway would be of insufficient depth and proportion to handle future traffic were proven correct, as the rise of container shipping (which involves standardized containers that can be moved between different transportation forms – for example, ships, truck, rail – without needing to be opened) and larger vessels meant that the majority of transoceanic vessels could not fit in the locks and canals.[41] When first opened, the seaway allowed somewhere in the neighbourhood of 80 percent of the world's merchant shipping fleet and had access to 60 percent of North America's population and 65 percent of its manufacturing capacity; the percentage of the shipping fleet that could use the seaway quickly declined after the first successful container ship made its maiden voyage in 1956 and the technology spread and moved toward standardization.[42] Ocean-going vessels that were of the requisite proportions moved awkwardly in the St. Lawrence canals, and draught restrictions meant that they generally could not be loaded to full capacity. In fact, the completed system was traversed chiefly by specialized intralake traffic that brought iron ore from the eastern half of the continent to the Great Lakes region, or grain from the Prairies. Thus, for all the technological progress it represented, in some key respects the seaway was already bordering on obsolescence when it was completed. Other transportation, technological, economic, and policy developments sometimes worked against the seaway. For example, although a great deal of iron ore moved through its locks, new processes for upgrading Mesabi ore meant that Ungava ore was not in as much demand as predicted. The US Congress prevented the SLSDC from advertising for the seaway, and the Mississippi River was maintained as a waterway alternative. Maintenance costs were also higher than expected.

Bulk iron ore and grain immediately became the dominant commodities, combining for well over half of the total cargo shipped, a status they have maintained to the present. Other important cargoes were coal, petroleum products, and various manufactured goods. Not much of the anticipated general cargo movement materialized, however, and traffic was not enough to warrant the remaining links in an all-Canadian seaway at Cornwall, notwithstanding the preparatory steps for a lock on the north side of the Moses-Saunders Power Dam that had been taken during construction.[43] Despite subsequent claims that the project was a federal government initiative foisted on Quebec, studies have tended to show that the seaway was in fact beneficial for that province's economic and port interests. In fact, the Montreal region may have actually benefited economically from the seaway more than any other area along the St. Lawrence.[44] Nonetheless, there are reasonable grounds for speculating that the seaway was linked to the rise of separatism in Quebec.

Its shortcomings notwithstanding, in both Canada and the United States, the St. Lawrence Seaway and Power Project has directly generated billions of dollars of commerce and created indirect benefits by virtue of further integrating the economies and industries of the two countries. The 2011-12 annual report by the St. Lawrence Seaway Management Corporation, the successor to the SLSA, states that marine shipping on the Great Lakes/Seaway system supports $34.6 billion in annual economic activity, saving cargo shippers $3.6 billion in transportation costs per year.[45] Nor should it be forgotten that the St. Lawrence project might have proven extremely valuable if the Cold War had turned hot. The seaway effectively carried the two products, iron ore and wheat, for which it was chiefly intended. It is fair to speculate that the St. Lawrence project's full potential would have been realized only if another major war had occurred. Indeed, the possibility of another conflict that would consume resources on a greater scale than the Second World War was an important motivation for the Canadian and American governments to seek out means of exploiting the St. Lawrence River. The wide-ranging benefits of hydroelectricity produced from the Moses-Saunders Power Dam – chiefly, providing power for manufacturing on both sides of the river – are often ignored in assessing the importance of the St. Lawrence dual project. American and Canadian planners knew that North America's power supplies had been vital in their previous war efforts and believed that power supplies would play a crucial role in determining the future of any protracted conflict, such as with the Communist bloc.

In St. Lawrence County, on the New York side, the majority of the industrial jobs in the Massena area would not have existed without the cheap industrial hydroelectricity. Nevertheless, although there was some economic gain for settlements along the upper St. Lawrence that were not dislocated, particularly during the construction phase, much of the local anticipated long-term prosperity did not materialize. The degree to which toll levels dissuaded users was a topic of debate. Gennifer Sussman contends that "if there had been institutional machinery for ensuring that the two governments increased tolls periodically to comply with the enabling legislation that made full recovery of costs mandatory, the [St. Lawrence] locks could have paid for themselves on schedule even if treated as a commercial venture." Alternatively, if the two national governments "had followed the more usual procedure of investing in the project, rather than of loaning capital at current rates of interest, they would have received a respectable return on their investment, even at prevailing toll rates."[46]

Nonetheless, since the self-amortization of the enterprise was predicated on expanding usage, the seaway never came close to paying for itself. The economic downturn of the 1980s hit the St. Lawrence Seaway hard, particularly the declining iron ore industry. The waterway has never fully recovered, today carrying about 60 percent of its capacity. Canada cancelled its portion of the seaway debt in 1977, and the United States did the same in 1983. In 1998, operational management of the Canadian portion was turned over to a not-for-profit corporation, the St. Lawrence Seaway Management Corporation, with the Government of Canada continuing to own the infrastructure and act as regulator. The St. Lawrence Seaway Management Corporation now oversees operations in tandem with the SLSDC.

Fractured Flows

The environmental repercussions of constructing the St. Lawrence Seaway and Power Project were enormous, and its legacy cannot be fully measured without an appreciation of the ecological impact. The reconfiguration of the local ecosystem inherent in the St. Lawrence project disrupted the aquaculture, which included circumscribing the success and mobility of many species. The American eel, historically an important part of the St. Lawrence biomass, could no longer travel down the St. Lawrence River, as it was blocked by dams, and it is now an endangered species (eel ladders were added to the Moses-Saunders dam in 1974 and Beauharnois dam in 1994). Brief consideration was given to fishways at the beginning of the planning for the power project but soon abandoned; though the Dominion Fisheries Act required all dams to provide a fishway subject to the responsible minister's interpretation, neither the federal government nor the provincial Department of Lands and Forests insisted on a fishway because of the cost of modifications to the St. Lawrence dams and the "general inefficiency" for the "presumed purpose."[47] Lands and Forests wanted a $40,000 study on problems relating to fish and wildlife on the Ottawa and St. Lawrence Rivers, but HEPCO was not willing to pay for it. Other government officials speculated that the conditions created by the lake might actually be beneficial to wildlife.[48]

Many fish had been trapped and relocated before the creation of the power pool. For example, in August 1955, two Kingston-area fishermen under the supervision of the district biologist with the Canadian government caught thousands of fish in a dewatered stretch between two cofferdams – the game fish were dumped back in at a different location; the coarse fish were packed in ice and marketed.[49] Restricted movement was not the only negative impact on fish, for the extensive dredging changed

FIGURE 6.12   Long Sault Dam and New York's Barnhart Island Beach. © *Stormont, Dundas, and Glengarry Historical Society, courtesy of Cornwall Community Museum*

spawning and feeding grounds, and conditions for aquatic life were very poor during the construction period. Numerous species, such as walleye, lake sturgeon, and different types of sucker, had used the now-removed Long Sault Rapids as spawning grounds. Modified flows also played a role, as the current was much slower than before the seaway; studies later showed that the changed water flows upstream from the Moses-Saunders dam destroyed fish habitat and caused cancerous tumours in bottom-dwelling fish.[50]

There are conflicting views as to whether the fishing quality returned to normal after 1959.[51] It does certainly appear that the populations of many of the most prominent fish species found in the river experienced a dramatic and extended decline because of the construction of the St. Lawrence Seaway and Power Project, though there has been a rebound in many stocks since the 1980s. Some cold-water species that were intentionally stocked in the Great Lakes have made their way into the St. Lawrence since the 1950s. A recent scientific investigation using a 1931 study as a baseline has concluded that the types of fish found in the upper St. Lawrence have changed little since the 1930s.[52]

Altered water speeds and flows also changed the relationship between the river and experienced fishers and boaters. Changed flow and ice patterns altered scouring and sedimentation processes. During construction, many islands were removed or reshaped, and new islands added. There was a significant increase in the nearshore aquatic habitat in Lake St. Lawrence on account of the greater surface area of the new lake, but much of this was shallow and subject to frequent fluctuations of up to 6.6 feet – brought on by both seasonal factors and the operation of the dams – that periodically turned these areas into mud flats unsuitable for fish.[53] Granted, these altered nearshore habitats benefited other types of flora and fauna.

The water was certainly more polluted after the creation of the seaway. Richard Carignan contends that the St. Lawrence at Montreal is essentially made up of three separate rivers or ecosystems because of differing water characteristics resulting from seaway channelization.[54] Michèle Dagenais describes how the seaway's construction contributed to the changing relationship between that city and the St. Lawrence, alienating Montreal from the river, increasing water pollution, and privatizing the shores.[55] In addition to the pollution caused directly by the construction process, it is likely that submerged infrastructure remains leeched all types of toxins and other contaminants into the water, and large amounts of decomposing plant life releases mercury. Gas stations, for example, were razed before the flooding, but some of their tanks were left underground, as were other types of chemicals. IJC studies leading to the Great Lakes water quality agreements in 1972 and 1978 confirmed that the international section of the St. Lawrence was already heavily polluted by the 1960s. The long-term impact on other types of wildlife is difficult to determine and based largely on anecdotal evidence: for example, some types of birds suffered, yet duck populations seem to have increased. The situation is complicated by a relative lack of baselines and evidence on which to base comparisons of the pre- and post-seaway conditions. There are some exceptions, such as the 1931 study mentioned above. In addition, during 1953 and 1954, botanists from the Canadian Department of Agriculture studied the plant life of the Canadian side of the St. Lawrence Valley but conjectured that the St. Lawrence project was not likely to eliminate any unique species, particularly as the International Rapids section "contained no species of specific floristic interest."[56] Testifying to the resiliency of nature, the authors

FIGURE 6.13　Saint-Lambert Lock and Victoria Bridge, Montreal. © *Library and Archives Canada / St. Lawrence Seaway Authority fonds*

of a recent study considering various effects of the St. Lawrence project conclude that

> based on the limited information available, it would seem that, in the long run, initial turbidity from the construction quickly declined, vegetation slowly recolonized banks of the new shoreline, especially in areas not disturbed by erosion and dredge[,] and both fish and animal life returned in new proportions to different locations along the shores and in the waters of the St. Lawrence.[57]

In one of the few efforts to intentionally conserve wildlife, PASNY, in conjunction with the New York State Department of Conservation and the federal Fish and Wildlife Service, cordoned off a shallow water area in the vicinity of Wilson Hill using dikes – what would have otherwise been marsh or mud flats – for "aquatic food plants and an environment suitable for nesting and forage for large numbers of wild fowl" and "anticipated that this treatment will also encourage abundant production of panfish."[58] A Canadian equivalent was the Upper Canada Migratory Bird Sanctuary.

The IJC's method of procedure gave precedence to navigation, riparian, and hydroelectric generation rights on the St. Lawrence over environmental and recreational rights. Peaking operations involving major water level fluctuations are hazardous for the riverine environment – they were not allowed under the initial IJC order, but the IJC and navigation interests quickly reversed this.[59] But regulating the river at too uniform a level, and not allowing for natural and seasonal variability, is also ecologically harmful. Immediately after the power pool had been raised, there were rumours that the major dike near the power dam leaked.[60] The extent may have been exaggerated, but there were small seepages that continued after the project was completed.[61] The claim that there has been shoreline erosion because of wave action from the seaway is supported by anecdotal evidence and IJC studies. For example, a former resident of the submerged community of Moulinette suggests there has been twenty feet of erosion.[62] Others claim that the St. Lawrence project changed the local weather conditions, possibly because of the increased amount of water surface area.[63]

Beyond the initial ecological impact of the massive reshaping of the St. Lawrence, one of the most prominent environmental concerns has been the charge that the seaway has increased the mobility of invasive and destructive foreign species of marine life, with zebra mussels as one of the most publicized examples. Jeff Alexander's *Pandora's Locks: The Opening of the Great Lakes–St. Lawrence Seaway* chronicles the environmental damage of these invasive species, and the various contributors to *Voices for the Watershed: Environmental Issues in the Great Lakes–St. Lawrence Drainage Basin* explore a wide range of environmental impacts.[64] These species hitch a ride in the ballast water of ocean-going vessels originating in foreign areas, and many create an ecological domino effect in the Great Lakes–St. Lawrence basin.[65] Despite knowledge that the older alignments of the Welland Canal had allowed exotic organisms such as lamprey and alewife to move between Lake Ontario and Lake Erie (particularly after the reworking of the water supply for the third iteration of the canal) and then to the upper Great Lakes, the involved governments and agencies apparently did not consider the possibility, or did not care, that the seaway would enable the infiltration of additional invasive species.[66] This despite the fact that in 1955 Canada and the United States had jointly established the Great Lakes Fishery Commission primarily to deal with the invasive sea lamprey. Out of the approximately 160 invasive species that have arrived in the Great Lakes–St. Lawrence basin since the early nineteenth century, it is estimated that about one-third have done so since the seaway's opening.[67] However, the invasive species were not the necessary result of the seaway, for if action had been taken earlier by the Canadian and American authorities to regulate the ballast water of foreign vessels, many of these invasions might have been prevented. Moreover, our understanding of invasive species and ecological change is complicated by the fact that some of these "invasive" species predated the seaway or were actually native to the Great Lakes.[68]

Seasonality was, and remains, an important consideration. From the earliest contemplations of a St. Lawrence development, ice formation in the winter – particularly frazil ice, which formed instead of solid ice cover when water flowed too swiftly – had concerned engineers.[69] After the seaway was complete, there were subsequent experiments and technological innovations to reduce ice formation and extend the winter navigation season. The seaway is now effectively closed for only about three months, starting at the end of December, and ice-jamming problems have been virtually removed.[70] Nevertheless, the removal of solid ice cover disrupts the river's chemical processes and ecosystem, and unnatural ice breakup can cause land scarring and other shoreline impacts.[71]

In 1978, the St. Regis (Akwesasne) Band expressed concern to the IJC about the impact of the St. Lawrence Seaway and Power Project on the International Rapids section. As Cornwall Island, part of Akwesasne territory, was directly downstream of the Moses-Saunders Power Dam, the changed water levels resulting from the power project had a noticeable and deleterious long-term impact on the ecological health of the area.

Furthermore, the General Motors and Reynolds industrial plants in New York had located adjacent to Akwesasne land, with resulting long-range environmental and health implications.[72] The adverse effects were grouped into three main categories: those caused by higher water levels, those caused by changes in flow, and those caused by environmental change.[73] A formal submission followed in 1981, and the St. Regis group made further inquiries about shoreline erosion and adverse effects on the muskrat population in the immediately following years. The IJC studied the issues using the limited evidence available, which included past gauge readings and aerial photography. The IJC's reports admitted a wide range of negative environmental impacts but deemed them generally not substantial enough, or their causes too uncertain, to require action or compensation. Aerial photography showed that littoral erosion and changes were occurring but were "localized in nature" at erosion-prone sites or had been one-time results of excavation or landfilling from the St. Lawrence project. Peaking and ponding fluctuations (controlled increase and decrease of the water flows) were not deemed a significant factor in the variability of the muskrat population in the St. Regis area.[74] The IJC concluded that water level fluctuations had decreased since the installation of the St. Lawrence development, and the contention that there had been a general increase in the mean level of the St. Lawrence since the 1950s was not supported by gauge data. At any rate, the wide range of changes identified in the St. Regis submission could not be fully evaluated, as they were beyond the IJC's limited time, sources, and evidence. Nevertheless, these reports serve as a summary of the ecological damage done by the St. Lawrence project: decreased water quality; loss of land by various types of erosion and island drowning; marshland flooding; weed growth; redistribution of pollutants; scouring and silting; fish and eels killed by powerhouse turbines; impediments to fish movement and spawning; and concentrated pollutants in fish.[75]

The experience of the St. Lawrence project probably contributed to the environmental awareness of the 1960s. It is likely no coincidence that the end of the project roughly coincided with the rise of an increased North American environmental consciousness and concomitant movement. Aided by public prompting, governments have in recent decades taken more notice of the ecological health of the St. Lawrence, recently passing legislation to require ships to flush ballast water in an effort to prevent invasive species.[76] Seaway advocates point out that water transportation is more environmentally friendly (because of less fuel consumption and emissions) and safer than road or rail transportation, as is hydroelectric development compared with other power-generating options such as coal or nuclear power. Currently, there are debates between proponents of an expanded seaway and those of a restored river. The American portion of the Moses-Saunders Power Dam, now named after Franklin D. Roosevelt, underwent a relicensing process between 1996 and 2003 that brought out the various sides of the debate.[77] In the early twenty-first century there seems to be some willingness by governmental and seaway agencies to address the negative repercussions of the St. Lawrence project, though it is far from sufficient.[78]

Conclusion

The completion of the St. Lawrence Seaway and Power Project was met with great fanfare, yet the subdued fiftieth anniversary of the opening of the seaway in 2009 demonstrates that the St. Lawrence project is not something that is celebrated in the Canadian national imaginary. Even the memoirs and recollections of Canadian officials who participated in the St. Lawrence negotiations either ignore the seaway or throw bromides at it as a testament to Canadian-American cooperation. There is a collective amnesia, or perhaps a subconscious desire to purge the failure of the all-Canadian option from memory. This stems in part from the economic and environmental problems of the seaway, but it can also perhaps be attributed to the fact that Canadians viewed the joint project, after appearing to be tantalizing close to an all-Canadian waterway, as a defeat and a reminder of living in the shadow of their southern neighbour.

Though the logic and methods of building the St. Lawrence Seaway and Power Project spread across the continent, and the world, it also served as a cautionary tale. The engineering plans for regulating the water levels of the St. Lawrence Seaway and Power Project were indicative of a faith in progress and technology, exemplified by the models the planners relied on, but the evolution of these plans equally reveal the contradictions and limits to this faith. The experts and engineers who planned the St. Lawrence were not as infallible as they, or the society that had produced them, liked to believe. Although the St. Lawrence Seaway and Power Project was an impressive achievement from an engineering perspective, and there were certainly economic benefits, in hindsight the project should be considered a mistake.

CONCLUSION

To the Heart of the Continent

On August 10, 1954, in a speech at the sod-turning ceremony marking the start of construction on the hydroelectricity phase of the St. Lawrence project, Prime Minister Louis St. Laurent waxed eloquent about the cooperative spirit shared by Canada and the United States, and the bond – rather than the barrier – that the St. Lawrence created between the two countries. However, several months later, St. Laurent was more forthright in a speech to a Montreal audience:

> It is more or less an open secret, I believe, that the Canadian Government would have much preferred to build all the canals in Canadian territory. Why did we not do it? Because in order to obtain the co-operation of the American Government before the International [Joint] Commission, we had stated that we would be willing to consider, at any time, any project of American participation provided it did not delay the beginning of operations, and in the meantime, that is between October 29, 1952, and June 7, 1954, the President and the Congress [approved the Wiley Bill] ... but the validity of the permit obtained by the New York Hydro could still have been contested before the American Courts or cancelled by the American Congress had we refused to discuss any project of agreement on the deepening of the waterway.[1]

The prime minister went on to stress the serious injury that would have been done to the bilateral North American relationship if Canada had insisted on unilateral action, but he had nonetheless indicated that the

Canadian government had hoped to build the seaway alone and was prevented from doing so by the United States.

The Montreal speech had originally been given in French. In it, the prime minister called the criticism his government had been receiving in the press and in public for the St. Lawrence agreement "*un mauvais marche.*" The English press translation quoted St. Laurent as saying that Ottawa "had got the best it could out of a bad bargain" in reference to the Canadian government's belief that it had little choice but to accept American involvement yet had still managed to obtain the best possible joint settlement under the circumstances.[2] The American embassy objected to this statement, and the Canadian Department of External Affairs countered that it was a mistranslation, the prime minister himself writing a letter to the editor of the *Ottawa Journal* claiming that the meaning of the phrase had been misrepresented.[3]

Whether or not St. Laurent had been misquoted, "the best of a bad bargain" was a fitting characterization not only for the August 1954 seaway agreement, from the perspective of the Canadian government and public, but equally for the decades of seaway negotiations that had preceded the agreement. The completion of the St. Lawrence Seaway and Power Project agreement was a mixed success for the Canadian government. It had achieved its goal, bringing about the desperately needed hydroelectric power and deep-draught waterway, after years of frustration. However, the seaway was to be a cooperative project, and after the idea of the all-Canadian seaway had captured the imagination of the Canadian public and government, American participation was a definite disappointment, if not a defeat.

The terms of the joint 1954 seaway agreement were decidedly less attractive for Canada than were those of the 1932 Great Lakes Waterway Treaty and the 1941 Great Lakes–St. Lawrence Basin Agreement. The 1954 agreement did not give Canada credit for the cost of the Welland Canal (over $130 million). The United States would have equal control of the seaway, including two of the three locks in the International Rapids section, despite contributing only about one-third of the total cost and with only a portion of the whole seaway within American territory. The St. Laurent government had attempted to atone for this loss by insisting that it build the Iroquois Lock on the Canadian side, installing infrastructure near Cornwall that could be used for a future Canadian waterway, and retaining the right to build this all-Canadian route at a future date.

The all-Canadian seaway appealed to Canadians on many levels, and the widespread public support for the Canadian project fed the St. Laurent

FIGURE C.1 Iroquois Lock, 2009. © *Daniel Macfarlane*

FIGURE C.2 Remains of the old fourteen-foot canal *(foreground)*, the Iroquois Control Dam behind it. © *Daniel Macfarlane*

government's attempt to go it alone. The extent to which Canadians embraced the all-Canadian concept was repeatedly substantiated by the available means of gauging public opinion as well as by every available report, memorandum, poll, and diplomatic dispatch produced by the Canadian and American governments. The economic, trade, and national security benefits that would accrue to Canada from a solely Canadian St. Lawrence enterprise can partially account for its popular embrace. But these benefits were not the main reason; in fact, many suspected that the advantages might be even greater if the seaway was undertaken in conjunction with the United States. Underlying the drive for the all-Canadian seaway was instead a blend of Canadian nationalism shaped by the early Cold War context, informed by the historic role and conceptions of the St. Lawrence River, contoured by the cultural interplay of technology and nature, and infused by ambivalent ideas about Canada's relationship with the United States. Although the drive for an all-Canadian seaway was in many ways a product of events that had transpired since the 1930s (the Depression, Second World War, and start of the Cold War), it also joined various strands of Canadian nationalism that stretched far back into the history of the country. The Laurentian thesis in particular sustained the conception of the St. Lawrence River as a fundamental and defining aspect of Canadian history and identity and, in turn, infused the notion of an all-Canadian seaway with the same nationalist importance and symbolism. As an American Department of State official recorded in the early 1950s, "Canada's decision to build the St. Lawrence Seaway as an all-Canadian project has seized the imagination of Canadians. It is a symbol of their new-found strength."[4] When added to the post-1945 context and mood – economic and technological ability, national self-assurance, and a growing resistance to perceived American domination – Canadian nationalists could not help but be captured by the notion of their country's *own* waterway.

As a transportation megaproject, the all-Canadian seaway offered a nation-building parallel to the transcontinental railways, promoting Canadian identity, national unity, progress, and prosperity while linking the country in an east-west orientation, in contrast to the north-south pull of the United States. The sense of identity with, and ownership of, the St. Lawrence led nationalists concerned about Canada's subservient role as a raw material exporter to the United States to fear American encroachment on the river. Conversely, the seaway represented Canada's ability to independently exploit, use, and control its natural resources via technological progress. Put a different way, Canada, which had been shaped

– and in many ways constrained – by its environment, could now thrive because of it.

Environmental Diplomacy

The St. Lawrence Seaway and Power Project is a vitally important subject for the modern political, diplomatic, and transnational history of North America. Prime Minister St. Laurent's Montreal statement about the ability of the St. Lawrence issue to disturb bilateral relations was no exaggeration. For example, an American official disclosed in 1951 that the St. Lawrence issue contained the potential to "probably injure our relations with Canada more than any other single incident which has occurred during this century."[5] Such a statement demonstrates the importance of the St. Lawrence issue in Canadian-American relations, for it is arguably the most underappreciated aspect of the bilateral relationship in the post-1945 years. Indeed, the seaway story indicates the nature of the Canadian-American relationship in the formative years of the Cold War. Both countries pursued their national interests, which often coincided but were sometimes in competition. The situation was also complicated by the St. Laurent government's belief that the greatest Canadian national interest, in terms of foreign policy, was the continuance of strong relations with Washington.

An all-Canadian seaway, although seemingly the fulfillment of the national interest, could prove its antithesis – the loss of a solid bilateral relationship – since the United States opposed a solely Canadian waterway. Despite Canada's stature as a trusted ally, the relationship extended only so far as it did not threaten American interests, for Washington was more than willing to override Ottawa's desires and sovereignty concerns if these created significant security or economic risks for the United States. The negotiations leading to the St. Lawrence project show Canada to be a self-interested and pragmatic international actor, and seaway diplomacy in the 1940s and 1950s shows the forging of Canadian foreign policy to be a complex process. Examining this process strips some of the lustre from the purported "golden age" of Canadian foreign policy.[6]

The seaway is often held up as a successful model of Canadian-American relations. It is generally believed that, in its negotiations, building, and then operation, the seaway is an apt metaphor for the harmonious relationship and integration of the two countries. The construction and operation of the St. Lawrence Seaway and Power Project may be an example of transnational and transboundary cooperation, but the hydropolitics of the St. Lawrence Seaway and Power Project better represent the asymmetry of the bilateral relationship in the early Cold War. It is tempting, particularly from a nationalist perspective, to speculate that the course of St. Lawrence negotiations shows that the United States regarded Canada as merely a great storehouse to be exploited in line with American wishes. There is some truth to such a view, but such speculations ignore the agency Canada had in pursuing the St. Lawrence project in the post-1945 years, and falsely attribute to the United States a more coherent and coordinated policy toward Canada in this period than was actually the case.

In retrospect, the political hurdles blocking the consummation of the St. Lawrence Seaway and Power Project proved more difficult than the natural obstacles. Punctuated issues, such as wars and crises, have of course profoundly shaped Canadian-American relations, though the general tenor of the relationship has also been predicated on persistent day-to-day and low-profile matters. Because it was debated and considered for more than half a century, the seaway was in many ways both a persistent and punctuated issue. Seaway diplomacy exacted a lasting, and not always positive, impact on Canadian-American relations in the early 1950s and likely contributed to the growth of anti-Americanism in the later 1950s and 1960s, as well as to the rise in tensions in the Canada-US relationship in subsequent years.

The political history of the St. Lawrence project suggests that, insofar as Canadian-American relations ran smoothly in the immediate post–Second World War period, this was the result of Canada's willingness to pre-emptively acquiesce in order to prevent major problems. Put another way, when it came to dealing with the American colossus, the Department of External Affairs was generally pragmatic enough to anticipate what policies would be unacceptable to the Americans and steer Canadian policy on a different tack. Thus, the United States had indirect means, which Washington itself might not have always realized, of shaping Canadian policy in advance. Of course, there were exceptions, and in some ways the St. Lawrence Seaway and Power Project is both the rule and the exception that proves the rule: from 1951 to 1954, Canada actively sought an all-Canadian seaway, cognizant of American preferences to the contrary, but once it became manifest that Washington was intent on participating, most Canadian officials accepted that a joint seaway was almost inevitable.

Linkage attempts were repeatedly threatened or employed. It is clear that, because of the enormous influence the United States exerted on the

Canadian economy and continental defence, the ramifications of Canada building alone would have been significant. The badly needed hydroelectric development would not have gone ahead if Canada had proceeded independently: as the available evidence indicates that certain American officials and interests would have, as they repeatedly warned Canada, done what was necessary with the Federal Power Commission licence to forestall the power project. The executive branch and bureaucracy (mainly the Department of State) of the US government tended to take Canadian sensitivities into account – or "humour" Canada – in large part because of Canada's geographic and trade significance, but the US Congress, with its powers over foreign policy, repeatedly showed itself more than willing to ignore or override Canadian interests. Statements by Canadian policy makers implied that negative repercussions would be felt in the loss of prestige or political capital that Canada would experience in the United States, which in turn would have a wide impact on many Canadian interests and policies. Regardless of the precise area where retribution would result, it is clear that Ottawa believed there would be severe repercussions if it had continued to seek an all-Canadian seaway. The nuclear weapons controversy during the Diefenbaker-Kennedy years demonstrates the means the United States had, both subtle and not so subtle, to influence Canadian policy that it disproved of and which threatened American national security.[7]

There were limits to asymmetry in the Canadian-American relationship, however, as evidenced by the Eisenhower administration's reluctant acceptance of Ottawa's determination to build the navigation works at Iroquois. In fact, Canada generally made out well in direct negotiations with the United States. But there were definite limits to American patience, and the St. Laurent government realized that to push ahead with an all-Canadian waterway after the enactment of the Wiley-Dondero Act would imperil the many benefits that Canada received from its close relationship with the United States and implicitly threaten Canadian sovereignty.

Why, then, did the St. Laurent government persist for so long in pursuing the all-Canadian seaway? New York would share in the hydroelectric power and the Ungava iron ore could still find its way to US steel mills on the Great Lakes even if Canada alone controlled the waterway, and the St. Laurent government hoped that these inducements would be sufficient to persuade Washington to let Canada proceed. Ottawa was thus banking – or perhaps gambling would be the more appropriate characterization – on the continued reluctance of Congress to authorize an American seaway role; the St. Laurent government hoped that, if no legislation was forthcoming, the United States would realize how important the project was for Canada, and North America in general, and allow Canada to begin construction. Moreover, the Canadian government believed that the United States, as a sensible and friendly neighbour, would recognize Canada's right to develop the St. Lawrence.

The St. Laurent government acted pragmatically in the end, but there were several leading officials in both the government and the bureaucracy intent on achieving the all-Canadian waterway, including C.D. Howe, Lionel Chevrier, Guy Lindsay, and General A.G.L. McNaughton. For these men, the all-Canadian St. Lawrence project represented a key piece in a nation-building agenda based on a confident Canadian postwar nationalism. Prime Minister St. Laurent and Secretary of State for External Affairs Pearson both favoured the all-Canadian alternative but realized the potentially negative ramifications for the Canadian-American relationship. They were joined, or perhaps preceded, in this view by some officials in the Department of External Affairs and by two of the Canadian ambassadors to Washington in the early Cold War period, Hume Wrong and A.D.P. Heeney, who tended to stress the political reality of a cooperative project.

The all-Canadian seaway was pushed by several key individuals with engineering expertise. In addition to having an engineering background, Howe had been involved with the St. Lawrence file since the 1930s and was likely St. Laurent's closest ally in the cabinet.[8] Within the cabinet, it was Howe, Pearson, and Brooke Claxton who held the primary policy reins on American relations; the prime minister also allowed the Department of External Affairs and a range of public servants – such as Norman Robertson and Hume Wrong, among others – considerable latitude in formulating foreign policy.[9] McNaughton proved to be a pivotal figure in the seaway diplomacy. In fact, McNaughton may be the most underappreciated Canadian foreign policy actor in the post-1945 period, for, in addition to his roles at the United Nations and on the Permanent Joint Board on Defence, he was instrumental in shaping Canada's approach to the border waters it shared with the United States. In that regard, McNaughton is best known for his role in the Columbia River negotiations of the 1950s and 1960s, which culminated in the Columbia River Treaty and subsequent hydroelectric projects, and it appears that his advocacy of an all-Canadian seaway was a formative influence on his resistance to sharing the benefits of developing the Columbia with the Americans.[10]

The St. Lawrence issue exerted an important influence on the subsequent patterns of other Canadian-American transborder water and environmental

FIGURE C.3 Contemporary view from the American side of the Moses-Saunders Power Dam, 2009. © *Daniel Macfarlane*

FIGURE C.4 Long Sault Control Dam, 2009. © *Daniel Macfarlane*

relations, such as the Columbia issue. The St. Lawrence power project was equally an extremely important step in Canadian dam building: out of the 613 large dams built in the country up to 1984, almost 60 percent were constructed between 1945 and 1975.[11] It was an age of grandiose hydro projects, and the St. Lawrence project was at the forefront of this movement. Tangentially, the St. Lawrence saga marked a low point in the history of the International Joint Commission, as the commission was used by its members – particularly the two chairmen, McNaughton and Roger McWhorter – for partisan and national purposes.

The clear public desire for a nationally controlled waterway helped drive the St. Laurent government on, as did electoral considerations, even when it received repeated rebuffs from the United States. The St. Lawrence development was an almost irresistible attraction from the perspective of the Canadian state. In retrospect, it is easy to see how – given the momentum generated by decades of planning and negotiating for the St. Lawrence project, and the nation-building, nationalist, technological, and political motivations for the project during the early Cold War – the logic of the seaway went virtually unquestioned. Moreover, from a pragmatic electoral perspective, the St. Laurent government hoped that the fulfillment of the St. Lawrence Seaway and Power Project would make it appear active, dynamic, and innovative, and it was the type of tangible achievement that would draw and focus voter attention (or, more cynically, distract the electorate from other issues). Such considerations were evident in Prime Minister St. Laurent's 1951 statement to President Truman that it was useful "to get the Canadian people talking about a constructive project of the magnitude of the St. Lawrence development, which might help to prevent overconcentration on their troubles over prices and short supplies."[12]

The St. Lawrence story therefore speaks to the issue of domestic sources of Canadian foreign policy. It shows the impact that domestic imperatives – in this case, the public desire for an all-Canadian seaway – had on the federal government's approach to seaway diplomacy. The public embrace of a national waterway pushed the St. Laurent government not only to pursue the all-Canadian seaway but also to exact concessions from the United States during the August 1954 negotiations in an attempt to appease a disappointed public. On the other hand, the Liberals also bowed to a joint project despite their firm belief that domestic opinion desired a national waterway. Thus, the St. Laurent government based its final policy choice on advancing what it determined to be the overriding national interest, maintaining harmonious relations with the United States, which also had major domestic implications because of the economic impact of the relationship. In extracting concessions on a future Canadian waterway and the Iroquois Lock, the St. Laurent government had attempted to chart a middle course between satisfying the United States and the public preference. Electoral calculations also factored into the equation. As there had just been a federal election in 1953, the St. Laurent government was more willing to concede to a joint project in 1954, thinking that the electorate might forget about the loss of the all-Canadian seaway by the next election. However, it appears that the abandonment of the all-Canadian route contributed to voter dissatisfaction with the St. Laurent Liberals, as did other issues connected to the United States, such as the TransCanada pipeline, and Canada's waning relationship with Britain.[13] Since Canadians were disappointed that an all-Canadian seaway was replaced by a joint waterway, it is reasonable to assume that the St. Lawrence development influenced some of those who voted against the Liberals in 1957 (and again in 1958), particularly those who were concerned about America's influence on Canada.[14]

There were other domestic political ramifications. In terms of Ottawa's relationship with Ontario, the federal-provincial agreement on the St. Lawrence project, and then joint construction of the St. Lawrence Seaway and Power Project, represented a positive shift in federal-provincial cooperation and went a long way toward erasing the acrimony of the pre-1945 years. The St. Lawrence project served to retain Ontario's position as the economic and communications nexus of Canada. If one invokes the macro theories of Canadian history – the Laurentian, metropolitan-hinterland, or staples theses – it would not be a stretch to speculate that the St. Lawrence project perpetuated a system by which central Canada, particularly Toronto and Montreal, remained the manufacturing metropoles at the expense of the hinterlands that supplied the raw resources.

Nature of the State

The St. Lawrence Seaway and Power Project has much to tell us about the interplay of state building, environment, technology, engineering, and modernism in North America during the early Cold War. The dream that eventually took shape in the 1950s as the St. Lawrence Seaway and Power Project stretched back several centuries. Starting with the initial European settlers and continuing up to the mid-point of the twentieth century, a deep navigational channel from the Atlantic Ocean to the heart of the continent had long captivated northern North Americans and their governments. In the process, the concept of the seaway and power project had undergone a noticeable transition in the minds of both the engineering experts and the Canadian public. A modernist ethos had underpinned the engineering plans for the seaway since the late nineteenth century or early twentieth century; however, in the decades after the First World War, with the concomitant change in scale, technology, and ambition, the St. Lawrence Seaway and Power Project planning became *high* modernist.

In the words of James C. Scott, to whom the concept is most prominently attributed, high modernism

> is best conceived as a strong, one might even say muscle-bound, version of the beliefs in scientific and technical progress that were associated with industrialization in Western Europe and in North America from roughly 1830 until World War I. At its core was a supreme self-confidence about continued linear progress, the development of scientific and technical knowledge, the expansion of production, the rational design of social order, the growing satisfaction of human needs, and, not least, an increasing control over nature.[15]

High modernist plans thus privilege bureaucratic and technocratic expertise over local knowledge, without recognition of the limitations of this top-down approach, in large-scale attempts to make "legible" – through simplification, standardization, and ordering – social and natural environments in order to control them and prescribe utilitarian plans for their betterment.[16]

Scott identifies three elements of what he calls "the most tragic episodes of state development in the late nineteenth and twentieth centuries": in addition to the actual high modernist ideology, he points to the unrestrained use of the immense powers wielded by modern states, and an incapacitated or prostrate civil society.[17] The latter element is not generally considered characteristic of mid-twentieth-century North American societies, though that may depend somewhat on one's definition of a "prostrate" society. But the first two elements of Scott's definition – which encompass a focus on the future and whitewashing of the past, a hubristic faith in technological and scientific progress and methods, and a drive to tame nature and reorder society on a large scale – are so readily apparent in the St. Lawrence project that, even though it was constructed by liberal democracies rather than authoritarian states, it should still be considered high modernist. Tina Loo, for example, has declared that in Canada "hydroelectric development was the most prominent manifestation of the high modernist impulse"[18] and Graeme Wynn specifically affirms the notion of the seaway as the epitome of a Canadian high modernist project.[19]

High modernism is an appropriate concept for understanding and characterizing the organizing logic and imperatives that drove plans for the St. Lawrence project, in terms of both the river and the people that resided near it. The St. Lawrence Seaway and Power Project was a nation-building exercise controlled by centralized bureaucracies with the aim of regimenting the natural environment for the sake of progress, and in turn attempting to organize, regulate, and improve Canadian society. But it was also reordered and attuned to liberal, capitalist, and democratic principles that were refracted through the prism of Cold War imperatives and modalities and further infused by Canadian St. Lawrence nationalism and American imperialism.

In many respects, the St. Lawrence Seaway and Power Project was a product of technological momentum or path dependencies, particularly those flowing from hydroelectric development, electrical distribution, and manufacturing technologies; but the completion of the St. Lawrence project was certainly not inevitable, as its long history demonstrates, for it took the historically contingent choices, ideas, and conditions of the postwar period to make it a reality. The creation of the St. Lawrence Seaway and Power Project was a consequence of its Cold War context. Richard P. Tucker reminds us that "much of the world's dammed rivers reflect Cold War zones of competition, and the concentration of fiscal and industrial resources at many dam sites in remote locations cannot be fully explained outside the framework of Cold War rivalries."[20] As an apolitical form of coercion, high modernism thrived in both capitalist and Communist countries, especially the United States and Soviet Union. The American desire to participate in the St. Lawrence undertaking emanated from the same strategic motivations that led the United States to undertake or assist with hydroelectric projects in a number of non-aligned countries during the Cold War, stretching from Egypt to the Philippines, situated along the Soviet Union's periphery.[21]

Similar American strategic motivations can be traced back to earlier massive waterways on foreign soil. The Panama Canal, for example, can be framed as an imperial project aimed at building an advanced, ordered, and technologically progressive American empire.[22] Within its own borders, the United States was a pioneer when it came to modernist water control projects, such as the Hoover/Boulder, Bonneville, and Tennessee Valley Authority projects.[23] Several large dams were also constructed in Canada in the first half of the twentieth century, though the majority were done privately, in contrast to the United States.[24] At the same time as the St. Lawrence project was under construction, the United States and Canada were about to finish another transborder water control megaproject, at Niagara Falls, and begin another on the Columbia River.

The Soviet Union was simultaneously undertaking large-scale hydroelectric and river basin industrialization projects, and there are striking parallels between the St. Lawrence project and the Communist dam developments on the Kuibyshev and Volga.[25] At the opening of the St. Lawrence project, the Hydro-Electric Power Commission of Ontario (HEPCO) chairman made an explicit comparison with the Kuibyshev hydro project, waxing poetic about the cooperative international nature of the St. Lawrence development and the example such collaboration projected to the world.[26] The Soviets further attempted to counter the West by assisting with Egypt's construction of the Aswan High Dam on the Nile River after capitalist support was withdrawn.[27]

I would argue that the high modernist wave in North America crested in the late years of the Second World War and early Cold War period (i.e., from approximately the early 1940s to the 1960s), with the St. Lawrence project a leading exemplar of the high modernist élan. In retrospect, such water control efforts represented part of an initial, or perhaps second, wave of large-scale dam building that swept the world in the last half of the twentieth century (the World Commission on Dams estimated that there were forty-five thousand dams worldwide by the end of the millennium). Dams represented modernity and power, both state and electric, for "the great blank wall of a dam was a screen on which [planners and governments] would project the future."[28] Although there has been in recent years a slowdown, and even reversal, of dam building in some countries, more recent hydro dam projects such as the Itaipu Dam on the Brazil-Paraguay border and Three Gorges Dam in China dwarf the early Cold War efforts. The transnational character of high modernism, applied across countries of all creeds and religions, demonstrates that the tendency toward dominating the environment was not the inherent result of a Christian ethic of nature, as some have argued; to be sure, the St. Lawrence project is emblematic of faith, but a faith in science, technology, and progress.[29]

The St. Lawrence Seaway and Power Project was the product of a global engineering fraternity and the transborder spread of engineering techniques and ideologies. The lead engineering firm for the American share of the power works, Uhl, Hall and Rich, had extensive experience in hydraulic projects stretching across the globe, including the Tennessee Valley Authority dams. Many American engineers and tradespeople had worked on one or more of the Bureau of Reclamation's hundreds of dams in the American West.[30] Canadian planners also had wide-ranging experiences, in addition to those projects in their own backyard like Niagara –

many Americans employed on the Moses generating station in the St. Lawrence went immediately to work on another power project named after Robert Moses, this one at Niagara Falls. During its construction, many officials and planners from other countries travelled to view the St. Lawrence complex, which was comparable to a graduate school for engineers. Those involved in the St. Lawrence project were afterward sought for their expertise on other megaprojects in North America and virtually every other continent.[31] The St. Lawrence project influenced subsequent Canadian hydroelectric megaprojects in British Columbia, Quebec, Labrador, and Manitoba, as well as the range of proposals for bulk transboundary water diversions from Canada to the United States, such as the NAWAPA and GRAND canal schemes.[32] Hydro nationalism was as apparent in Quebec's postwar projects as it had been in the case of the St. Lawrence project.[33] However, compared with other Canadian hydro megaprojects, the St. Lawrence Seaway and Power Project was close to major urban centres, rather than in predominantly rural or isolated areas. It is also worth pointing out that planning for such grandiose projects was an exclusively male domain.[34]

Nothing could stand in the way of progress, not even whole towns and communities. Mass displacement in the St. Lawrence Valley was a small price to pay for the production of electricity and the increased accessibility of iron ore deposits. Flooding out thousands of people in the Lost Villages and surrounding rural areas (including the Mohawk reserves) was justified in the name of progress and for the benefit of the wider nation. The reorganization and resettlement of those affected by the power development would be for their own benefit, as they would be placed in consolidated new towns – instead of scattered about in inefficient villages, hamlets, and farms – with modern living standards and services. To replace the inundated communities and farms, HEPCO designed the relocated towns on the latest planning principles: new homes with basements; paved streets that looped around instead of following a grid pattern; modern sewers, water and hydro facilities, and sewage treatment plants. As high modernist project are wont to do, history was literally erased (as was the case with the War of 1812 site of the Battle of Crysler's Farm) in favour of a heroic future. The transportation networks along and across the river were also reconfigured and upgraded. Such organizational and settlement plans imposed state-defined political, economic, and social values, and enabled Ottawa (and Toronto), as well as Washington (and Albany), to control how these communities fit into the emerging postwar order so that they could be more fully integrated into the wider political and industrial capitalist

Figure C.5 Old Highway 2 running into Lake St. Lawrence. © *Daniel Macfarlane*

Figure C.6 Old road *(centre)* and building foundations *(upper left, in the distance)* visible at the former site of Aultsville, in Lake St. Lawrence, 2012. © *Daniel Macfarlane*

structures. In this sense, the Canadian state was at least partially motivated by its role as "a client of the business community," to borrow from H.V. Nelles's argument about Ontario and hydroelectric development in an earlier context, with this community dominated in the early Cold War period by large corporations often controlled by American capital.[35]

The state used the St. Lawrence project as a spectacle to demonstrate its power and prove its legitimacy to its citizens. Sampling, polling, surveying, testing, and modelling were extensively used; as fundamental techniques of a high modernist approach, they allowed the state to control information, set the terms of debate, and manufacture consent – if people knew the facts, the rationality of the project would inevitably compel them to accept its logic.[36] There was a societal deference to government, which in turn reflected a deference to experts and engineers. For the involved governments, as well as the general public, the idea that it was all a sacrifice worth making was pervasive. There were certainly those who resisted in various ways, but for many the project carried an aura of inevitability. Moreover, those dislocated by the power pool expected that the St. Lawrence project would bring with it great prosperity. There were a range of responses to the Lost Villages experience that resists easy generalizations. Living standards were improved in the new model towns, but people also lost a way of life for which there was no clear compensation. HEPCO was more responsive to the preferences of those it dislocated, compared with its American counterpart and relative to other previous megaprojects, but the utility was ultimately unable to see beyond the momentum and high modernist logic of the St. Lawrence Seaway and Power Project. Although relocation looked different on the American side, the Power Authority of the State of New York's head, Robert Moses, was every bit the populist high modernist prophet.

The experts, and by extension, the state and society, viewed nature as something to be controlled and ordered through technology with little to no consideration of the wider environmental impact. The rhetoric used by experts and governments focused on defeating, dominating, exploiting, and mastering the river. A megaproject ethos is also evinced by the type of language that was not used: namely, acknowledgment of the environmental limits and repercussions inherent in a project on the scale of the St. Lawrence Seaway and Power Project. This is supported by a study surveying the environmental impact of manipulating the St. Lawrence: "With limited exceptions ... environmental concerns were of little interest to either the engineers who designed the project or the general public."[37]

Implicit in the engineering and governmental planning was an arrogance regarding their ability to control and remake the St. Lawrence, one of the world's largest rivers based on water volume, down to the most precise water levels and flows. The International Joint Commission was responsible for developing a river profile so that construction could proceed. Engineers spent the duration of the construction phase trying to establish a satisfactory range of minimum and maximum water levels for both the St. Lawrence project and Lake Ontario. This involved ascertaining the "natural" levels of the river and lake, which was a self-defeating endeavour, since "natural" was subjective and difficult to determine because of a lack of information and previous modifications to the Great Lakes–St. Lawrence basin. The engineers kept producing methods of regulation to prescribe the water levels, but because of faulty approaches and lack of full knowledge they were forced to repeatedly revise them over the span of a few years. Despite their public claims of preciseness and expertise, among themselves they continuously qualified their water level prescriptions with the phrase "as nearly as may be," and the engineering process revealed many errors, assumptions, guesses, and subjective assessments. Nevertheless, they were able to overcome these engineering miscalculations, and we should not overlook the fact that, in the end, the construction and operation of the St. Lawrence project went according to plan. From the perspective of the experts, there were relatively few unintended environmental consequences, since the more pernicious (e.g., invasive species) could likely have been prevented, and other negative repercussions were considered by planners to be tolerable by-products. This is not to excuse the damage wrought by the St. Lawrence project but rather to highlight that environmental repercussions were assumed and accepted by planners, thus suggesting that high modernist megaprojects do not inevitably fail wholesale. The one major area where the seaway did not live up to its planners' expectations was in the amount of cargo moved annually and the seaway's resulting inability to be self-liquidating. However, this inadequacy was, ironically, because the navigation elements were not ambitious enough: bigger locks and canals could have hosted the larger vessels that increasingly plied ocean waters.

There is a strong temptation to adopt a declensionist narrative and portray the St. Lawrence solely as a river that has been completely lost, corrupted into a seaway and power pool. But stressing only this perspective ignores the power of nature in general and the St. Lawrence in particular. Scholars have variously conceived of the blending of artificial and mechanical with the natural and the organic manifest in water control projects, proposing useful concepts such as socio-technological systems, second nature, and organic machine.[38] Several historians have recently sought to further erase what they identify as an artificial divide between the natural and the industrial: Liza Piper identifies an assimilative and integrative industrial process in the Canadian subarctic where "nature, economy, and society each adapted to one another," and Sarah B. Pritchard and Thomas Zeller seek to "naturalize industrialization" by stressing "the ways in which industrial processes were embedded within, and thus ultimately dependent upon, natural resources, environmental processes, and ecosystems."[39]

Building on the idea that the industrial remaking of a river involves changing, rather than severing, the connections between an ecosystem, technology, and society, it is clear that the waters of the St. Lawrence were, on the one hand, certainly commodified and abstracted; on the other hand, engineers and planners also engaged with and knew the St. Lawrence environment in great detail. On many levels, those living along the river were separated and alienated from their past ways of knowing and using its waters, but the hybrid envirotechnical system that was the seaway also forged new relationships, dependencies, and ways of knowing between the St. Lawrence and industrialized society. Dams and canals changed the river, but they nonetheless became established parts of the St. Lawrence environment, which was ultimately more natural than industrial, even if only slightly. Even though the hydrological regime of the St. Lawrence was radically altered, and lengthy sections converted to a lake and a shipping channel, it is also a very resistant and fluid river. Outside of a few concentrated locations, the impact of human intervention is no longer readily apparent along the St. Lawrence – even the changed contours wrought by the flooding in the International Rapids section (a title that now has no real applicability since the rapids have been drowned) are now essentially natural, the abrupt farmers' field shorelines of the 1950s shaped by wave action and colonized by flora and fauna native to littoral zones. Trees quickly replaced bulldozer tracks, vegetation covered the spoil piles, animals returned, and fish found new spawning grounds. Newcomers to the water's edge are unaware that the river was so substantially reshaped, and it is likely that the majority of the residents in the communities near the river, or the weekend campers on the Long Sault Parkway, have never known the St. Lawrence to be anything but its current form.

The high modernist concept can gain additional purchase as an explanatory tool when it is recalibrated and nuanced to take into account specific locations and cultures. Scott does appreciate that high modernism changed, or had staggered developments, across eras and locations: for example, the author ascribes a "softer" form of authoritarian high modernism to Tanzanian villagization in the 1970s, speaks of Soviet collective farms and the Brazilian creation of a new capital city from scratch as "ultra-high modernism," and in a separate study detailed how high modernist aspirations for the development of the Tennessee Valley Authority were derailed by its liberal-democratic setting.[40] Scott not only proposes that high modernist projects are bound to fail because of their intrinsic contradictions but suggests that such projects cannot even get fully off the ground in liberal political economies because of three major barriers: the private sphere of activity, the private sector of the economy (i.e., the free market), and effective democratic institutions.[41] At the same time, he identifies as high modernist luminaries several American figures, including one of the lead St. Lawrence hydroelectric planners, Robert Moses.

In his study of state-led reform in the context of US New Deal agriculture techniques – which he identifies as "low modernism" because of the emphasis on citizen participation and betterment – Jess Gilbert takes this democratic factor into account, but the participatory democratic aims he ascribes to New Deal authorities were generally not shared by St. Lawrence planners.[42] In the context of the seaway, the argument that private sectors make the economic reins too complicated to manage does not necessarily hold up, nor does the idea that democratic representative institutions invariably thwart high modernist schemes. True, consent for high modernist schemes had to be continuously – or, at least, every four years or so – earned or manufactured. But when consent was achieved, as was the case with the seaway, it gave high modernist plans greater legitimacy. At least in the case of the St. Lawrence project, one must conclude that democratic institutions and the free market served only to complicate and partially modify, rather than block, high modernist interventions.[43]

Tina Loo and Meg Stanley have convincingly shown there was actually an intimate engagement with place in Canadian postwar dam-building efforts, a "high modernist local knowledge" defined by detailed and intimate awareness of specific environmental locales.[44] Building on Loo and Stanley, I would argue that the St. Lawrence River was reduced and abstracted to simplified schematics in order to make the riverine environment legible, but on-the-ground conditions were not ignored. In fact, the two power entities, the involved governments, and transboundary agencies had spent decades closely scouring, studying, and analyzing the St. Lawrence bioregion in order to ascertain specific place information. These micro-level details were then translated and projected into macro-level plans. St. Lawrence planners repeatedly modified and adapted their engineering plans based on what they faced on the ground. But even though the engineers were repeatedly made aware of their fallibility and the limits of their expertise, they continued to ignore the underlying flaws in their methods and maintained full trust in their models and technological expertise. When certain aspects of the project, such as ice formation, seemed beyond expert control, planners would ignore the problem and disclaim responsibility (e.g., label it an "act of God"). In the end, it was in the engineers' belief that they could actually know and master every square inch of the environment, rather than ignorance of the river basin which they were manipulating, that the hubris of the St. Lawrence undertaking was perhaps most apparent.

What is needed, then, is a more appropriate conceptualization of the form that the high modernist ethos took in regard to the St. Lawrence project. The Canadian and American federal governments in the 1950s would not be considered by most definitions as authoritarian, but calling the St. Lawrence case a "softer" form of high modernism does not adequately capture its unique qualities and variations. We see what I have termed *negotiated* high modernism: high modernist in conception, but negotiated politically in practice. Although the St. Lawrence project was fundamentally about control, the governments still needed to continually mediate and legitimate their authority, which was accomplished through involving people in the process (depending on their levels of political, cultural, and economic power) as participants and admirers, as workers and participants (i.e., those being relocated), and the various means of manufacturing consent. Lacking the centralized and autocratic authority to simply impose schemes without some measure of approval from civil society and other levels of government, the involved states had to repeatedly adapt, negotiate, and legitimize themselves – in relation to both the specific natural environments and the societies they aimed to control – and their high modernist St. Lawrence vision. Moreover, the St. Lawrence project can be distinguished by the need to formally negotiate not only between the different levels of government within one nation but between those of two countries, a necessity that undoubtedly exerted a profound influence on the final form.

Although I believe that negotiated high modernism is applicable to the United States, I am more confident in ascribing it to the Canadian state. This stems from my analysis of Canada's approach to the St. Lawrence Seaway and Power Project, which was manifestly a product of the uniquely Canadian cultural conceptions of the St. Lawrence River; national links between identity, technology, and the natural environment; Canada's relationship with the United States; the very existence of Canada as a negotiated state (i.e., a compromise between ethnicities and cultures); and the distinctive Canadian relationship between the state and civil society.

The seaway served as a lightning rod for many expressions of Canadian nationalism: geographic, environmental, technological, political, and economic. The focusing of these various sources of Canadian identity, particularly technological nationalism, on the waters of the St. Lawrence can be thought of as a form of hydro nationalism, which can be conceptualized as a combination of hydrological and hydraulic nationalism since Canadians identified with the river in both its "natural" state (hydrological) and as an "improved" seaway/power project (hydraulic).[45] The significance of the St. Lawrence is apparent when the role that water and rivers in general have played in fostering Canadian nationalism and mythology is taken into account. Canadians "consider water part of their natural identity,"[46] as "rivers are Canadian cultural icons; they have consistently communicated the idea of Canada, its meta-narrative of nation-building and collective identity."[47] The power of the river narrative and hydrological (and hydraulic) symbolism in Canadian (and American) historiography is outlined by the authors of an environmental history of the Bow River in Alberta, for example, who illustrate the use of the St. Lawrence as a central organizing metaphor in Canadian history.[48]

The connection in Canada between water and nationalism can appear contradictory, particularly in relation to the United States.[49] The St. Lawrence River had much greater purchase in the Canadian national imaginary. Moreover, the St. Lawrence saga suggests that there were, at least at the time, important cultural differences between Canadian and American views of the natural environment, particularly where it intersected with technology and national development.[50] Noted historian W.L. Morton, among other prominent cultural commentators such as Northrop Frye and Margaret Atwood, identified the importance to the national psyche of Canada of a small population struggling against the Canadian Shield and the North – a vast, foreboding, cold, and hostile landscape. Similar sentiments about the link between Canadians and the natural world are reflected in the paintings of the Group of Seven and Emily Carr.

This study suggests there is some truth to this hostility thesis, though it is fraught with oversimplifications and potential contradictions. Other geographical and environmental factors, such as the abundance of untamed wilderness in Canada, the country's historic reliance on extracting staple resources, and the concentration of the population in certain areas, can help account for different Canadian and American views of nature.[51] I think there is something to the notion that a uniquely Canadian conception of the natural world was produced by Canadians' greater exposure to a harsher and more frigid environment; the historic development of economic, technological, and natural resource extraction activities associated with this environment; and relationships with more powerful countries (primarily Britain and the United States) seeking to acquire these resources.

The seaway saga signals the hubris to which twentieth-century governments, experts, and planners were prone when dealing with plans to radically reshape water systems. The litany of major water transfer schemes that have appeared since the St. Lawrence Seaway and Power Project was completed indicates the persistence of these attitudes. Contemporary megaprojects generally require canvassing public input and lengthy environmental impact assessments, yet it cannot be assumed that projects like the seaway will not happen today; development of the western Canadian tar sands, for example, indicate that large-scale undertakings with a deleterious impact on the natural environment can happen when there are sufficient financial and political interests at stake, particularly when they are distant from large population centres. The future of the St. Lawrence River is intimately connected to wider questions about the future of North American water supplies, and Canadian-American sharing of water resources, particularly in the Great Lakes basin. There have been calls to close the seaway because of the invasive species it made possible, and the host of concerns about long-terms impacts on the riverine system, as well as the threat of spills or other accidents. Proponents of expanding shipping capabilities, such as the US Army Corps of Engineers, have proposed enlarging the waterway so that it can accept larger ships, whereas others draw attention to water shipping as an environmentally friendly form of bulk transport compared to road and rail.

The St. Lawrence Seaway and Power Project stands as one of the longest-running issues in the history of bilateral North American relations, and one with a tremendous impact on the economic, political, social, and environmental structures of the two countries. This volume demonstrates that the St. Lawrence project was one of the pre-eminent issues in Canadian-American relations during the early Cold War period, and in revealing the labyrinth-like twists and turns that characterized the genesis of the project, provides a comprehensive account of the process. With the completion of the St. Lawrence Seaway and Power Project, the empire of the St. Lawrence, the dream that had captivated so many throughout Canadian history, had been fulfilled, albeit in a form that few of its early exponents could have foreseen: "Above all, the river remained, the river which cared not whether it was valued or neglected, the river which would outlast all the ships that sailed upon it and survive all the schemes which it could possibly inspire."[52]

Notes

FOREWORD: NATIONAL DREAMS

1 Harold A. Innis, *The Fur Trade in Canada: An Introduction to Canadian Economic History* (Toronto: University of Toronto Press, 1930); Pierre Berton, *The National Dream: The Great Railway, 1871-1881* (Toronto: McClelland and Stewart, 1970); Marc Raboy, *Missed Opportunities: The Story of Canada's Broadcasting Policy* (Montreal and Kingston: McGill-Queen's University Press, 1990), 45; Michel Filion, "Broadcasting and Cultural Identity: The Canadian Experience," *Media, Culture and Society* 18, 3 (1996): 447-67.
2 Graeme Wynn, "Dignity and Power," foreword to *Home Is the Hunter: The James Bay Cree and Their Land*, by H.M. Carlson (Vancouver: UBC Press, 2008), xiii; and Graeme Wynn, *Canada and Arctic North America: An Environmental History* (Santa Barbara, CA: ABC-Clio, 2006), 273-321.
3 Lowell J. Thomas, *The Story of the St. Lawrence Seaway* (Buffalo, NY: Stewart, 1957), 13.
4 The St. Lawrence Seaway Management Corporation, "Seaway History," *Great Lakes St. Lawrence Seaway System*, 2013, http://www.greatlakes-seaway.com/en/seaway/history/; Claire Puccia Parham, *The St. Lawrence Seaway and Power Project: An Oral History of the Greatest Construction Show on Earth* (Syracuse, NY: Syracuse University Press, 2009).
5 Norm Tufford, *Tommy Trent's ABC's of the Seaway* (Cornwall, ON: St. Lawrence Seaway Management Corp., 1988) and subsequent editions.
6 CBC Digital Archives, "The St. Lawrence Seaway: Let the Flooding Begin," 2013, http://www.cbc.ca/archives/.
7 See "The Lost Villages," 2013, http://www.ghosttownpix.com/lostvillages/; *Ottawa Citizen* online feature, "The Lost Villages: Upper Canada's Heritage Beneath the Waves," http://www2.canada.com/ottawacitizen/features/lostvillages/. Songs by James Gordon about the Lost Villages with audio recordings are available at "The Lost Villages" *Ottawa Citizen* website under "Gone but Not Forgotten"; Louis Helbig's Sunken Villages exhibition, a series of aerial photographs of submerged sites (made possible, ironically, by the work of

otherwise ecologically troublesome zebra mussels introduced into the St. Lawrence by the seaway, which have helped to clarify Great Lakes waters) was on display at the Marianne van Silfhout Gallery on the Brockville campus of St. Lawrence College in fall 2013 (see http://www.sunkenvillages.ca/).

8 Johanna Skibsrud, *The Sentimentalists* (Kentville, NS: Gaspereau Press, 2009); Maureen Corrigan, "'The Sentimentalists': Submerged Emotions Surface," NPR Books, http://www.npr.org/.

9 Anne Michaels, *The Winter Vault* (Toronto: McClelland and Stewart, 2009), 35, 50.

10 H.D. Thoreau, "The Scenery of Quebec; and the River St. Lawrence," in *A Yankee in Canada, with Anti-Slavery and Reform Papers* (Boston: Ticknor and Fields, 1866), 83. *A Yankee in Canada* was first published serially in 1853, in *Putnam's Monthly*, as "An Excursion to Canada."

11 John Ogilby (translated from Dutch, original author Arnoldus Montanus) *America: being an accurate description of the New World: containing the original of the inhabitants; the remarkable voyages thither: the conquest of the vast empires of Mexico and Peru, their ancient and later wars. With their several plantations, many, and rich islands; their cities, fortresses, towns, temples, mountains, and rivers: their habits, customs, manners, and religions; their peculiar plants, beasts, birds, and serpents* (London: 1671).

12 P. Veyret, "Un cas d'isolement: Les Canadiens français," in *Mélanges géographiques offerts à Ph. Arbos*, Institut de géographie de Clermont-Ferrand (Paris, 1953), 293-99, cited in Cole Harris, "The St. Lawrence: River and Sea," *Cahiers de géographie du Québec* 11, 23 (1967): 171-79.

13 J. Bouchette, *The British Dominions in North America, or, A topographical and statistical description of the provinces of lower and upper Canada, New Brunswick, Nova Scotia, the islands of Newfoundland, Prince Edward, and Cape Breton: including considerations on land granting and emigration; and a topographical dictionary of Lower Canada; to which are annexed, statistical tables and tables of distance I* (London: Colburn and Bentley, 1831) vii, 126.

14 Thoreau, *Yankee in Canada*.

15 Bouchette, *British Dominions*, 154-57.

16 Marion I. Newbigin, *Canada: The Great River, the Lands and the Men* (New York: Harcourt Brace, 1926), 300; H.P. Biggar, review of *Canada: The Great River, the Lands and the Men*, by Marion I. Newbigin, *Geographical Review* 69, 5 (1927): 474-75.

17 D.G. Creighton, *The Commercial Empire of the St. Lawrence, 1760-1850* (Toronto: Ryerson Press, 1937); J.M.S. Careless, "Frontierism, Metropolitanism, and Canadian History," *Canadian Historical Review* 35, 1 (1964): 14; D.G. Creighton, *John A. Macdonald: The Young Politician* (Toronto: Macmillan, 1952) and *John A. Macdonald: The Old Chieftain* (Toronto: Macmillan, 1955). See also for this discussion generally, C.C. Berger, *The Writing of Canadian History: Aspects of English-Canadian Historical Writing, 1900-1970* (Toronto: Oxford University Press, 1976, 2nd ed., 1986).

18 *Canadian Encyclopedia*, s.v. "Historiography in English," by A.B. McKillop, http://www.thecanadianencyclopedia.com/.

19 John Bartlet Brebner, *North Atlantic Triangle: The Interplay of Canada, The United States and Great Britain* (New Haven, CT: Yale University Press, 1945), xi. For Lower and Underhill, see Berger, *The Writing of Canadian History*.

20 J.W. Dafoe, *Canada: An American Nation* (New York: Columbia University Press, 1935); George Grant, *Lament for a Nation: The Defeat of Canadian Nationalism* (Toronto:

Macmillan, 1965); Ian Lumsden, ed., *Close the 49th Parallel Etc.: The Americanization of Canada* (Toronto: University of Toronto Press, 1970). For a wide-ranging examination of this question and a useful guide to the literature, see Allan Smith, *Canada – An American Nation? Essays on Continentalism, Identity and the Canadian Frame of Mind* (Montreal and Kingston: McGill-Queen's University Press, 1994).

21 J.R. McNeill and Corinna Unger, eds., *Environmental Histories of the Cold War* (Cambridge: Cambridge University Press, 2010); Kurk Dorsey, "Perhaps I Was Mistaken: Writing about Environmental Diplomacy over the Last Decade," *Passport: The Society for Historians of American Foreign Relations Review* 44, 2 (2013): 37-41.

22 Kurkpatrick Dorsey, *The Dawn of Conservation Diplomacy: US-Canadian Wildlife Protection Treaties in the Progressive Era* (Seattle: University of Washington Press, 1998). See also Kurk Dorsey, "International Environmental Issues," in *A Companion to American Foreign Relations*, ed. Robert Schulzinger (Malden, MA: Wiley-Blackwell, 2006), 31-47; Kurk Dorsey, "Dealing with the Dinosaur (and Its Swamp): Putting the Environment into Diplomatic History," *Diplomatic History* 29, 4 (2005): 573-87; and "Forum: New Directions in Diplomatic and Environmental History," eds. Kurk Dorsey and Mark Lytle, special issue, *Diplomatic History* 32, 4 (2008).

23 Dorsey, "Perhaps I Was Mistaken."

24 By way of passing example, one might reference here work on the "Trail Smelter dispute," perhaps the most storied case in international environmental law. As Martijn van de Kerkhof writes in the first sentence of his article "The Trail Smelter Case Re-Examined: Examining the Development of National Procedural Mechanisms to Resolve a Trail Smelter Type Dispute," "To study international environmental law without being confronted with the Trail Smelter case is like studying literature without ever coming across the works of William Shakespeare" (*Merkourios* 27, 73 [2011]: 68-83). The case has been much more studied in the United States than in Canada. See J.D. Wirth, *Smelter Smoke in North America: The Politics of Transborder Pollution* (Lawrence: University Press of Kansas, 2000); J.D. Wirth, "The Trail Smelter Dispute: Canadians and Americans Confront Transboundary Pollution, 1927-41," *Environmental History Review* 1, 2 (1996): 34-51; and R.M. Bratspies and R.A. Miller, eds., *Transboundary Harm in International Law: Lessons from the Trail Smelter Arbitration* (New York: Cambridge University Press, 2006), and several articles cited in van de Kerkhof (2011). In Canada, most of one of the main contributions to the study of this case remains unpublished: James R. Allum, "Smoke across the Border: The Environmental Politics of the Trail Smelter Dispute" (PhD diss., Queen's University, 1995). See also James R. Allum, "'An Outcrop of Hell': History, Environment, and the Politics of the Trail Smelter Dispute," in Bratspies and Miller, *Transboundary Harm*, chap. 1. One might similarly note work on the International Joint Commission and the Boundary Waters Treaty of 1909, which has perhaps drawn a little more Canadian interest (see the bibliography at http://bwt.ijc.org/uploads/docs/Bibiliography.pdf). See also Shannon Stunden Bower, *Wet Prairie: People, Land, and Water in Agricultural Manitoba* (Vancouver: UBC Press, 2011) for her treatment of the implications of the transborder flow of water into Manitoba.

25 G.F. Kennan, *American Diplomacy, 1900-1950* (Chicago: University of Chicago Press, 1984), 66.

26 Dorsey, "Dealing with the Dinosaur (and Its Swamp)."

27 Terry Pedwell, "Stephen Harper 'Won't Take No for an Answer' from U.S. on Keystone XL Pipeline," *Financial Post*, September 26, 2013.

28 Michaels, *Winter Vault*, 45.

Introduction: River to Seaway

1 Hugh MacLennan, "By Canoe to Empire," *American Heritage* 12, 6 (1961): 71, quoted in Neil S. Forkey, "'Thinking like a River': The Making of Hugh MacLennan's Environmental Consciousness," *Journal of Canadian Studies* 41, 2 (Spring 2007): 48-49.
2 The Thousand Islands section extends from Lake Ontario to just below Prescott, a distance of about 69 miles. The International Rapids section then runs 46.5 miles to the head of Lake St. Francis, over which the water level falls 93 feet. The Lake St. Francis section then covers a distance of 26 miles to the Soulanges section, which extends from the foot of Lake St. Francis to the head of Lake St. Louis, a total of approximately 18 miles, with a fall of about 82 feet. The Lachine section runs 24 miles from the head of Lake St. Louis to Montreal Harbour, with a total difference of water level of almost 50 feet. An accessible overview of the abiotic and biotic features of the St. Lawrence ecoregions can be found in Gordon Nelson et al., *The Great River: A Heritage Landscape Guide to the Upper St. Lawrence Corridor*, Environments Publication (Waterloo, ON: University of Waterloo, 2005).
3 James Eayrs, *The Art of the Possible: Government and Foreign Policy in Canada* (Toronto: University of Toronto Press, 1961), 157.
4 John Herd Thompson and Stephen J. Randall, *Canada and the United States: Ambivalent Allies*, 4th ed. (Montreal: McGill-Queen's University Press, 2008), 213.
5 Gordon T. Stewart, "America's Canada Policy," in *Partners Nevertheless: Canadian-American Relations in the Twentieth Century*, ed. Norman Hillmer (Toronto: Copp Clark Pittman, 1989), 21.
6 Annette Baker Fox, Alfred O. Hero Jr., and Joseph S. Nye Jr., *Canada and the United States: Transnational and Transgovernmental Relations* (New York: Columbia University Press, 1976); Carl E. Beigie and Alfred O. Hero Jr., *Natural Resources in US-Canadian Relations*, vols. 1 and 2 (Boulder, CO: Westview Press, 1980); John E. Carroll, *Environmental Diplomacy: An Examination and a Prospective of Canadian-US Transboundary Environmental Relations* (Ann Arbor: University of Michigan Press, 1983); John Kenneth M. Curtis and John E. Carroll, "Transboundary Environmental and Resource Issues," in Hillmer, *Partners Nevertheless*; G. Bruce Doern, *Green Diplomacy: How Environmental Policy Decisions Are Made* (Toronto: C.D. Howe Institute, 1993); Dorsey, *The Dawn of Conservation Diplomacy*; Dorsey, "Dealing with the Dinosaur"; Philippe G. Le Prestre and Peter Stoett, eds., *Bilateral Ecopolitics: Continuity and Change in Canadian-American Environmental Relations* (London: Ashgate, 2006); Joseph Taylor III, *Making Salmon: An Environmental History of the Northwest Fishery Crisis* (Seattle: University of Washington Press, 1999).
7 A 1951 State Department review of Canadian-American relations concluded that the two countries had a "unique relationship." Lawrence Aronsen, *American National Security and Economic Relations with Canada, 1945-1954* (Westport, CT: Praeger, 1997), xvi.
8 The most prominent proponents of the North American school include Norman Hillmer, Robert Bothwell, J.L. Granatstein, and Greg Donaghy. This school is in many ways an outgrowth of "continentalist" interpretations and the Carnegie Endowment for International Peace series on Canadian-American relations edited by James Shotwell and capped by John Bartlet Breber's *North Atlantic Triangle: The Interplay of Canada, the United States, and Great Britain*, which are less anti-American than "imperialist" approaches. For a summary and retrospective of the North Atlantic triangle concept, see the special edition of *London Journal of Canadian Studies* 20 (2004/2005). For a discussion of continentalist and imperialist interpretations, see Damien-Claude Bélanger, *Prejudice and Pride:*

Canadian Intellectuals Confront the United States, 1891-1945 (Toronto: University of Toronto Press, 2011).

9 Norman Hillmer and J.L. Granatstein, *For Better or for Worse: Canada and the United States into the Twenty-First Century* (Toronto: Nelson Thomson, 2007).

10 Reginald Stuart's *Dispersed Relations* makes the case for seeing more similarities than difference in the general Canadian-American relationship, and Victor Konrad and Heather Nicol's *Beyond Walls: Reinventing the Canadian-United States Borderlands* is one of the few works dedicated to the northern North American borderlands. An excellent example of a recent borderlands approach focused on the Great Lakes region is the multi-authored *Permeable Border*, and Claire Puccia Parham provides a comparative history of two St. Lawrence settlements, Cornwall, Ontario, and Massena, New York. Reginald Stuart, *Dispersed Relations: Americans and Canadians in Upper North America* (Washington, DC: Woodrow Wilson Center Press and Johns Hopkins University Press, 2007); Claire Puccia Parham, *From Great Wilderness to Seaway Towns: A Comparative History of Cornwall, Ontario, and Massena, New York, 1784-2001* (Albany, NY: State University of New York Press, 2004); John J. Bukowczyk et al., *Permeable Border: The Great Lakes Basin as Transnational Region, 1650-1990* (Pittsburgh: University of Pittsburgh Press, 2005); Victor Konrad and Heather N. Nicol, *Beyond Walls: Reinventing the Canada–United States Borderlands* (London: Ashgate, 2008).

11 Scholars such as William Appleman Williams and Walter LeFeber pointed to cultural sources of America's international policies and imperialism, dating back to the beginnings of the New Left. This has been articulated more recently in what has been termed the new cultural history, including postmodern and poststructuralist approaches (e.g., the works of Akira Iriye, Andrew Johnston, and Walter Hixson). See William Appleman Williams, *The Tragedy of American Diplomacy* (New York: Dell, 1962); Walter LeFeber, *The New Empire: An Interpretation of American Expansion, 1860-1898* (Ithaca, NY: Cornell University Press, 1963); Akira Iriye, *Power and Culture: The Japanese-American War, 1941-1945* (Boston: Harvard University Press, 1981); Andrew Johnston, *Hegemony and Culture in the Origins of NATO First-Use, 1945-1955* (New York: Palgrave Macmillan, 2005); Walter Hixson, *The Myth of American Diplomacy: National Identity and US Foreign Policy* (New Haven, CT: Yale University Press, 2008). A number of studies of Canadian external relations history have recently examined race and culture: Robert Teigrob, *Warming Up to the Cold War: Canada and the United States' Coalition of the Willing, from Hiroshima to Korea* (Toronto: University of Toronto Press, 2009); David Webster, *Fire and the Full Moon: Canada and Indonesia in a Decolonizing World* (Vancouver: UBC Press, 2010); Ryan Touhey, "Dealing in Black and White: The Diefenbaker Government and the Cold War in South Asia 1957-1963," *Canadian Historical Review* 92, 3 (2011): 429-54; John Price, *Orienting Canada: Race, Empire, and the Transpacific* (Vancouver: UBC Press, 2011).

12 A prominent recent example is Brian J. Bow, *The Politics of Linkage: Power, Interdependence, and Ideas in Canada-US Relations* (Vancouver: UBC Press, 2009), 3. John Holmes makes an argument for linkage on a personal but not institutional level: John W. Holmes, *Life with Uncle: The Canadian-American Relationship* (Toronto: University of Toronto Press, 1981), 55. For a further exploration of a contrary view about linkage and the St. Lawrence project see Daniel Macfarlane, "'Caught between Two Fires': St. Lawrence Seaway and Power Project, Canadian-American Relations, and Linkage," *International Journal* 67, 2 (Summer 2012): 465-82.

13 An example of this charge comes from Donald Creighton, *The Forked Road: Canada, 1939-1957* (Toronto: McClelland and Stewart, 1976).
14 Lionel Chevrier, *The St. Lawrence Seaway* (Toronto: Macmillan, 1959); Theo L. Hills, *The St. Lawrence Seaway* (London: Methuen, 1959); William Willoughby, *The St. Lawrence Waterway: A Study in Politics and Diplomacy* (Madison, WI: University of Wisconsin Press, 1961); Carleton Mabee, *The Seaway Story* (New York: Macmillan, 1961).
15 Claire Puccia Parham produced a very useful oral history, titled *The St. Lawrence Seaway and Power Project;* Jeff Alexander's *Pandora's Locks: The Opening of the Great Lakes–St. Lawrence Seaway* focuses on the invasive species and environmental damage in the Great Lakes that resulted from the seaway, primarily in the post-1959 period. Timothy Heinmiller provides a chapter about the conceptual change necessary for the St. Lawrence to go from a "river" to a "seaway." Ronald Stagg's recent *The Golden Dream: A History of the St. Lawrence Seaway* is a fine popular survey, covering the history of St. Lawrence navigation from its earliest canals to recent decades. Continuing a tradition of popular historical writing on the seaway, D'Arcy Jenish's *The St. Lawrence Seaway: Fifty Years and Counting,* commissioned by the St. Lawrence Seaway Management Corporation to commemorate the fiftieth anniversary of the opening of the seaway, is a brief but informative account of the seaway's history since 1959. Some of these works also deal with aspects of the construction and rehabilitation phase of the St. Lawrence project. Robert Passfield's article and William Becker's book on the role of the US Army Corps of Engineers provide excellent information about the technical aspects of construction, whereas J.B. Bryce's study for HEPCO is a little known but very useful explanation of the engineering process. A number of local histories, and a small selection of academic studies, have been published about the Lost Villages. On this score, a recent contribution of note is Joy Parr's chapter on perceptions of the changed St. Lawrence environment at Iroquois, Ontario. Gennifer Sussman, *The St. Lawrence Seaway: History and Analysis of a Joint Water Highway* (Montreal: C.D. Howe Research Institute, 1978); Gennifer Sussman, *Quebec and the St. Lawrence Seaway* (Montreal: C.D. Howe Institute, 1979); J.B. Bryce, *A Hydraulic Engineering History of the St. Lawrence Power Project with Special Reference to Regulation of Water Levels and Flows* (Toronto: Ontario Hydro, 1982); William H. Becker, *From the Atlantic to the Great Lakes: A History of the US Army Corps of Engineers and the St. Lawrence Seaway* (Washington, DC: US Army Corps of Engineers, 1984); Gary Pennanen, "Battle of the Titans: Mitchell Hepburn, Mackenzie King, Franklin Roosevelt, and the St. Lawrence Seaway," *Ontario History* 89, 1 (March 1997): 1-21; Gregory Witol, ed., *The St. Lawrence Seaway and Quebec* (Nepean, ON: Naval Officers Association of Canada, 1997); Robert W. Passfield, "Construction of the St. Lawrence Seaway," *Canal History and Technology Proceedings* 22 (2003): 1-55; Claire Puccia Parham, "The St. Lawrence Seaway: A Bi-National Political Marathon," *New York History* 85, 4 (Summer 2004): 359-85; Timothy Heinmiller, "The St. Lawrence: From River to Marine Superhighway," in *Canadian Water Politics: Conflicts and Institutions,* ed. Mark Sproule-Jones, Carolyn Johns, and B. Timothy Heinmiller (Montreal: McGill-Queen's University Press, 2008); Claire Puccia Parham, *The St. Lawrence Seaway and Power Project: An Oral History of the Greatest Construction Show on Earth* (Syracuse, NY: Syracuse University Press, 2009); Jeff Alexander, *Pandora's Locks: The Opening of the Great Lakes–St. Lawrence Seaway* (East Lansing, MI: Michigan State University Press, 2009); Joy Parr, *Sensing Changes: Technologies, Environments, and the Everyday, 1953-2003* (Vancouver: UBC Press, 2009); Ronald Stagg, *The Golden Dream: A History of the St. Lawrence Seaway*

(Toronto: Dundurn, 2010); D'Arcy Jenish, *The St. Lawrence Seaway: Fifty Years and Counting* (Manotick, ON: Penumbra Press, 2009).

16 For example, Willoughby, *The St. Lawrence Waterway*; Harry R. Mahood, "The St. Lawrence Seaway Bill of 1954: A Case Study of Pressure Groups in Conflict," *Southwestern Social Science Quarterly* 53 (September 1966): 141-49; Rudolph S. Comstock, "The St. Lawrence Seaway and Power Project: A Case Study in Presidential Leadership" (PhD diss., Ohio State University, 1956).

17 Two of the major exceptions include C.P. Stacey, *Canada and the Age of Conflict: A History of Canadian External Policies*, vol. 2, *1921-1948, The Mackenzie King Era* (Toronto: University of Toronto Press, 1981); Aronsen's *American National Security*.

18 Holmes, *Life with Uncle*, 123.

19 According to Creighton, the seaway was the "only great new undertaking of the period; but the division of construction and ownership between Canada and the United States reduced its significance as a national Canadian achievement," for "Canada's sense of national self-reliance and self-sufficiency suffered another serious injury when, at the eleventh hour, and on the ungenerous terms of its own choosing, the United States decided to participate in the St. Lawrence scheme, and the hope of an all-Canadian Seaway was gone forever." Donald Creighton, *Canada's First Century, 1867-1967* (Toronto: St. Martin's Press, 1970), 290-91.

20 William Kilbourn, "The 1950s," in *The Canadians, 1867-1967*, ed. J.M.S. Careless and R. Craig Brown (Toronto: Macmillan, 1967), 319.

21 "Laurentian Thesis," *The Canadian Encyclopedia* (Historica Foundation), http://www.thecanadianencyclopedia.com. For the classic exposition of the Laurentian thesis, see Donald Creighton, *The Empire of the St. Lawrence: A Study in Commerce and Politics* (Toronto: Macmillan, 1956) and the earlier version, Creighton, *The Commercial Empire*; Creighton, *Dominion of the North: A History of Canada* (Toronto: Houghton Mifflin, 1944).

22 Donald Creighton, "The Decline and Fall of the Empire of the St. Lawrence," *Canadian Historical Association, Historical Papers*, 1969: 14-25; Jean-Claude Robert, "The St. Lawrence and Montreal's Spatial Development in the Seventeenth through the Twentieth Century," in *Urban Rivers: Remaking Rivers, Cities, and Space in Europe and North America*, ed. Stéphane Castonguay and Matthew Evenden (Pittsburgh: University of Pittsburgh Press, 2012), 145-59.

23 Creighton, "Decline and Fall," 18.

24 Simon Schama, *Landscape and Memory* (Toronto: Random House, 1995), 363; Mark Cioc, *The Rhine: An Eco-Biography, 1815-2000* (Seattle: University of Washington Press, 2002); Christof Mauch and Thomas Zeller, eds., *Rivers in History: Perspectives on Waterways in Europe and North America* (Pittsburgh: University of Pittsburgh Press, 2008); Sara B. Pritchard, *Confluence: The Nature of Technology and the Remaking of the Rhône* (Cambridge, MA: Harvard University Press, 2011).

25 Donald F. Davis charts the history of metropolitanism and identifies five variants, of which Donald Creighton represents the "entrepreneurial" approach. See Donald F. Davis, "The 'Metropolitan Thesis' and the Writing of Canadian Urban History," *Urban History Review* 14, 2 (October 1985): 95-114. W.H. New also analyzes the links in Canadian writing between identity and landscape in *Land Sliding*, and took up these themes concerning the St. Lawrence, and Laurentian thesis, in more detail in "The Great River Theory: Reading MacLennan and Mulgan." W.H. New, "The Great River Theory: Reading MacLennan and

26 Mulgan," in *Essays on Canadian Writing* 56 (Fall 1995): 162-82; W.H. New, *Land Sliding: Imagining Space, Presence, and Power in Canadian Writing* (Toronto: University of Toronto Press, 1997).
26 Christopher Armstrong, Matthew Evenden, and H.V. Nelles, *The River Returns: An Environmental History of the Bow* (Montreal: McGill-Queen's University Press, 2009), 11.
27 Creighton, *Empire of the St. Lawrence*, 6-7. Carl Berger provides an excellent historiographical examination of the works of Innis and Creighton, and also notes that Scottish geographer Marion Newbigin's stress on the St. Lawrence predated and influenced both of them, and suggests that Innis's concern with the centrality of the St. Lawrence to Canada's historical development may have stemmed from contemporary debates about creating the deep waterway. George Brown's doctoral dissertation also adumbrated some of the Laurentian themes. Historian Donald Wright is working on a biography of Creighton that promises to examine the evolution of the ideas that led to Creighton's conception of the St. Lawrence. See Berger, *The Writing of Canadian History*, 2nd ed., 22-23; 91-93; 213-23; George Brown, "The St. Lawrence as a Factor in International Trade and Politics, 1783-1854" (PhD diss., University of Chicago, 1924); Newbigin, *Canada, the Great River;* Harold A. Innis, *The Fur Trade in Canada: An Introduction to Canadian Economic History* (Toronto: University of Toronto Press, 1956).
28 W.L. Morton, "Clio in Canada: The Interpretation of Canadian History," *University of Toronto Quarterly* 15, 3 (1946); W.L. Morton, *The Kingdom of Canada: A General History from Earliest Times* (Toronto: McClelland and Stewart, 1963); J.M.S. Careless, *Canada: A Story of Challenge* (Toronto: Macmillan, 1963).
29 William Toye, *The St. Lawrence* (Toronto: Oxford University Press, 1959); Jean L. Gogo, ed., *Lights on the St. Lawrence: An Anthology* (Toronto: Ryerson Press, 1958); Henry Beston, *The St. Lawrence* (Toronto: Rinehart, 1945).
30 Stéphane Castonguay and Darin Kinsey, "The Nature of the Liberal Order: State Formation, Conservation, and the Government of Non-Humans in Canada," in *Liberalism and Hegemony: Debating the Canadian Liberal Revolution*, ed. Jean-François Constant and Michel Ducharme (Toronto: University of Toronto Press, 2009), 223.
31 Janice Cavell, "The Second Frontier: The North in English-Canadian Historical Writing," *Canadian Historical Review* 83, 3 (September 2002): 4. A.B. McKillop also affirms the dominance of the Laurentian thesis in "Historiography in English," *The Canadian Encyclopedia* (Historica Foundation), http://www.thecanadianencyclopedia.com/.
32 J.M.S. Careless makes this point. See Davis, "The 'Metropolitan Thesis,'" 96.
33 A well-attended roundtable at a recent Canadian Historical Association annual meeting organized by the author suggests there is a hunger for a re-engagement with these metatheories: "A Roundtable on the Macro-Theories of Canadian History (staples, metropolitan-hinterland, Laurentian theses)" (participants: Christopher Dummit, Sean Kheraj, Daniel Macfarlane, Doug Owram, Shirley Tillotson), annual conference of the Canadian Historical Association, Waterloo, Ontario, May 28, 2012.
34 Joseph Bouchette, *The British Dominions*, 1 (1831): 126-70, quoted in John Warkentin, ed. *So Vast and Various: Interpreting Canada's Regions in the Nineteenth and Twentieth Centuries* (Montreal: McGill-Queen's University Press, 2011).
35 For example, see Raoul Blanchard, *Le Canada francais: Province de Quebec* (Montreal: Fayard, 1960); Jean-Claude Lasserre, *Le Saint-Laurent, grande porte de l'Amérique* (LaSalle, QC: Hurtubise, 1980); Gilles Matte and Gilles Pellerin, *Carnets du St-Laurent*

(Montreal: Heures Blues, 1999); Marie-Claude Ouellet, *Le Saint-Laurent-Fleuve à découvrir* (Montreal: Éditions de l'Homme, 1999); Jean Gagne, *À la découverte du Saint-Laurent* (Montreal: Éditions de l'Homme, 2005); Alain Franck, *Naviguer sur le fleuve au temps passé 1860-1960* (Quebec City: Publications Québec, 2000); Robert, "Montreal's Spatial Development," 145-47. Henry Beston, though an English-speaking American, was also attuned to the links between nature, the St. Lawrence, and French-Canadian culture and identity.

36 Ralph Heintzman, "Political Space and Economic Space: Quebec and the Empire of the St. Lawrence," *Journal of Canadian Studies* 29, 2 (1994): 19-63. On the links between Quebec nationalism, water, and hydroelectric development in relation to the James Bay projects see Caroline Desbiens, "Producing North and South: A Political Geography of Hydro Development in Quebec," *Canadian Geographer* 48, 2 (2004): 101-18.

37 Christopher Armstrong and H.V. Nelles, *Wilderness and Waterpower: How Banff National Park Became a Hydro-Electric Storage Reservoir* (Calgary: University of Calgary Press, 2013), 215.

38 The need to avoid the "nature" and "culture" dualism is discussed in Terje Tvedt and Terje Oestigaard, "A History of the Ideas of Water: Deconstructing Nature and Constructing Society," in *Ideas of Water from Ancient Societies to the Modern World*, A History of Water 2, 1 (London: I.B. Tauris, 2009). This monograph is part of the eminently useful multivolume series A History of Water, which boasts an array of contributions to water history and water studies from various disciplinary backgrounds (see the Bibliography for the various volumes).

39 On hydroelectric development in Canada see John Dales, *Hydroelectricity and Industrial Development: Quebec, 1898-1940* (Cambridge, MA: Harvard University Press, 1957); Ronald Richardson, Walter G. Rooke, and George McNevin, *Developing Water Resources: The St. Lawrence Seaway and the Columbia/Peace Power Projects* (Toronto: Ryerson Press, 1969); James W. Wilson, *People in the Way: The Human Aspect of the Columbia River Project* (Toronto: University of Toronto Press, 1973); H.V. Nelles, *The Politics of Development: Forests, Mines, and Hydro-Electric Power in Ontario*, 2nd ed. (Montreal: McGill-Queen's University Press, 2005); Neil Swainson, *Conflict over the Columbia: The Canadian Background to an Historic Treaty* (Montreal: McGill-Queen's University Press, 1979); Christopher Armstrong, *The Politics of Federalism: Ontario's Relations with the Federal Government, 1867-1942* (Toronto: University of Toronto Press, 1981); Sean McCutcheon, *Electric Rivers: The Story of the James Bay Project* (Montreal: Black Rose, 1991); James Waldrum, *As Long as the Rivers Run: Hydroelectric Development and Native Communities* (Winnipeg: University of Manitoba Press, 1993); Neil B. Freeman, *The Politics of Power: Ontario Hydro and Its Government, 1906-1995* (Toronto: University of Toronto Press, 1996); Ronald J. Daniels, ed., *Ontario Hydro at the Millennium: Has Monopoly's Moment Passed?* (Montreal: McGill-Queen's University Press, 1996); Jeremy Mouat, *The Business of Power: Hydro-Electricity in South Eastern British Columbia, 1897-1997* (Victoria: Sono Nis, 1997); Karl Froschauer, *White Gold: Hydroelectric Power in Canada* (Vancouver: UBC Press, 1999); Jean L. Manore, *Cross-Currents: Hydroelectricity and the Engineering of Northern Ontario* (Waterloo, ON: Wilfrid Laurier University Press, 1999); David Massell, *Amassing Power: J.B. Duke and the Saguenay River, 1897-1927* (Montreal: McGill-Queen's University Press, 2000); Jamie Swift and Keith Stewart, *Hydro: The Fall and Decline of Ontario's Electric Empire* (Toronto: Between the Lines, 2004); Matthew Evenden, *Fish versus Power: An Environmental History of the Fraser River* (New York: Cambridge University Press, 2004);

Tina Loo, "People in the Way: Modernity, Environment, and Society on the Arrow Lakes," *BC Studies* 142/143 (Summer/Autumn 2004): 161-96; Desbiens, "Producing North and South"; Stéphane Castonguay, "The Production of Flood as Natural Catastrophe: Extreme Events and the Construction of Vulnerability in the Drainage Basin of the St. Francis River (Quebec), Mid-Nineteenth to Mid-Twentieth Century," *Environmental History* 12, 4 (October 2007): 816-40; Alexander Netherton, "The Political Economy of Canadian Hydro-Electricity: Between Old 'Provincial Hydros' and Neoliberal Regional Energy Regimes," *Canadian Political Science Review* 1, 1 (2007): 107-24; Tina Loo, "Disturbing the Peace: Environmental Change and the Scales of Justice on a Northern River," *Environmental History* 12, 4 (October 2007): 895-919; Armstrong, Evenden, and Nelles, *The River Returns*; Matthew Evenden, "Mobilizing Rivers: Hydro-Electricity, the State, and World War II in Canada," *Annals of the Association of American Geographers* 99, 5 (2009): 845-55; Thibault Martin and Steven M. Hoffman, eds., *Power Struggles: Hydro-Electric Development and First Nations in Manitoba and Quebec* (Winnipeg: University of Manitoba Press, 2009); James L. Kenny and Andrew G. Secord, "Engineering Modernity: Hydroelectric Development in New Brunswick, 1945-1970," *Acadiensis* 39, 1 (Winter/Spring 2010): 3-26; Philip Van Huizen, "Building a Green Dam: Environmental Modernism and the Canadian-American Libby Dam Project," *Pacific Historical Review* 79, 3 (August 2010): 418-53; Meg Stanley, *Voices from Two Rivers: Harnessing the Power of the Peace and Columbia* (Vancouver: Douglas and McIntyre, 2011); David Massell, *Quebec Hydropolitics: The Peribonka Concessions of the Second World War* (Montreal: McGill-Queen's University Press, 2011); Armstrong and Nelles, *Wilderness and Waterpower*; Caroline Desbiens, *Power from the North: Territory, Identity, and the Culture of Hydroelectricity in Quebec* (Vancouver: UBC Press, 2013).

40 Martin Reuss and Stephen Cutcliffe, eds., *The Illusory Boundary: Environment and Technology in History* (Charlottesville: University of Virginia Press, 2010). On the Envirotech group see http://envirotechweb.org/what-is-envirotech/.

41 Josephson, *Industrialized Nature*; David E. Nye, *American Technological Sublime* (Cambridge, MA: MIT Press, 1996). On the theme of the link between nature and technology in the United States, see also Leo Marx, *The Machine in the Garden: Technology and the Pastoral Idea in America* (New York: Oxford University Press, 1964).

42 James C. Scott, *Seeing Like a State: How Certain Schemes to Improve the Human Condition Have Failed* (New Haven, CT: Yale University Press, 1998). For earlier usage of "high modernism" see David Harvey, *The Condition of Postmodernity: An Enquiry into the Origins of Cultural Change* (Cambridge, MA: Wiley-Blackwell, 1990). See the various collections of reviews of Scott and high modernism in "Review Essays: *Seeing Like a State*," *American Historical Review* 196 (February 1, 2001): 106-29; "Seeing like a State: A Conversation with James C. Scott," *Cato Unbound* (September 2010), http://www.cato-unbound.org/archives/.

43 Timothy Mitchell engages many of the same issues of technology, environment, and modernism as Scott, though Mitchell focuses on "the kinds of social and political practices that produce simultaneously the powers of science and the powers of modern states" rather than the ways that modern states have misused the powers of science relative to the "proper" use of science. Indeed, Scott perhaps mistakenly assumes that an ideal "proper" scientific way of thinking exists. Timothy Mitchell, *Rule of Experts: Egypt, Techno-Politics, Modernity* (Los Angeles: University of California Press, 2002), fn 77, 312.

44 Donald Worster, "Water in the Age of Imperialism – and Beyond," in *The World of Water, A History of Water* 1, 3, ed. T. Tvedt and T. Oestigaard (London: I.B. Tauris, 2006). In his seminal *Rivers of Empire,* Worster contends that the American West was "a modern hydraulic society, which is to say, a social order based on the intensive, large-scale manipulation of water and its products in an arid setting." Donald Worster, *Rivers of Empire: Water, Aridity, and the Growth of the American West* (New York: Pantheon, 1985), 7. Worster borrowed his dialectical approach in part from Karl Wittfogel's classic *Oriental Despotism: A Comparative Study of Total Power* (New Haven, CT: Yale University Press, 1957). See also Theodore Steinberg, "'That World's Fair Feeling': Control of Water in Twentieth Century America," *Technology and Culture* 34 (April 1993), 401-9; Erik Swyngedouw, "Modernity and Hybridity: Nature, *Regeneracionismo,* and the Production of the Spanish Waterscape, 1890-1930," *Annals of the Association of American Geographers* 89, 3 (September 1999): 443-65; David Biggs, *Quagmire: Nation-Building and Nature in the Mekong Delta* (Seattle: University of Washington Press, 2010); Erik van der Vleuten and Cornelius Disco, "Water Wizards: Reshaping Wet Nature and Society," *History and Technology* 20, 3 (2004): 291-309; François Molle, Peter P. Mollinga, and Philipp Wester, "Hydraulic Bureaucracies: Flows of Water, Flows of Power," *Water Alternatives* 2, 3 (October 2009): 328-49. For a recent overview of the different approaches and historiography on conceptions of water and society, see Jamie Linton, *What Is Water? The History of a Modern Abstraction* (Vancouver: UBC Press, 2009).

45 Quoted in "Stresses Canada's Need of Scientists, Engineers," *Globe and Mail,* January 29, 1951, 12.

46 I put a heavier emphasis on the Canadian perspective in large part because previous studies have focused to a greater extent on the American side of things, and because the majority of the St. Lawrence Seaway and Power Project lies within Canada. I primarily employ sources from the federal governments of Canada and the United States, presidential archives, state and provincial governments, other involved organizations (such as the International Joint Commission, Hydro-Electric Power Commission of Ontario, and Power Authority of the State of New York), local records, and oral interviews.

Chapter 1: Accords and Discords

1 For an overview of canal developments in and around the St. Lawrence River and Great Lakes up to the twentieth century, see Chapter 1 of the author's doctoral dissertation, as well as several other works: Daniel Macfarlane, "To the Heart of the Continent: Canada and the Negotiation of the St. Lawrence Seaway and Power Project, 1921-1954" (PhD diss., University of Ottawa, 2010); Stagg, *The Golden Dream*; Normand Lafrenière, *Canal Building on the St. Lawrence: Two Centuries of Work, 1779-1959* (Ottawa: Parks Canada, 1983).

2 The Williamsburg canal system was made up of several grouping of canals: Farran's Point, Rapide Plat, Iroquois-Galop. The 1871 Treaty of Washington established that the St. Lawrence River west of Cornwall would serve as the international boundary between Canada and the United States. However, it also stipulated that American access to the Canadian canals could be terminated on two years' notice.

3 Willoughby, *The St. Lawrence Waterway,* 64.

4 Stagg, *The Golden Dream,* 102.

5 Willoughby, *The St. Lawrence Waterway,* 68.

6 On the origins of the IJC see N.F. Dreiziger, "The International Joint Commission of the United States and Canada, 1895-1920: A Study in Canadian-American Relations" (PhD

diss., University of Toronto, 1974); William R. Willoughby, *The Joint Organizations of Canada and the United States* (Toronto: University of Toronto Press, 1979); Robert Spencer, John Kirton, Kim Richard Nossal, eds., *The International Joint Commission Seventy Years On* (Toronto: University of Toronto Centre for International Studies, 1981).

7   Hydro-Electric Power Commission of Ontario (hereafter HEPCO), St. Lawrence Power Project (hereafter SPP) Series, 91.123, Memorandum to Howard Ferguson from HEPCO Chairman re: St. Lawrence River Power Development, January 9, 1924.

8   Arthur V. White, *Long Sault Rapids, St. Lawrence River: An Enquiry into the Constitutional and Other Aspects of the Project to Develop Power Therefrom* (Ottawa: Mortimer, 1913), Appendix 30: Proceedings before the International Waterways Commission, February 8-9, 1910.

9   See, for example, ibid., 317-19.

10 HEPCO, SPP Series, 91.123, Memorandum to Howard Ferguson from HEPCO Chairman re: St. Lawrence River Power Development, January 9, 1924.

11 White, *Long Sault Rapids*, 35-38. The author was assistant chairman to Clifford Sifton, who was chairman of the Commission of Conservation. This study was part of the commission's extensive efforts to inventory hydroelectric sites in Canada.

12 Ibid., 31. Emphasis in original. This report includes what may be among the earliest studies of frazil ice on the St. Lawrence. Included in the appendices are studies on ice conditions in the St. Lawrence: Appendix 25, "Ice Conditions on the St. Lawrence," by Prof. H.T. Barnes; Appendix 26, "Ice Jams between Morrisburg and Cornwall," by J.W. Rickey; Appendix 27, "Effect in Canadian Waters of a Power-House Built in the Long Sault Channel," by J.W. Rickey; Appendix 28, "Reasons for Opposing the Project to Dam the Long Sault Channel," by J. Wesley Allison.

13 Stagg, *The Golden Dream*, 110.

14 There are many examples in the congressional record to draw on. For a summary of the traditional opposition to legislation to improve the St. Lawrence see Government of the United States, US Congress, Senate, S.J. Res. 111, 80th Congress, 2nd session, 525-30. The Lester K. Sillcox Series in the St. Lawrence Seaway Collection in the St. Lawrence University Archives is a particular strong repository of anti-seaway sentiment. Sillcox reputedly coined the term "iceway" to refer to the limitations that winter conditions would impose on a deep waterway.

15 Hills, *St. Lawrence Seaway*, 58.

16 Ibid., 83.

17 HEPCO, SPP Series, 91.123, box 68, Memorandum (Chief Engineer) re: Power Possibilities at Morrisburg, October 3, 1918.

18 See George Washington Stephens, *The St. Lawrence Waterway Project: The Story of the St. Lawrence River as an International Highway for Water-Borne Commerce* (Montreal: L. Carrier, 1930), 168-79.

19 US Senate, *St. Lawrence Waterway: Report of the United States and Canadian Government Engineers on the Improvement of the St. Lawrence River from Montreal to Lake Ontario made to the International Joint Commission (Wooten-Bowden Report)*, Supplementary to Senate Document 114, 67th Congress (Washington, DC: Government Printing Office, 1922).

20 Stacey, *Age of Conflict*, vol. 2, *1921-1948, The Mackenzie King Era*, 110; Peter G. Oliver, *Howard Ferguson: Ontario Tory* (Toronto: University of Toronto Press, 1977), 174-76.

21 Pennanen, "Battle of the Titans," 2. Stacey suggests King's resistance might also have stemmed from the fact that the IJC reference had been authorized by the Borden government. Stacey, *Age of Conflict*, 110.

22 Willoughby, *The St. Lawrence Waterway*, 94.
23 Ibid., 99-100.
24 *Toronto Globe*, July 19, 1921, quoted in ibid., 99.
25 Bernard L. Vigod, *Quebec before Duplessis: The Political Career of Louis-Alexandre Taschereau* (Montreal: McGill-Queen's University Press, 1986), 123-25.
26 HEPCO, SPP Series, 91.123, Memorandum to Howard Ferguson from HEPCO Chairman re: St. Lawrence River Power Development, January 9, 1924.
27 Armstrong, *Politics of Federalism*, 160.
28 As part of the 1922 approval, the IJC also established an International Massena Board of Control. In 1928, the St. Lawrence River Power Company, a subsidiary of the Aluminum Company of America, applied for the authority to increase the level of the weir, but this was abandoned because of Canadian opposition. See L.M. Bloomfield and Gerald F. Fitzgerald, *Boundary Waters Problems of Canada and the United States: The International Joint Commission, 1912-1958* (Toronto: Carswell, 1958), dockets 15, 17, and 24.
29 H. Blair Neatby, *William Lyon Mackenzie King, 1924-1932*, vol. 2, *The Lonely Heights* (Toronto: University of Toronto Press, 1963), 257-58.
30 HEPCO, The St. Lawrence Development: "An Outline of the Events Leading up to the Selection of the Controlled 238-242 Single Stage Project Forming the Basis of the 1941 Agreement and a Comparison of That Scheme with the C-217 Two Stage Project Proposed in the 1932 Treaty," December 15, 1943.
31 Armstrong, *Politics of Federalism*, 178.
32 See Armstrong, *Politics of Federalism;* Freeman, *The Politics of Power*.
33 Gordon T. Stewart, *The American Response to Canada since 1776* (East Lansing, MI: Michigan State University Press, 1992), 138.
34 For example, see Government of Canada, Department of External Affairs, "Prime Minister to Secretary of State of United States, July 12, 1927," *Documents on Canadian External Relations*, Alex I. Inglis, ed., 4, 1926-1930 (Ottawa: Minister of Public Works and Government Services Canada, 1971), 420-26.
35 University of Toronto Archives, Vincent Massey Papers, B87-0082, box 385, file 006, Skelton to Massey, February 14, 1928.
36 Some senators thought it granted too much to private power interests, whereas others wanted to keep separate hydro development and remedial works affecting the scenic appeal of the cataract. See Daniel Macfarlane, "Creating a Cataract: The Transnational Manipulation of Niagara Falls to the 1950s," in *Urban Explorations: Environmental Histories of the Toronto Region*, ed. Colin Coates, Stephen Bocking, Ken Cruikshank, and Anders Sandberg (Hamilton, ON: L.R. Wilson Institute for Canadian Studies-McMaster University, 2013); Daniel Macfarlane, "'A Completely Man-Made and Artificial Cataract': The Transnational Manipulation of Niagara Falls," *Environmental History* 18, 4 (October 2013): 759-84.
37 This megaproject featured a gigantic combined power channel (with the infrastructure for an eventual deep navigation canal) boasting a length of about 15 miles that displaced vast tracts of agricultural land on the south shore of the St. Lawrence River. The Beauharnois power station was initially allowed to divert 40,000 cfs, which was increased to 53,000 cfs in 1932, with another 30,000 permitted in 1942 because of wartime needs. See Louis-Raphael Pelletier's doctoral dissertation, which provides an analysis of the creation of the Beauharnois project: "Revolutionizing Landscapes: Hydroelectricity and the Heavy Industrialization of Society and Environment in the Comté de Beauharnois, 1927-1948" (PhD diss., Carleton University, 2005). Richard Kottman discusses the political, diplomatic, and federal-

provincial water rights issues involved in the approval of the Beauharnois project in "The Diplomatic Relations of the United States and Canada, 1927-1941" (PhD diss., Vanderbilt University, 1958), 63-68.
38 It also featured three consecutive flight locks to pass ships over the Niagara Escarpment. The route remained much the same from Port Colborne to Thorold, though after crossing the Niagara Escarpment it ran fairly straight north to a new Lake Ontario connection, Port Weller. Roberta M. Styran and Robert R. Taylor, *This Great National Object: Building the Nineteenth-Century Welland Canals* (Montreal: McGill-Queen's University Press, 2012).
39 Kottman, "Diplomatic Relations," 63-68.
40 As a result of such projects, by the time the St. Lawrence project was constructed the United States had amassed a great deal of experience in multipurpose dam projects. See David P. Billington and Donald C. Jackson, *Big Dams of the New Deal Era: A Confluence of Engineering and Politics* (Norman: University of Oklahoma Press, 2006).
41 Sarah Phillips, "FDR, Hoover, and the New Rural Conservation, 1920-1932," in *FDR and the Environment*, ed. Henry L. Henderson and David B. Woolner (New York: Palgrave Macmillan, 2009), 142.
42 Neatby, *William Lyon Mackenzie King*, vol. 2, 284-85; Hillmer and Granatstein, *For Better or for Worse*, 107; Lawrence Martin, *The Presidents and the Prime Ministers: Washington and Ottawa Face to Face; The Myth of Bilateral Bliss, 1867-1982* (Toronto: Doubleday, 1982), 103.
43 Government of Canada, Library and Archives Canada (hereafter LAC), MG 26 J4 (King fonds), Memorandum: St. Lawrence Waterway, Draft of points for consideration, 1930.
44 Ibid.
45 Government of the United States, National Archives and Records Administration II (hereafter NARA II), RG 59, box 4043, telegram 1283, Ottawa Embassy to Department of State, January 23, 1930.
46 LAC, MG 26 J13 (King diaries), June 30, 1930.
47 LAC, MG 26 J4 (King fonds), letter from Premier George Ferguson to King, February 24, 1930; LAC, MG 26 J4 (King fonds), letter from King to Taschereau, March 8, 1930; LAC, MG 26 J4 (King fonds), letter from St. Laurent to Deputy Attorney General, April 26, 1930. The author of the latter letter was future prime minister Louis St. Laurent, a lawyer at the time.
48 Willoughby, *The St. Lawrence Waterway*, 137.
49 NARA II, RG 59, file 711.42157 SA 29/765, box 4044, Memorandum to the Acting Secretary, July 1, 1931.
50 Completed in 1900, this canal reversed the Chicago River and facilitated the Illinois (or Chicago) diversion that provided sewage disposal and navigation to the Mississippi watershed. In addition to Canada, other US states complained about this diversion, and in 1930 the US Supreme Court capped the amount of water that could be diverted. See John W. Larson, *Those Army Engineers: A History of the Chicago District* (Washington, DC: US Army Corps of Engineers, 1980); Libby Hill, *The Chicago River: A Natural and Unnatural History* (Chicago: Lake Claremont Press, 2000).
51 Willoughby, *The St. Lawrence Waterway*, 138.
52 Canadian documentation on the 1931-32 negotiations is sparse, apparently because Bennett preferred to keep the record in his head and between himself and William Herridge, the new Canadian minister to the United States. The Americans also left a scanty record: a State Department memorandum from the 1940s maintains that many of the US pre-1933 St. Lawrence files were probably destroyed. Ibid., 147; Government of the United States,

President Franklin D. Roosevelt's Office Files, 1933-1945, pt. 1, Safe and Confidential Files, Memorandum from the Under Secretary of State to the President, July 22, 1942, ed. William E. Leuchtenburg (Washington: University Publications of America, 1990).

53 NARA II, RG 59, file 711.42157 SA 29/1165, box 4045, Department of State Memorandum, November 16, 1931.

54 In 1929, the Canadian and Ontario governments had created the Conference of Canadian Engineers to study what type of hydroelectric development best suited Canadian interests; it recommended the use of two separate gravity power dams (known as the C-217 dual-stage scheme), the Canadian at Crysler Island and the American at Barnhart Island. HEPCO, The St. Lawrence Development: "An Outline of the Events Leading up to the Selection of the Controlled 238-242 Single Stage Project Forming the Basis of the 1941 Agreement and a Comparison of That Scheme with the C-217 Two Stage Project Proposed in the 1932 Treaty," December 15, 1943. It is worth noting that all plans since the St. Lawrence power project's first conception involved gravity dams, which relied on mass and were dominant in the United States. Gravity dams exemplified the "massive" tradition of dam building, in contrast to the "structural" tradition, characterized by thinner arch dams. See Billington and Jackson, "Introduction," in *Big Dams of the New Deal Era.*

55 Richard Kottman, "Herbert Hoover and the St. Lawrence Treaty of 1932," *New York History* 56 (July 1975): 337.

56 With a treaty agreement pending, Roosevelt sent Hoover a telegram offering to return from his boat trip to work out the power details. Hoover, however, declined his offer. See *President Franklin D. Roosevelt's Office Files, 1933-1945,* pt. 1, Safe and Confidential Files, Hoover to Roosevelt, July 10, 1932.

57 The contributors to a recent collection contend that FDR's environmental legacy has not been fully appreciated. Indeed, there is precious little on the creation of PASNY. Henderson and Woolner, *FDR and the Environment.*

58 The Chicago diversion was effectively limited by the 1930 US Supreme Court decision to 3,200 cfs on an annual basis. The United States appealed for an extension because of worries that low water levels would threaten public health conditions in Chicago, as financial difficulties stemming from the Depression had caused work to cease on sewage disposal work. Stepping down the diversion amount was supposed to be done by 1938, although Canada later granted the United States an extension. NARA II, RG 59, file 711.42157 SA 29/1048, box 4045.

59 L. Martin, *The Presidents and the Prime Ministers,* 107.

60 Willoughby, *The St. Lawrence Waterway,* 148.

61 *President Franklin D. Roosevelt's Office Files, 1933-1945,* pt. 4, Subject File, Memorandum (no author), February 22, 1934; Kottman, "Herbert Hoover," 345.

62 NARA II, RG 59, file 711.42157 SA 29/1296, box 4048, Memorandum: All-Canadian Route, January 26, 1934.

63 NARA II, RG 59, file 711.42157 SA 29/1288-1/2, box 4047, Memorandum (by Hickerson), June 23, 1934.

64 Memorandum by the Under Secretary of State (Phillips), 711.42157SA29/1291½, July 11, 1934, Government of the United States, *Foreign Relations of the United States (FRUS), 1934* (Washington, DC: US Government Printing Office, 1934), vol. 1, Canada, 973-74.

65 Nelles, *The Politics of Development,* 484; Pennanen, "Battle of the Titans," 5, 16.

66 LAC, MG 26 J13 (King diaries), June 27, 1939.

67 NARA II, RG 59, file 711.42157 SA 29/1273, box 4047, Extract from the President's Press Conference, March 14, 1934.

68 These included alterations such as the sovereignty of Lake Michigan, diversion of water from Georgian Bay, expenditure of American funds in the International Rapids section, the deepening of the draught to thirty feet instead of twenty-seven, single-stage instead of dual-stage power development, and giving the United States an additional share of the water for hydroelectricity.
69 *President Franklin D. Roosevelt's Office Files, 1933-1945*, pt. 4, Subject File, Under Secretary of State to President, January 4, 1935.
70 See Hillmer and Granatstein, *For Better or for Worse*, chap. 5.
71 Pennanen, "Battle of the Titans," 9.
72 Armstrong, *Politics of Federalism*, 190.
73 Felix Belair Jr., "Aims at Dictators," *New York Times*, August 19, 1938, 1 and 3.
74 For a collection of letters between Hepburn and King see *Correspondence and Documents Relating to the Great Lakes–St. Lawrence Basin Development, 1938-1941* (Ottawa: Edmond Cloutier, 1941).
75 LAC, MG 26 J4, Memorandum for File re: Meeting of Cabinet with Representatives of the Ontario Legislature, October 3, 1939.
76 Beatrice Bishop Berle and Travis Beal Jacobs, eds., *Navigating the Rapids, 1918-1971: From the Papers of Adolf A. Berle* (New York: Harcourt Brace Jovanovich: 1973), 309; Freeman, *The Politics of Power*, 85.
77 LAC, MG 26 J4, Memorandum for the Prime Minister: Niagara-St. Lawrence Discussions, October 12, 1939.
78 HEPCO, SPP Series, "The St. Lawrence Development: An Outline of the Events Leading up to the Selection of the Controlled 238-242 Single Stage Project Forming the Basis of the 1941 Agreement and a Comparison of That Scheme with the C-217 Two Stage Project Proposed in the 1932 Treaty," December 15, 1943.
79 US Department of Commerce (N.R. Danielian), *The St. Lawrence Survey*, 7 vols. (Washington, DC: US Government Printing Office, 1941).
80 LAC, RG 25, file 1268-B-40, pt. 2, St. Lawrence–Niagara River Treaty Proposals (International Correspondence) (1940/41), Memorandum from Robert H. Jackson (US Attorney General) to Secretary of State, March 13, 1941.
81 *President Franklin D. Roosevelt's Office Files, 1933-1945*, pt. 1, Safe and Confidential Files, Secretary of War to the President, August 5, 1942.
82 Roosevelt Archives, Olds Papers, Federal Power Commission, National Defense and War Production Power Activities: 1939-1947, Canada, box 106, Cordell Hull to Josiah W. Bailey, April 10, 1944.
83 Ontario's use of the additional diverted water at Niagara Falls also received the necessary approval of the International Joint Commission in 1940. See Freeman, *The Politics of Power*; J.C. Day, "Canadian Interbasin Diversions" (Ottawa: Government of Canada, 1985); Peter Annin, *The Great Lakes Water Wars* (Washington, DC: Island Press, 2006).
84 HEPCO, SPP Series, Memorandum: St. Lawrence Project: Benefits Which Accrue to Ontario under the Canada–United States Agreement, March 5, 1941.
85 Roosevelt Archives, Olds Papers, Federal Power Commission, National Defense and War Production Power Activities: 1939-1947, Canada, box 106, J. Dibblee to H.J. Symington, March 17, 1941.
86 Norman D. Wilson, *The Rehabilitation of the St. Lawrence Communities: A Report on the Factors in the Rehabilitation of the St. Lawrence Communities Partly or Wholly Inundated in the Development for Power and Navigation of the International Rapids Section of the St. Lawrence River* (Ottawa: Canadian Advisory Committee on Reconstruction, 1943).

Notes to pages 47-53

87 Bruce Hutchison, *The Unknown Country* (1942), quoted in Warkentin, *So Vast and Various*, 129-31.

Chapter 2: Watershed Decisions

1 Thompson and Randall, *Canada and the United States*, 189.
2 Evenden, "Mobilizing Rivers."
3 Harry S. Truman Library and Archives, Papers of Eben A. Ayers, Subject File, 1898-1971, box 12, St. Lawrence Seaway, Message to Congress, October 3, 1945. In general, Truman favoured the internationalization of key waterways, such as the Rhine-Danube, Suez, Panama, and Bosporus Strait, which he mentions in his diary entry of July 30, 1945, and in an unsent correspondence with James F. Byrnes on January 5, 1946. See Robert H. Ferrell, ed., *Off the Record: The Private Papers of Harry S. Truman* (New York: Harper and Row, 1980), 58, 80.
4 NARA II, RG 59, file 711.42157 SA 29/1296, box 4048, Summary of Reports and Data Relative to the Great Lakes Project, January 16, 1934.
5 LAC, RG 2, Cabinet Conclusions, March 21, 1947.
6 LAC, RG 25, file 1268-D-40C, St. Lawrence River-Niagara River Treaty Proposals – General Correspondence, pt. 7 (June 1, 1947-January 16, 1948), 3560, Minutes from Informal Meeting to discuss the St. Lawrence Seaway Project, January 16, 1948.
7 LAC, RG 25, file 1268-Y-40, St. Lawrence Deep Waterway – Importance of Quebec Labrador Iron Ore Deposits (November 6, 1948-April 8, 1952), 4169, Summary of "The Market for Labrador Ore: A Study of the Iron Ore Situation," prepared by H.G. Cochrane (Department of Reconstruction) and W.M. Goodwin (Bureau of Mines), November 16, 1948; Hills, *St. Lawrence Seaway*, 71.
8 See Maeva Marcus, *Truman and the Steel Seizure Case: The Limits of Presidential Power* (New York: Columbia University Press, 1977); Truman Archives, Papers of Clarence H. Osthagen, box 22, Subject Files – US Department of Commerce: Steel Seizure, 1952.
9 Graham D. Taylor and Peter A. Baskerville, *A Concise History of Business in Canada* (Toronto: Oxford University Press, 1994), 401.
10 Willoughby, *The St. Lawrence Waterway*, 208-9.
11 Aronsen, *American National Security*, 151.
12 HEPCO, 91.123, Thomas Hogg (HEPCO chairman) to Francis Wilby (PASNY chairman), April 30, 1946.
13 LAC, RG 25, file 1268-D-40, pt. 14 (FP. 1), vol. 6345, Report: "The St. Lawrence Waterway and the Canadian Economy," Department of Trade and Commerce (Economic Research Division), Government of Canada, January 1951, 18.
14 For example, Robert Saunders, chairman of HEPCO, intimated to C.D. Howe that "Ontario's real motive in pressing for separate power development was to hasten the two Federal Governments toward implementation of the combined power-waterway project," as the province anticipated that its expenses under a jointly integrated project would be lower. Conversely, interactions between a HEPCO official and the acting secretary of the Canadian Section of the International Joint Commission left the impression that all the pressure for the scheme was coming from New York. LAC, RG 25, file 1268-U-40C, St. Lawrence–New York–Ontario–Power Priority Plan – Application to International Joint Commission (3/3/48-n.d.), 3563, Memorandum, May 15, 1948; LAC, RG 25, file 1268-U-40C, St. Lawrence–New York–Ontario–Power Priority Plan – Application to

International Joint Commission (3/3/48-n.d.), 3563, Memorandum re: St. Lawrence Waterway Project and the Ontario–New York Power Scheme, June 14, 1948.
15 For example, HEPCO, 91.123, Draft Minutes of a joint meeting of the Power Authority of the State of New York and the Hydro-Electric Power Commission of Ontario, New York, April 8, 1948.
16 LAC, RG 25, file 1268-D-40C, St. Lawrence River-Niagara River Treaty Proposals – General Correspondence, pt. 8 (January 14-December 31, 1948), 3560, Memorandum for the Acting Under Secretary of State for External Affairs, June 4, 1948.
17 NARA II, RG 59, file 711.42157 SA/5-2848, Stone to Foster, May 28, 1948.
18 According to Danielian, a memorandum he had written, which John Blatnik passed along, helped convince Truman to stick to the dual project. See Dwight D. Eisenhower Library and Archives, Columbia University Oral History Project, OH 177, Oral History Interview with N.R. Danielian (1972), 15.
19 Charles Ritchie, *Diplomatic Passport: More Undiplomatic Diaries, 1946-1962* (Toronto, Canada: Macmillan, 1981), 70.
20 NARA II, RG 59, file 711.42157 SA 29/11-148 to 711.4216/10-1447, box 3304, Memorandum from Harry Truman to George C. Marshall, December 3, 1948.
21 LAC, RG 25, file 1268-U-40C, St. Lawrence–New York–Ontario–Power Priority Plan – Application to International Joint Commission (3/3/48-n.d.), 3563, Note for file 1268-U-40C, December 21, 1948.
22 Interview with Dennis Dack, Toronto, May 2, 2011. According to Dack, the use of this person, whose last name was Hunt, may have been known to only a few people.
23 See Macfarlane, "Creating a Cataract," in Coates et al., *Urban Explorations*.
24 Ibid.
25 NARA II, RG 59, file 711.4216/1-2549, box 3304, despatch 54: US Embassy to Department of State: Canadian Attitudes toward the St. Lawrence Waterway and Power Project, January 25, 1949.
26 LAC, RG 25, file 1268-H-40, St. Lawrence River Waterway Project – Defence Aspects (April 1, 1941-October 31, 1941), 3335, Speech by Brooke Claxton at Sault Ste. Marie, January 11, 1949. The standard work on Claxton is David J. Bercuson, *True Patriot: The Life of Brooke Claxton, 1898-1960* (Toronto: University of Toronto Press, 1993).
27 Aronsen, *American National Security*, 155.
28 LAC, RG 25, file 1268-D-40C, St. Lawrence River-Niagara River Treaty Proposals – General Correspondence, pt. 9 (January 5-December 30, 1949), 3560, Status of the St. Lawrence Project (March 17, 1949), appended to Memorandum for the Under Secretary, March 22, 1949.
29 A prime minister with a parliamentary majority in Canada's responsible government system, such as St. Laurent enjoyed, could pass the necessary enabling legislation with little difficulty; the US separation of powers and congressional system meant that a St. Lawrence project supported by the administration could be repeatedly frustrated by sectional interests that were regionally or geographically concentrated.
30 LAC, RG 25, file 1268-Q-40, St. Lawrence Waterway Project-Interdepartmental Committee – General File, pt. 1.2, 6184, Memorandum: An All-Canadian St. Lawrence Waterway, May 17, 1949.
31 LAC, RG 25, file 1268-K-40C, St. Lawrence–Niagara River Treaty Between Canada and United States – Additional Diversion of Water at Niagara Falls, pt. 4 (January 1, 1948-November 30, 1949), 3561, Draft letter from Prime Minister to President, May 25, 1949.

32 LAC, RG 2, file W-10-1 (vol. 1), filed separately 1951, St. Lawrence Waterway – Possible Future Action, February 17, 1950.
33 Truman Archives, Papers of Oscar L. Chapman, Correspondence File, box 63, St. Lawrence Seaway, undated memorandum.
34 Karl Boyd Brooks, "Introduction," in *The Environmental Legacy of Harry S. Truman*, ed. Karl Boyd Brooks (Kirksville, MO: Truman State University Press, 2009).
35 On this topic, see Elmo Richardson, *Dams, Parks and Politics: Resource Development and Preservation in the Truman-Eisenhower Era* (Lexington: University of Kentucky Press, 1973).
36 Truman Archives, Papers of Harry S. Truman, David E. Bell Subject File, box 1, Memorandum for Mr. Murphy, Subject: St. Lawrence Project, November 26, 1949; Truman Archives, Dean Acheson Papers, Memoranda of Conversations File – December 1949, box 65, Notes on Cabinet Meeting, December 22, 1949.
37 Truman Archives, Papers of Eben A. Ayers, Subject File, 1898-1971, box 12, St. Lawrence Seaway.
38 Truman Archives, Dean Acheson Papers, Memoranda of Conversations File – December 1949, box 65, Notes on Cabinet Meeting, December 22, 1949.
39 LAC, RG 2, file W-10-1 (1), Waterways; Water Development Projects; etc. St. Lawrence Waterway and Power Project, 1950, 207, St. Lawrence Waterway – Possible Future Action, February 17, 1950.
40 Ibid.
41 LAC, RG 25, file 1268-H-40, St. Lawrence River Waterway Project – Defence Aspects (April 1, 1941-October 31, 1951), vol. 3335, Report: "Development Prospects for Labrador Iron Ore," Department of Trade and Commerce, Government of Canada, May 5, 1950.
42 LAC, RG 25, file 1268-D-40, pt. 14 (FP. 1), vol. 6345, Report: "The St. Lawrence Waterway and the Canadian Economy," Department of Trade and Commerce (Economic Research Division), Government of Canada, January 1951. Figures for American traffic in this report were based on two studies by the US Department of Commerce: *An Economic Appraisal of the St. Lawrence Seaway Project* (November 1947) and *Potential Traffic on the St. Lawrence Seaway* (December 1948). It is worth noting that American projections for the potential annual traffic (i.e., 57-84 million tons) on a seaway were consistently higher than those given by Canadian counterparts (i.e., about 45 million tons). American traffic estimates were predicated in large part on Canada figures for existing traffic on the Welland Canal and St. Lawrence fourteen-foot canals, and estimates of the iron ore deposits in the Ungava region depended on Canadian studies. United States Senate, 83rd Congress, 1st Session, "Notes re St. Lawrence Seaway," Hearings before the Subcommittee of the Committee on Foreign Relations, "Economics and Self-Liquidation of Navigation Phase (Enclosure 2a)," April 14-16 and May 20-21, 1953.
43 Louis St. Laurent, "Our North American Partnership," address at St. Lawrence University, Canton, New York, June 11, 1950, *Statements and Speeches* (Canadian Department of External Affairs), 50/23.
44 NARA II, RG 84, Canada, US Embassy, Ottawa, Classified General Records, 1950, box 9, Memorandum of Meeting: Niagara Diversion Treaty and the St. Lawrence Seaway and Power Project, June 21, 1950.
45 Chevrier, *The St. Lawrence Seaway,* 42.
46 LAC, RG 25, file 1268-D-40, pt. 10.2, St. Lawrence and Niagara River Treaty Proposals – General Correspondence (January 21-December 12, 1950), 6344, Secretary of State to Canadian Ambassador, September 16, 1950.

47 Dale Thomson, *Louis St. Laurent: Canadian* (Toronto: Macmillan, 1967), 307.
48 Sussman, *Quebec and the St. Lawrence Seaway*, 8; Susan Mann Trofimenkoff, *The Dream of Nation: A Social and Intellectual History of Quebec* (Toronto: Macmillan, 1982), 269.
49 Sussman, *Quebec and the St. Lawrence Seaway*, 2, 30-32. See also, for example, speeches and reports by municipal officials, such as C.E. Campeau, the assistant director of the Montreal City Planning Department. St. Lawrence University Archives, St. Lawrence Seaway Series, Mabee Series (D. Canadian Materials), box 71, file 6, C.E. Campeau, Assistant Director of the Montreal City Planning Department, "The Saint-Lawrence Seaway and Its Effects on the Montreal Region" (lecture, Canadian Progress Club of St-Laurent, December 9, 1954).
50 Mabee, *The Seaway Story*, 158.
51 Stephen Azzi, *Walter Gordon and the Rise of Canadian Nationalism* (Montreal: McGill-Queen's University Press, 1999); Hillmer and Granatstein, *For Better or for Worse*, 187-88. Cultural nationalism manifested itself in other interesting forms, as Steven High has shown: Steven High, "The 'Narcissism of Small Differences': The Invention of Canadian English, 1951-1967," in *Creating Postwar Canada: Community, Diversity, and Dissent, 1945-1975*, ed. Magda Fahrni and Robert Rutherdale (Vancouver: UBC Press, 2008).
52 See the American embassy's summary of Canadian newspapers on that topic: Harry S. Truman Library and Archives, Papers of Harry S. Truman, David E. Bell Subject File, box 1, US Embassy Ottawa to State Department: St. Lawrence Project, November 2, 1950.
53 NARA II, RG 59, file 611.42321-SL/12-1650, box 2795, Memorandum of Conversation – St. Lawrence Seaway and Power Project, December 16, 1950.
54 LAC, RG 25, file 1268-D-40, pt. 11.1. St. Lawrence and Niagara River Treaty Proposal – General Correspondence (January 4-March 31, 1951), 6344, teletype from Canadian Ambassador to Secretary of State, WA-54, January 5, 1951.
55 See Truman Archives, RG 220: President's Water Resources Policy Commission; in particular, see "A Water Policy for the American People: Summary of Recommendations from the Report of the President's Water Resources Policy Commission," RG 220, box 58, file 592, Final report: "Summary of Recommendations."
56 Hillmer and Granatstein, *For Better or for Worse*, 180.
57 Robert Bothwell, *Alliance and Illusion: Canada and the World, 1945-1984* (Vancouver: UBC Press, 2007), 89. The standard history of Canada-US relations during the Korean War remains Denis Stairs, *The Diplomacy of Constraint: Canada, the Korean War, and the United States* (Toronto: University of Toronto Press, 1974).
58 Aronsen, *American National Security*, xvi.
59 NARA II, RG 84, US Embassy, Ottawa – Classified General Records, box 9, External Affairs Minister Pearson's Views on Canadian-American Relations, Toronto Consulate, 260, by Orson N. Nielsen, May 4, 1951.
60 Ahead of issues such as a delay in implementing the civil air agreement of June 1949, the inability of Canadians to effect military procurement in the United States and vice versa, the "Buy-American Act," recent immigration and customs incidents, and Newfoundland bases. Memorandum by the Director of the Office of British Commonwealth and Northern European Affairs (Labouisse) to the Assistant Secretary of State for European Affairs (Perkins), 842.00/11-849, November 8, 1949, *FRUS, 1949*, vol. 2, Canada, 402.
61 NARA II, RG 59, 611.42321-SL/5-151, Memorandum of Conversation, St. Lawrence Project, Raynor and Ignatieff, May 1, 1951.

62 NARA II, RG 84, file 322.2, St. Lawrence Seaway, Canada and US (1951), box 14, Memorandum of Conversation, St. Lawrence Seaway, August 4, 1951.
63 Government of the United States, *President Harry S. Truman's Office Files, 1945-1953*, pt. 3, Subject File, State Department Memorandum for the President: Visit of Prime Minister St. Laurent on September 27 to Discuss the St. Lawrence Seaway and Power Project, September 27, 1951, ed. William E. Leuchtenburg (Washington: University Publications of America, 1990); Memorandum by the Under Secretary of State (Webb) to the President, Subject: Visit of Prime Minister St. Laurent on September 28 to Discuss the St. Lawrence Seaway and Power Project, file 611.42321 SL/9-2751, September 28, 1951, *FRUS, 1951*, vol. 2, Canada, 916-22.
64 Memorandum by the Under Secretary of State (Webb) to the President: Subject: Visit of Prime Minister St. Laurent on September 28 to Discuss the St. Lawrence Seaway and Power Project, file 611.42321 SL/9-2751, September 28, 1951, *FRUS, 1951*, vol. 2, Canada, 916-22. Emphasis in original.
65 NARA II, RG 84, file 322.2, St. Lawrence Seaway, Canada and US (1951), box 14, Ottawa Embassy, St. Lawrence Navigation and Power Project, despatch 1451, May 10, 1951.
66 Ibid.
67 "Canada's Newspapers Look at the Seaway," *Ottawa Citizen*, August 3, 1951.
68 NARA II, RG 84, file 322.2, St. Lawrence Seaway, Canada and US (1951), box 14, Press Comment on St. Lawrence Seaway and Power Project, despatch 153, August 15, 1951.
69 Ibid.
70 In August 1951, these newspapers included *Quebec Chronicle Telegraph, Halifax Chronicle Herald, Victoria Daily Colonist, Moncton Transcript*, and *Saint John Telegraph Journal*. See NARA II, RG 84, file 322.2, St. Lawrence Seaway, Canada and US (1951), box 14, Press Comment on St. Lawrence Seaway and Power Project, despatch 153, August 15, 1951.
71 NARA II, RG 84, file 322.2, St. Lawrence Seaway, Canada and US (1951), box 14, Transmittal of Text of Gallup Poll on St. Lawrence Seaway, Enclosure: Gallup Poll of Canada (of June 27), July 28, 1951.
72 LAC, RG 2, file W-10-1 (1), Waterways; Water Development, Waterworks, Projects, etc. St. Lawrence Waterway and Power Project, 1951 (January-August 31), 207, Memorandum to Cabinet – prepared by the Interdepartmental Committee on Great Lakes–St. Lawrence Development re: the All-Canadian Waterway, June 1951.
73 LAC, RG 2, file W-10-1 (vol. 1), Waterways; Water Development, Waterworks, Projects, etc. St. Lawrence Waterway and Power Project, 1951 (September 1-November 15, 1951), 207 – Note on Cabinet Conclusions, September 21, 1951.
74 Chevrier, *The St. Lawrence Seaway*, 46.
75 LAC, RG 25, file 1268-D-40, pt. 13.1, St. Lawrence and Niagara River Treaty Proposals – General Correspondence (July 4-October 13, 1951), 6344, Memorandum, Meeting between the President and the Prime Minister of Canada on the St. Lawrence Project on September 28, 1951, September 28, 1951.
76 Ibid.; NARA II, RG 59, 611.42321-SL/9-2851, box 2795, Department of State, Notes on the Meeting between the President and Prime Minister St. Laurent of Canada, October 5, 1951; "Announcement Subsequent to a White House Conference between Prime Minister Louis St. Laurent of Canada and President Truman," in US Congress, *St. Lawrence Seaway Manual: A Compilation of Documents on the Great Lakes Project and Correlated Power Development* (Washington, DC: US Government Printing Office, 1955), 112-13.
77 Thomson, *Louis St. Laurent*, 320.

78 Ibid.
79 LAC, RG 25, file 1268-D-40, pt. 13.2, St. Lawrence and Niagara River Treaty Proposals – General Correspondence (July 4-October 13, 1951), 6344, Memorandum: St. Lawrence Seaway Legislation, October 25, 1951. Emphasis in original.
80 Mabee, *The Seaway Story*, 158.
81 Chevrier, *The St. Lawrence Seaway*, 47-48.
82 NARA II, RG 84, file 322.2, St. Lawrence Seaway (1952), US Embassy, Ottawa, Classified General Records, 1950-1961, Perkins to Secretary, St. Lawrence Seaway and Power Project, November 1, 1952.
83 "Exchange of Notes between Canada and the United States, 35, January 11, 1952," in R.R. Baxter, *Documents on the St. Lawrence Seaway* (New York: Frederick A. Praeger, 1960), 33-34.
84 Willoughby, *The St. Lawrence Waterway*, 235.
85 LAC, RG 25, file 1268-D-40, pt. 14.2, vol. 6345, St. Lawrence and Niagara River Treaty Proposals – General Correspondence, Address by the Hon. Lionel Chevrier, Minister of Transport, Over the CBC network, "The Nation's Business," January 8, 1952.
86 For examples see "Canada's Plan for Building Seaway Alone," *Charleston News and Courier*, December 2, 1951, 4H; Editorial, "No Bluff on the Seaway," October 25, 1951, *Ottawa Citizen*.
87 NARA II, RG 84, file 322.2, box 14, St. Lawrence Seaway (1952), US Embassy, Ottawa – Classified General Records, 1950-61, Memorandum for Economic Section: Views of Lt. Col. Lucien Dansereau on St. Lawrence Waterway, from Consulate General in Montreal, February 7, 1952.
88 Truman Archives, Dean Acheson Papers, box 70, Memoranda of Conversations – March 1952, Memorandum of Conversation: Status of the St. Lawrence Project, March 26, 1952.
89 Truman Archives, Dean Acheson Papers, box 70, Memoranda of Conversations – April 1952, Memorandum of Conversation with the President, Item 2, St. Lawrence Seaway, April 7, 1952.
90 Truman Archives, Harry S. Truman Papers, Charles S. Murphy Files, Correspondence and General File, 1945-1953, box 33, St. Lawrence: 1952, Memorandum of Conversation: St. Lawrence Seaway and Power Project, April 14, 1952.
91 NARA II, RG 84, file 322.2, St. Lawrence Seaway (1952), Minutes of Discussions with the President on the St. Lawrence Project Held at the White House (April 14, 1952), April 14, 1952.
92 LAC, RG 2, W-10-1 (May) 1952, 240, Record of Cabinet Decisions, 193 (June 18, 1952), including Item 3: St. Lawrence Development Project; modification in financial basis of Canadian proposal, June 18, 1952.

Chapter 3: Caught between Two Fires

1 This information about the book *Canada at War* is noted in Michael Bliss, *Northern Enterprise: Five Centuries of Canadian Business* (Toronto: McClelland and Stewart, 1987), 463.
2 MacEachern makes this statement in regard to the reissue of Janet Foster's *Working for Wildlife: The Beginning of Preservation in Canada*, 2nd ed. (Toronto: University of Toronto Press, 1998). Alan MacEachern, "Voices Crying in the Wilderness: Recent Works in Canadian Environmental History," *Acadiensis* 31, 2 (Spring 2002): 215-26.

3 Christopher Armstrong and H.V. Nelles, *Monopoly's Moment: The Organization and Regulation of Canadian Utilities, 1830-1930* (Philadelphia: Temple University Press, 1986), 237-38.
4 Memorandum to the Assistant Secretary of State for European Affairs (Perkins) to the Secretary of State, Subject: Material on Canada, 955, 611.42/11-1952, November 19, 1952, *FRUS, 1951-1952*, vol. 6, Canada, 2051-53.
5 See Marco Adria, *Technology and Nationalism* (Montreal: McGill-Queen's University Press, 2010); R. Douglas Francis, *The Technological Imperative in Canada: An Intellectual History* (Vancouver: UBC Press, 2009); Cole Harris, "The Myth of the Land in Canadian Nationalism," in *Nationalism in Canada*, ed. Peter Russell (Toronto: McGraw-Hill, 1966); Bélanger, *Prejudice and Pride*.
6 Adria, *Technology and Nationalism*, 22.
7 Ibid., 45. On technological nationalism see also Maurice Charland, "Technological Nationalism," *Canadian Journal of Political and Social Theory* 10, 1 (1986): 196-220; Robert McDougall, "The All-Red Dream: Technological Nationalism and the Trans-Canada Telephone System," in *Canadas of the Mind: The Making and Unmaking of Canadian Nationalisms in the Twentieth Century*, eds. Norman Hillmer and Adam Chapnick (Montreal: McGill-Queen's University Press, 2007).
8 Francis, *Technological Imperative*, 2. This view of the impact of technology on North American integration was also promulgated by philosopher George Grant. See his *Lament for a Nation;* George Grant, *Technology and Empire: Perspectives on North America* (Toronto: House of Anansi, 1969).
9 A.A. den Otter, *The Philosophy of Railways: The Transcontinental Railway Idea in British North America* (Toronto: University of Toronto Press, 1997). Jeremy Mouat provides an interesting analysis of the reality of railways and nation building in the British Columbia context: Jeremy Mouat, "Nationalist Narratives and Regional Realities: The Political Economy of Railway Development in Southeastern British Columbia, 1895-1905," in *Parallel Destinies: Canadian-American Relations West of the Rockies*, ed. John M. Findlay and Kenneth S. Coates (Seattle: University of Washington Press, 2002). On the links between organized capitalism, technological innovation, megaprojects, and nation building, also see Philip Resnick, *The Masks of Proteus: Canadian Reflections on the State* (Toronto: University of Toronto Press, 1990), particularly chap. 8, "'Organized Capitalism' and the Canadian State."
10 John Swettenham, *McNaughton*, vol. 3, *1944-1966* (Toronto: Ryerson Press, 1969), 212.
11 Government of Canada, LAC, RG 25, file 1268-D-40, pt. 17, St. Lawrence and Niagara River Treaty, St. Lawrence Project – General File (July 2-August 30, 1952), vol. 6345, Secretary of State to Canadian Ambassador (Pickersgill to Wrong), July 8, 1952.
12 LAC, RG 25, file 1268-D-40, pt. 17, St. Lawrence and Niagara River Treaty, St. Lawrence Project – General File (July 2-August 30, 1952), vol. 6345, Secretary of State to Canadian Ambassador (Wrong to Pickersgill), WA-1814, July 10, 1952.
13 LAC, RG 25, file 1268-D-40, pt. 17, St. Lawrence and Niagara River Treaty, St. Lawrence Project – General File (July 2-August 30, 1952), vol. 6345, Canadian Ambassador to Secretary of State, July 15, 1952.
14 LAC, RG 25, file 1268-D-40, pt. 17, St. Lawrence and Niagara River Treaty, St. Lawrence Project – General File (July 2-August 30, 1952), vol. 6345, Memorandum of Interview with Dr. N.R. Danielian, August 1, 1952.

15 LAC, RG 25, file 1268-D-40, pt. 17, St. Lawrence and Niagara River Treaty, St. Lawrence Project – General File (July 2-August 30, 1952), vol. 6345, Memorandum on the St. Lawrence Project, August 1, 1952.
16 LAC, RG 25, file 1268-D-40, pt. 17, St. Lawrence and Niagara River Treaty, St. Lawrence Project – General File (July 2-August 30, 1952), vol. 6345, Memorandum of Interview with Dr. N.R. Danielian, August 1, 1952.
17 NARA II, RG 84, file 320, Canada–United States, US Embassy Ottawa, Classified General Records, 1950-1961 (1950-1952: 050 to 1950-1952: 310), Perkins to the Secretary: Material on Canada, November 19, 1952.
18 LAC, RG 25, file 1268-D-40, pt. 17.2, St. Lawrence and Niagara River Treaty, St. Lawrence Project – General File (August 1- 30, 1952), 6346, Memorandum for the Minister: St. Lawrence Project; Recent Developments, August 27, 1952.
19 LAC, RG 25, file: St. Lawrence Seaway 1952-1954, 3175, DEA Minute on the St. Lawrence Project, September 4, 1952.
20 LAC, RG 25, file 1268-D-40, pt. 17.2, St. Lawrence and Niagara River Treaty, St. Lawrence Project – General File (August 1, 1952-August 30, 1952), 6346, Memorandum for the Minister: St. Lawrence Project, August 26, 1952.
21 LAC, RG 25, file 1268-D-40, pt. 19.1, St. Lawrence and Niagara River Treaty, St. Lawrence Project – General File (November 1-December 13, 1952), 6346, Note from the Canadian Ambassador in Washington to the Secretary of State of the United States, November 4, 1952.
22 LAC, RG 25, file 1268-D-40, pt. 19.1, St. Lawrence and Niagara River Treaty, St. Lawrence Project – General File (November 1-December 13, 1952), 6346, letter from Wrong to Acheson, November 4, 1952.
23 Ibid.
24 George Adams and Charlie Wilson counted among the strong supporters within the cabinet; others, such as White House chief of staff Sherman Adams, were opposed to the project. Dwight D. Eisenhower Library and Archives, Columbia University Oral History Project, OH 177, Oral History Interview with N.R. Danielian (1972); LAC, RG 25, file 1268-D-40, pt. 19.2, St. Lawrence and Niagara River Treaty, St. Lawrence Project – General File (November 1-December 13, 1952), 6346, Canadian Ambassador to Secretary of State, December 8, 1952.
25 Government of the United States, Eisenhower Presidential Library, Online Documents, "Confidential Memorandum from Senator Wiley for General Eisenhower, December 30, 1952."
26 LAC, RG 25, file 1268-D-40, pt. 20, St. Lawrence and Niagara River Treaty, St. Lawrence Project – General File (December 16, 1952-January 31, 1953), 6346, Secretary of State to Canadian Ambassador, December 17, 1952.
27 NARA II, RG 84, file 322.2, St. Lawrence Seaway (1952), US Embassy, Ottawa, Classified General Records, 1950-1961, Woodward to Charles S. Murphy (Special Counsel to the President), December 18, 1952. The draft letter from Truman to St. Laurent is attached.
28 LAC, RG 25, file 1268-D-40, pt. 20, St. Lawrence and Niagara River Treaty, St. Lawrence Project – General File (December 16, 1952-January 31, 1953), 6346, Memorandum to the Under Secretary: St. Lawrence Project, December 22, 1952.
29 *President Harry S. Truman's Office Files, 1945-1953*, pt. 3: Subject File, Stanley Woodward to Truman, January 2, 1952; LAC, RG 25, file 1268-D-40, pt. 20, St. Lawrence and Niagara River Treaty, St. Lawrence Project – General File (December 16, 1952-January 31, 1953), 6346, Canadian Ambassador to Secretary of State, January 2, 1953.

30 Truman Archives, Papers of Harry S. Truman, David E. Bell Subject File, box 1, Memorandum of Conversation: St. Lawrence Project, January 2, 1953.
31 Ibid.
32 LAC, RG 25, file 1268-D-40, pt. 20, St. Lawrence and Niagara River Treaty, St. Lawrence Project – General File (December 16, 1952-January 31, 1953), 6346, Canadian Ambassador to Secretary of State, January 2, 1953.
33 LAC, RG 25, file 1268-D-40, pt. 20, St. Lawrence and Niagara River Treaty, St. Lawrence Project – General File (December 16, 1952-January 31, 1953), 6346, Memorandum to the United States Ambassador, January 8, 1953.
34 Willoughby, *The St. Lawrence Waterway*, 247.
35 Eisenhower Archives, Dwight D. Eisenhower: Papers as President of the United States, 1953-61 (Ann Whitman Files), St. Lawrence Seaway, Memorandum for Legislative Meeting: St. Lawrence Seaway, March 30, 1953.
36 Evan Thomas, *Ike's Bluff: President Eisenhower's Secret Battle to Save the World* (New York: Little, Brown, 2012). Other revisionist works include Jean Edward Smith, *Eisenhower in War and Peace* (New York: Random House, 2011); Jim Newton, *Eisenhower: The White House Years* (New York: Doubleday, 2011).
37 Richardson, *Dams, Parks and Politics*.
38 Eisenhower Archives, OH 354, Oral History Interview with Len B. Jordan, August 18, 1975.
39 Ibid.
40 LAC, RG 25, file 1268-D-40, pt. 21.2, St. Lawrence and Niagara River Treaty, St. Lawrence Project – General File (February 2-May 20, 1953), 6347, Memorandum for the American Division: St. Lawrence Project, March 16, 1953. Apparently, in December 1952, Pearson had dined with Dulles and stated that it would be most unfortunate if the incoming administration did nothing to facilitate the granting of the FPC licence. RG 25, St. Lawrence Seaway – 1954-1964, vol. 3175, Memorandum: The St. Lawrence and Seaway Project, June 1, 1955.
41 Eisenhower Archives, Online Documents, letter from Pearson to Dulles, March 24, 1953; LAC, RG 25, vol. 6641, file 11513-40, pt. 3.2, Pearson to Dulles, March 21, 1953.
42 LAC, RG 25, file 1268-D-40, pt. 21.2, St. Lawrence and Niagara River Treaty, St. Lawrence Project – General File (February 2-May 20, 1953), 6347, Secretary of State to Canadian Ambassador, April 7, 1953.
43 NARA II, RG 59, file 611.32321-SL/4-853, box 2800 (January-September 1953), Memorandum of Conversation: Remarks in New York by the Hon. C.D. Howe regarding the St. Lawrence Waterway, April 8, 1953.
44 The day after his New York speech, Howe told an American diplomat that he had inside information that the soon-to-be appointed FPC chairman would be friendly to the PASNY application. NARA II, RG 59, file 611.32321-SL/4-853, box 2800 (January-September 1953), Memorandum of Conversation: Remarks in New York by the Hon. C.D. Howe regarding the St. Lawrence Waterway, April 8, 1953.
45 Ibid.
46 Eisenhower Archives, White House Office, Office of the Special Assistant for National Security Affairs: Records, 1952-1961, box 5, National Security Council (hereafter NSC) 150/1 – St. Lawrence Seaway, a Report to the National Security Council, by the NSC Planning Board on National Security Interests in the St. Lawrence–Great Lakes Seaway Project, April 16, 1953. The report was NSC 150. NSC 150/1 was released shortly thereafter, on April 23:

Eisenhower Archives, White House Office, Office of the Special Assistant for National Security Affairs: Records, 1952-1961, box 1, President's Papers 1953 (6), a Report to the National Security Council by the Executive Secretariat on National Security Interests in the St. Lawrence–Great Lakes Seaway Project, April 23, 1953.

47 Eisenhower Archives, White House Central Files, Official File, 1953-1961, box 696, OF 155-D-1, Inland Waterways-Navigation, St. Lawrence Waterway, 1952-53 (4), letter from Eisenhower to Wiley, April 24, 1953. A few days later, the chairman of the FPC responded, stating that he was in "hearty accord" with the findings and recommendations of the National Security Council. Eisenhower Archives, White House Central Files, Official File, 1953-1961, box 696, OF 155-D-1, Inland Waterways-Navigation, St. Lawrence Waterway, 1952-53 (4), Thomas C. Buchanan to Eisenhower, April 27, 1953; Eisenhower Archives, Columbia University Oral History Project, OH 177, Oral History Interview with N.R. Danielian (1972), 21.

48 J.W. Pickersgill, *My Years with Louis St. Laurent: A Political Memoir* (Toronto: University of Toronto Press, 1975), 186.

49 Eisenhower Archives, Dwight D. Eisenhower: Records as President, White House Central Files (Confidential File) 1953-61, Subject Series, box 69, Department of State (June 1954) (1), Report of Staff Committee on St. Lawrence–Great Lakes Seaway Project, May 6, 1953.

50 Hugo Spalinksi, from the Public Power and Water Corporation, was a licence competitor with an unworkable alternative plan for a deep waterway. The available evidence suggests that he was serving as a proxy for US interests opposed to the seaway, particularly the railroads: LAC, RG 25, file 1268-D-40, pt. 22.2, 6347: The St. Lawrence Project (May 20-August 14, 1953), Item for Weekly Divisional Notes: St. Lawrence Project, July 23, 1953.

51 Ibid.

52 Canadian officials sought American help in leap-frogging the Court of Appeals by petitioning the US Supreme Court for *certiorari* (requesting a lower court to forward its decision for review by a higher court) immediately upon the filing of the notices of appeal. LAC, RG 25, file 1268-D-40, pt. 23, 6347: The St. Lawrence Project (August 17-October 21, 1953), Burling to LePan, September 14, 1953.

53 LAC, RG 25, file 1268-D-40, pt. 23, 6347: The St. Lawrence Project (August 17-October 21, 1953), Secretary of State to Canadian Ambassador, October 8, 1953.

54 NARA II, RG 59, file 611.42321-SL/10-2053, box 2801 (October 1953-August 1954), Memorandum of Conversation: Canadian Request for Assistance in Expediting Handling of Anticipated Legislation re: St. Lawrence Power Development, October 20, 1953.

55 Oral arguments on the substance of the PASNY licence appeals were subsequently moved up to December 23, a situation that the Canadian ambassador deemed "quite satisfactory." LAC, RG 25, file 1268-D-40, pt. 24.2, 6347: The St. Lawrence Project (October 22-November 23, 1953), Canadian Ambassador to Secretary of State, November 19, 1953.

56 NARA II, RG 59, file 611.42321-SL/10-2253, box 2801 (October-August 1954), letter from Dulles to Herbert Brownell Jr., Attorney General, November 2, 1953.

57 Eisenhower Presidential Library, Online Documents, "Letter from Arthur Radford (JCS) to Senator Homer Ferguson, January 18, 1954."

58 For example, a Canadian memorandum noted the US authorities had explicitly asked for Canadian views on February 16, 1953, and March 19, 1953. LAC, RG 25, file: St. Lawrence Seaway 1952-54, Department of External Affairs Minute on the St. Lawrence Project, 3175, Memorandum for Mr. LePan: Notes on Canadian Attitude towards United States Participation in the St. Lawrence Seaway Project, January 14, 1954.

59 Ibid.
60 LAC, RG 25, file 1268-D-40, pt. 25.1, 6348: St. Lawrence General Correspondence (November 23, 1953-January 29, 1954), Note for file 1268-D: St. Lawrence Project: Proposal for United States Participation in the Seaway, January 8, 1954.
61 LAC, RG 25, file: St. Lawrence Seaway 1952-54, Department of External Affairs Minute on the St. Lawrence Project, 3175, Memorandum for Mr. LePan: Notes on Canadian Attitude towards United States Participation in the St. Lawrence Seaway Project, January 14, 1954.
62 Eisenhower Archives, White House Central Files, Official File, 1953-1961, box 696, OF 155-D-1, Inland Waterways-Navigation, St. Lawrence Waterway, 1954 (3), Memorandum for Governor Sherman Adams, May 17, 1954.
63 LAC, RG 25, file 1268-D-40, pt. 29, 6348, St. Lawrence Project: General Correspondence (May 17-June 10, 1954), Memorandum: St. Lawrence Seaway, May 24, 1954.
64 LAC, MG 32 B-39, 112, "Canadian Seaway Preferable: Howe," *Globe and Mail,* May 18, 1954.
65 Ibid.
66 LAC, RG 25, file 1268-D-40, pt. 25.2, 6348: St. Lawrence General Correspondence (November 25, 1953-January 29, 1954), Canadian Ambassador to Secretary of State (Walter Lippmann, "A Project Outgrown," *New York Herald Tribune*), January 21, 1954.
67 Ibid.
68 LAC, RG 25, file 1268-D-40, pt. 30.2, 6349, St. Lawrence Project: General Correspondence (June 7-30, 1954), St. Lawrence Seaway Meeting (June 28, 1954), June 29, 1954.
69 NARA II, RG 84, US Embassy, Ottawa – Classified and Unclassified General Records, 1938-1963, despatch 681, Ottawa Embassy, Comment on United States Participation in St. Lawrence Seaway, February 5, 1954. For an excellent collection of newspaper views and reports on the St. Lawrence project, see box 76 in the St. Lawrence Series (3) in the St. Lawrence Seaway Collection in the St. Lawrence University Archives.
70 LAC, RG 25, file 1268-D-40, pt. 26.2, 6348: St. Lawrence General Correspondence (February 1-March 11, 1954), Advertisement in the *Ottawa Citizen* by Tim Buck, National Leader, Labor-Progressive Party, titled "An Open Letter to Members of Parliament," February 23, 1954.
71 LAC, RG 25, file 1268-D-40, pt. 26.1, 6348: St. Lawrence General Correspondence (February 1-March 11, 1954), Clark Davey, "All-Canada Seaway Bartered for Power," *Globe and Mail,* February 19, 1954.
72 LAC, RG 25, file 1268-D-40, pt. 29, 6348, St. Lawrence Project: General Correspondence (May 17-June 10, 1954), Meeting of the Interdepartmental Committee on the St. Lawrence Project (May 3, 1954), May 27, 1954.
73 Ibid.
74 LAC, RG 25, file 1268-D-40, pt. 28, 6349, St. Lawrence Project: General Correspondence (April 9-May 14, 1954), Canadian Ambassador to Secretary of State, May 7, 1954.
75 Ibid.
76 "Exchange of Notes between the United States and Canada, June 7 and 16, 1954," *St. Lawrence Seaway Manual,* 168-70.
77 LAC, RG 25, file: St. Lawrence Seaway 1952-54, Department of External Affairs Minute on the St. Lawrence Project, 3175, DEA Minute on St. Lawrence Seaway, July 20, 1954.
78 NARA II, RG 59, file 611.42321-SL/6-1954, box 2802, Memorandum from Merchant to Robert B. Anderson (Dep. Sec. of Defense), June 19, 1954; NARA II, RG 59, file 611.42321-

SL/6-2254, box 2802, Memorandum from Phleger to Merchant: St. Lawrence Seaway – New Arrangements with Canadian Government, June 22, 1954.
79 LAC, RG 2, Cabinet Conclusions, July 6, 1954; LAC, RG 25, 1268-D-40, pt. 31.2, 6349, St. Lawrence Project: General Correspondence (July 1-20, 1954), Memorandum for Mr. MacKay: St. Lawrence Seaway, July 10, 1954.
80 John Swettenham also gives an account of the McNaughton-Anderson talks: Swettenham, *McNaughton,* vol. 3, *1944-1966,* 227-28.
81 Ibid.
82 LAC, RG 2, Cabinet Conclusions, July 6, 1954.
83 LAC, RG 25, file: St. Lawrence Seaway – 1954-1964, 3175. DEA Memorandum: The St. Lawrence Seaway and Power Project – Preliminary Note, June 1, 1955.
84 LAC, RG 25, file 1268-D-40, pt. 32, 6349, St. Lawrence Project: General Correspondence (July 21-31, 1954), Memorandum: St. Lawrence Seaway, Draft Report of Discussion Between United States–Canadian Legal Officers, July 26, 1954.
85 A detailed 1955 retrospective of the evolution of the St. Lawrence project by two members of the American Division of DEA, G.C. Cox and S.A. Friefeld, states that building at Iroquois was Howe's idea for recouping the $15 million contribution to the power entities. LAC, RG 25, file B-58, The St. Lawrence Seaway and Power Project, June 1, 1955.
86 LAC, RG 25, file 1268-D-40, pt. 33.2, 6349, St. Lawrence Project: General Correspondence (August 3-14, 1954), Pelletier to St. Laurent (at St. Patrick), August 5, 1954.
87 LAC, RG 25, file 1268-D-40, pt. 34.2, 6350, St. Lawrence Project: General Correspondence (August 16-19, 1954), Memorandum from American Division: St. Lawrence Seaway Talks August 12 and 13, August 20, 1954. According to a later recollection, Pearson had obtained US permission to rearrange the order of agenda items, which he did, so that those points on which the United States was unlikely to be able to satisfy Canadian desires would be first, thus providing stronger grounds for Canada's claim that it would need to build the Iroquois works as compensation. LAC, RG 25, The St. Lawrence Seaway – 1954-1964, vol. 3175, Memorandum – The St. Lawrence Seaway and Power Project, June 1, 1955.
88 Ibid.
89 NARA II, RG 59, 611.42321-SL/8-1954, box 2802, despatch 122, Ottawa Embassy, St. Lawrence Seaway Discussions, Ottawa (August 1954), August 19, 1954.
90 Ibid.
91 "Exchange of Notes between Canada and the United States of America, Modifying the Exchange of Notes of June 30, 1952: Concerning the Construction of the St. Lawrence Seaway; Signed at Ottawa, August 17, 1954. X-214," in Baxter, *Documents on the St. Lawrence Seaway,* 50-53.
92 International Joint Commission (hereafter IJC), Canadian Section, 68-2-5:3-3: St. Lawrence Power Application, Chairman's Meeting, 1954-1957, Memorandum of Meeting by E.M. Sutherland, March 25, 1957.
93 LAC, RG 25, file 1268-D-40, pt. 34, 6349, St. Lawrence Project: General Correspondence (August 16-19, 1954), Secretary of State to Canadian Ambassador, August 18, 1954; LAC, RG 25, file 1268-D-40, pt. 34, 6349, St. Lawrence Project: General Correspondence (August 16-19, 1954), press release 50, August 18, 1954.
94 NARA II, RG 84, US Embassy, Ottawa – Classified and Unclassified General Records, 1938-1963, despatch 140, Canadian Reaction to the St. Lawrence Seaway Agreement, August 27, 1954.

95 LAC, RG 25, file 1268-D-40, pt. 34.2, 6350, St. Lawrence Seaway Project: General Correspondence (August 16-19, 1954), Pearson to Marler, August 27, 1954; LAC, RG 25, file 1268-D-40, pt. 36.2, 6350, St. Lawrence Seaway Project: General File (August 26-30, 1954), Address by the Rt. Hon. Louis S. St-Laurent on the Celebration of the Tenth Anniversary of Le Syndicat des Journalistes de Langue Francaise de Montreal (October 23, 1954), October 23, 1954.

96 LAC, RG 25, 1268-D-40, pt. 34, 6349, St. Lawrence Seaway Project: General Correspondence (August 16-19, 1954), Editorial, "A Mess of Pottage," *Globe and Mail,* August 16, 1954.

97 NARA II, RG 84, US Embassy, Ottawa – Classified and Unclassified General Records, 1938-1963, despatch 140, Canadian Reaction to the St. Lawrence Seaway Agreement, August 27, 1954.

98 Editorial, "Call It 'A Bad Deal'?" *Ottawa Journal,* October 26, 1954.

99 LAC, RG 25, file 1268-D-40, pt. 34.2, 6350, St. Lawrence Seaway Project: General Correspondence (August 16-19, 1954), Draft reply to persons and organizations urging the Government to construct the St. Lawrence Seaway entirely in Canada, August 24, 1954.

100 "Appendix B. Memorandum on St. Lawrence Seaway by Department of External Affairs, August 26, 1954," *St. Lawrence Seaway Manual,* 201-2. On the "national interest" and Canadian foreign policy see also the various contributions to Greg Donaghy and Michael K. Carroll, eds., *In the National Interest: Canadian Foreign Policy and the Department of Foreign Affairs and International Trade, 1909-2009* (Calgary: University of Calgary Press, 2011).

101 LAC, RG 25, file 1268-D-40, pt. 35, 6350, St. Lawrence Seaway Project: General Correspondence (August 26-September 30, 1954), Extract from an Address Entitled "Canada and the United States: Our Area of Economic Co-operation," delivered by the Hon. L.B. Pearson, Secretary of State for External Affairs, at the University of Rochester, Rochester, New York, September 2, 1954.

102 LAC, RG 25, file 1268-D-40, pt. 34.2, 6350, St. Lawrence Project: General Correspondence (August 16-19, 1954), Draft Memorandum on St. Lawrence Seaway, August 25, 1954.

103 See the Bruce Hutchison quote in the concluding paragraph of Chapter 1.

Chapter 4: Fluid Relations

1 On "hybrid" waterscapes see Richard White, *The Organic Machine: The Remaking of the Columbia River* (New York: Hill and Wang, 1995); Cioc, *The Rhine.*

2 Some 197 million cubic yards of material were excavated and dredged; 6.5 million yards of concrete were poured. Robert W. Passfield, "Construction," 47.

3 "The greatest construction show on earth" is part of the subtitle of Claire Parham's book *St. Lawrence Seaway,* on the oral history of the construction of the St. Lawrence project, which she attributes to M.W. Oettershagen, deputy administrator of the Seaway Development Corporation in 1959.

4 Timothy Heinmiller effectively makes this point with his chapter "The St. Lawrence: From River to Marine Superhighway."

5 LAC, RG 2, W-10-1, 1, Waterways: Water Development, Waterworks, Projects, etc., St. Lawrence Waterway and Power Project (January 1-August 31, 1951), Memorandum to Cabinet, Prepared by the Interdepartmental Committee on Great Lakes–St. Lawrence Development re: the All-Canadian Waterway, n.d. [July 1951].

6 The values given here are in American dollars.

7 For a detailed history of the Corps of Engineers' role in constructing the seaway see Becker, *From the Atlantic.* Also see chap. 7, "US Army Engineers and the Rise of Cost-Benefit

Analysis," in Theodore M. Porter, *Trust in Numbers: The Pursuit of Objectivity in Science and Public Life* (Princeton, NJ: Princeton University Press, 1995).

8   Parham, *The St. Lawrence Seaway and Power Project*, 129.
9   J.B. Bryce provides a history focused on the hydraulic engineering aspects of the St. Lawrence power project: Bryce, *Hydraulic Engineering History*.
10 A ceremonial groundbreaking was originally planned for the seaway portion of construction but was eventually called off. Canada was opposed to these ceremonies because the works were entirely in US territory and would put focus on only the American portion of the seaway and dampen enthusiasm for the 1959 celebration. Dwight D. Eisenhower Presidential Library and Archives, Staff Files: Records of I. Jack Martin as Administrative Assistant to the President, 1953-58, box 3, file: St. Lawrence Seaway, Memorandum to General Wilton B. Persons (by Gruenther), April 25, 1958; LAC, RG 25, 6778, file 1268-D-40, pt. 43.2, St. Lawrence Seaway and Power Project – General File, March 29-April 29, 1955, DEA Memorandum: St. Lawrence Seaway: Ground-breaking Ceremonies, April 29, 1955.
11 Canada awarded its first contract in October of that year, and the United States did the same in January 1955.
12 Parham, *The St. Lawrence Seaway and Power Project*, 121.
13 LAC, RG 25, 6778, file 1268-D-40, pt. 47.2, St. Lawrence Seaway and Power Project – General File, June 16, 1955, to July 19, 1955, Memorandum (by Cote), June 17, 1955.
14 Chevrier, *The St. Lawrence Seaway*, 60.
15 Overtures to the Quebec government accelerated in the late 1940s, and in 1951 the federal government suggested that the cost to Quebec for the development of approximately 1.2 million horsepower was $235 million. LAC, RG 25, 2636, file: 1268-D-40C: St. Lawrence River-Niagara River Treaty Proposals (General Correspondence) (January 4, 1938-December 21, 1940); LAC, RG 25, 6344, file 1268-D-40, pt. 13.1, St. Lawrence and Niagara River Treaty Proposals – General Correspondence (July 4-October 13, 1951).
16 LAC, RG 25, file: 11513-40: Copies of Minutes, Meetings, Telegrams on the St. Lawrence Project Circulated to Legal Division, pt. 7, July 5-September 29, 1954, Note of Cabinet Decision: St. Lawrence Power Project; special customs and immigration arrangements, July 23, 1954. These "international works" also raised complicated questions surrounding issues such as workers compensation and unemployment insurance. LAC, RG 25, 6642, file: 11513-40: Copies of Minutes, Meetings, Telegrams on the St. Lawrence Project Circulated to Legal Division, pt. 7, July 5-September 29, 1954, DEA Memorandum: St. Lawrence Power Project: Unemployment Insurance and Workmen's Compensation, September 28, 1954.
17 There was a fault line running under the original foundation. Engineers found that the new location had greater discharge capacity, better ice control potential, and cost less. IJC, Canadian Section, 68-3-V2 – St. Lawrence Power Application, Correspondence from January 1-December 21, 1954, Memorandum to St. Lawrence River Joint Board of Engineers (from HEPCO): Relocation of Iroquois Control Dam, August 3, 1954; Bryce, *Hydraulic Engineering History*, 67.
18 During January 1955 discussions, a SLSDC official also stated that the United States intended to shift the location of one of its locks so that it could build twin locks in the Wiley-Dondero Canal south of Barnhart Island. This caught the Canadian representatives by surprise and seemed to undercut Canadian hopes of eventually constructing the connecting links of an all-Canadian seaway. Although that may have been a consideration, it turned out that the main US motivation was to site its locks in such a way as to allow for duplication at some undetermined future point. The St. Laurent government concluded that this was simply prudent planning.

19 The lock at Iroquois was constructed with two sets of steel sector gates at each end, rather than the conventional mitre gates, to enable the lock gates to serve as an emergency dam in case any one pair of gates were destroyed. Passfield, "Construction," 16; Government of Canada, Department of External Affairs, *Documents on Canadian External Relations*, no. 20, chap. 5, "Relations with the United States," pt. 4, "Economic Issues," Section D: St. Lawrence Seaway, Memorandum from Under Secretary of State for External Affairs to Secretary of State for External Affairs, November 22, 1954, DEA/1268-D-40 (Ottawa: Minister of Public Works and Government Services Canada, 1971), 585.
20 Anderson was replaced by Wilber Brucker in September 1955.
21 LAC, RG 25, 6352, file 1268-AD-40, pt. 1.2, St. Lawrence Project: Dredging at Cornwall Island, December 1, 1954-March 25 1955, St. Lawrence Seaway: Non-Duplication of Navigation Facilities, February 21, 1955.
22 Eisenhower Archives, Robert B. Anderson Papers, box 33, file: St. Lawrence Seaway (2), Comments on Memorandum from the State Department (February 11, 1955), February 14, 1955.
23 Eisenhower Archives, Robert B. Anderson Papers, box 33, file: St. Lawrence Seaway (1), Abbreviated Minutes of Meeting in SLSDC, May 18, 1955.
24 Power Authority of the State of New York (hereafter PASNY), Minutes of Trustee Meetings, Appendix: Memorandum from Moses, February 23, 1956.
25 LAC, RG 25, 6784, file 1268-AD-40, pt. 5.3, January 29, 1962, Memorandum to Douglas Turner. Attached is DEA Memorandum: Cornwall Island Dredging, November 7, 1956.
26 HEPCO and PASNY had in July indicated that the additional 1.25 feet of head at the power dam that would result from dredging the tailrace was worth between $8.5 million and $9 million to them. LAC, RG 25, 6783, file 1268-AD-40, pt. 2, St. Lawrence Project: Dredging and Facilities at Cornwall Island 1956, DEA Memorandum: St. Lawrence Project: Dredging North and South of Cornwall Island, May 14, 1956.
27 LAC, RG 25, 6784, file 1268-AD-40, pt. 5.3, 29 January 1962, Memorandum to Douglas Turner. Attached is DEA Memorandum: Cornwall Island Dredging, November 7, 1956; "Draft Source-Book Paper: St. Lawrence Project: Cornwall Island Dredging, January 17, 1957."
28 The original name "Long Sault Canal" was changed to "Wiley-Dondero Canal" in February 1958. The "Grasse River" Lock was changed to the "Bertrand H. Snell" Lock at the same time. The "Robinson Bay" Lock had been changed to the "Dwight D. Eisenhower" Lock in 1956.
29 The standard work on Moses is the Pulitzer Prize-winning tome by Robert Caro. Robert Caro, *The Power Broker: Robert Moses and the Fall of New York* (New York: Vintage, 1975).
30 LAC, RG 25, 6778, file 1268-D-40, pt. 45, St. Lawrence Seaway and Power Project – General File, June 16-July 19, 1955, Notes of Meeting with US Officials on St. Lawrence Seaway and Power Project, June 17, 1955.
31 LAC, RG 25, 7735, file 12279-40, pt. 1. St. Lawrence Seaway – Roosevelt Bridge, May 15-September 28, 1955, Notes of Meeting with US officials on St. Lawrence Seaway, June 20, 1955.
32 Chevrier, *The St. Lawrence Seaway*, 103.
33 LAC, RG 12, 1475, "Report of 1947 Board of Engineers on Lachine Section of the St. Lawrence Waterway."
34 LAC, RG 25, 6778, file 1268-D-40, pt. 46, St. Lawrence Seaway and Power Project – General File, July 21-September 30, 1955, SLSA press release 62, August 17, 1955.

35 LAC, RG 12, 3746, file 11300, Various Claims by South Shore Municipalities re: the St. Lawrence Seaway; LAC, RG 12, 3746, file 8122-58, St. Lawrence Waterway, Studies re: Lands and Services of the South Shore Municipalities as Affected by the St. Lawrence Seaway, 2. April 30, 1959.
36 Chevrier, *The St. Lawrence Seaway*, 107.
37 Government of Ontario, Ontario Provincial Archives (hereafter Ontario), RG 19-61-1, Municipal Affairs, Research Branch – Special Studies, St. Lawrence Seaway Study, box 21, file 14.1, St. Lawrence Seaway – General, West to Bunnell re: Township of Cornwall, Ontario, April 5, 1955; Ontario, RG 19-61-1, Municipal Affairs, Research Branch – Special Studies, St. Lawrence Seaway Study, box 21, file 14.1, St. Lawrence Seaway – General, "Expropriation Starts on Key Seaway Acres," *Globe and Mail*, April 7, 1955.
38 See Laurence M. Hauptman, *The Iroquois Struggle for Survival: World War II to Red Power* (Syracuse, NY: Syracuse University Press, 1986); Gerald R. Alfred (Taiaiake), *Heeding the Voices of Our Ancestors: Kahnawake Mohawk Politics and the Rise of Native Nationalism* (Toronto: Oxford University Press, 1995); Omar Z. Ghobashy, *The Caughnawaga Indians and the St. Lawrence Seaway* (New York: Devin-Adair, 1961); Kallen Martin, "Akwesasne Environments, 1999: Relicensing a Seaway after a Legacy of Destruction," *Native Americas* 16, 1 (1999): 24-27; Stephanie K. Phillips, "The Kahnawake Mohawks and the St. Lawrence Seaway" (MA thesis, McGill University, 2000); "Kahnawà:ke Revisited: The St. Lawrence Seaway," directed by Kakwirano:ron Cook, Kakari:io Pictures and Millennium Productions, 1999. The Kahnawake community also hired legal counsel, Joan Holmes and Associates, to appeal historical grievances arising from the seaway.
39 LAC, RG 25, 7735, file 12279-40, pt. 1., St. Lawrence Seaway – Roosevelt Bridge, DEA Memorandum: St. Lawrence Project: Relocation of Roosevelt Bridge, July 14, 1955; LAC, RG 25, 7735, file 12279-40, pt. 1., St. Lawrence Seaway – Roosevelt Bridge, DEA Memorandum, St. Lawrence Project: Relocation of Roosevelt Bridge, July 15, 1955.
40 Alfred, *Heeding the Voices*, 64-67.
41 Ibid., 64.
42 Mabee, *The Seaway Story*, 50.
43 Chevrier, *The St. Lawrence Seaway*, 104.
44 Ibid.
45 Alfred, *Heeding the Voices*, 65-67; Phillips, "The Kahnawake Mohawks," 12.
46 LAC, RG 25, 6779, file 1268-D-40, pt. 50, St. Lawrence Seaway and Power Project – General File, October 1, 1956-March 18, 1957, SLSA press release 139, November 28, 1956.
47 The seaway channel cut into the river a thousand feet from the south shore, with a channel two hundred feet wide. At Kahnawake, several islands were joined together as part of this dike formation.
48 LAC, RG 25, 6779, file 1268-D-40, pt. 50, St. Lawrence Seaway and Power Project – General File, SLSA press release 157, November 11, 1957. The ten bridges and their bridge types are as follows: Saint-Lambert: two vertical lift spans, railway and highway; Côte Ste-Catherine: one rolling lift, highway; Caughnawaga: two vertical lift, railway; Melocheville: one swing span, railway; St. Louis: one vertical lift, railway and highway; Valleyfield: one vertical lift, railway and highway; Iroquois: one rolling lift, highway; Cornwall: one high-level suspension bridge, highway.
49 Fifty feet of this amount was provided by actually raising the bridge itself. The remaining 30 feet resulted from changing the 250-foot span crossing the seaway channel (between the ninth and tenth pier) from a deck span to a through span. LAC, RG 25, 6780, file

1268-D-40, 52, St. Lawrence Project General File, September 18, 1957-April 21, 1958, SLSA press release 176: New Seaway Span Goes into Jacques Cartier Bridge, October 20, 1957.
50 PASNY officials decided that they could afford only the Ogdensburg bridge or the Iroquois bridge, and the former was more useful and better connected with existing traffic routes on both sides of the border. PASNY, Minutes of Trustee Meetings, October 11, 1954.
51 According to Stagg, at the height of construction there were 9 dredges, 140 shovels and draglines, 80 scrapers, 400 tractors, and 730 trucks working in the IRS. Stagg, *The Golden Dream*, 188.
52 From Prescott to Cornwall the river falls about ninety-two feet (Galop Rapids: nine feet; Rapide Plat Rapids: twelve feet; Long Sault Rapids: thirty feet).
53 Mabee, *The Seaway Story*, 182.
54 On the Long Sault Rapids as a tourist destination in the nineteenth century, see Chapter 3 of Patricia Jasen, *Wild Things: Nature, Culture, and Tourism in Ontario, 1790-1914* (Toronto: University of Toronto Press, 1995).
55 St. Lawrence University Archives, St. Lawrence Seaway Series, Mabee Series (D. Canadian Material), box 66, Power Authority of the State of New York, "Land Acquisition on the American Side of the St. Lawrence Seaway and Power Projects," July 18, 1955.
56 HEPCO, SPP Series, Minutes of Meeting Held in the Uhl, Hal and Rich Office, Boston, January 13, 1955.
57 This involved work on Iroquois Point, such as improving the cemetery and its surroundings, providing a park area for visitors, and rounding the slopes of the spoil piles so they would blend more attractively with the general landscape. This cost $65,000, rather than the estimated $1 million it would cost to remove and dispose of the spoil, which HEPCO was not willing to spend "in the light of results that would be obtained and, further, there was no record of commitment in this respect." LAC, RG 52, vol. 113, file C5-1-4-2, pt. 3, Memorandum to J.E. Devine, May 16, 1963.
58 Grasse River Lock (Snell Lock): lift of forty-six feet; Robinson Bay Lock (Eisenhower Lock): lift of forty-two feet.
59 Becker, *From the Atlantic*, 62-64.
60 Parham, *St. Lawrence Seaway*, 128-32.
61 Eisenhower Archives, Robert B. Anderson Papers, 1933-89, box 58, file: Saint Lawrence Seaway, Annual Report of the Saint Lawrence Seaway Development Corporation – 1955.
62 The Canadian powerhouse alone involved the excavation of 1,796,000 cubic yards of earth and 171,000 cubic yards of rock; the structure required 947,000 cubic yards of concrete and 32,100 tons of structural and reinforcing steel.
63 An average of sixty people perished during the construction of the Panama Canal, Golden Gate Bridge, Hoover Dam, and Empire State Building. "Seaway Statistics in Meters, Cubic Yards, and Dollars," *Toronto Star Weekly*, 27 June 1959, reference in Stephen Doheny-Farina, *The Grid and the Village: Losing Electricity, Finding Community, Surviving Disaster* (New Haven, CT: Yale University Press, 2001), 109.
64 There were smoother labour relations on the Canadian side. Parham, *St. Lawrence Seaway*, 121.
65 By 1962, just over $3.5 million of the almost $12.5 million Canadian claims had been settled. LAC, MG 32 B-39, 127, file 9-6-1-1, Claims by Contractors – General, St. Lawrence Development Corporation – Resume of Major Construction Claims, January 1962.
66 HEPCO, SPP Series, Moses to Duncan (Chairman HEPCO), August 9, 1957.

67 SLSA built a series of observation towers and enclosures between Montreal and Iroquois for the public to see, including towers at the Saint-Lambert Lock, Côte Ste-Catherine Lock, Caughnawaga, Iroquois Point, and at two towers at Melocheville (just west of Beauharnois powerhouse and near upper lock).
68 Parham, *St. Lawrence Seaway*, 27.
69 The Canadians kept pouring concrete in the winter, whereas the Americans stopped. See Passfield, "Construction," 35-44.
70 Ibid., 23.
71 Scott M. Campbell, "Backwater Calculations for the St. Lawrence Seaway with the First Computer in Canada," *Canadian Journal of Civil Engineering* 36 (2009): 1164-69.
72 Stagg, *The Golden Dream*, 216.
73 Parham, *St. Lawrence Seaway*, 40.
74 See Parham, "Construction Dilemmas," in *The St. Lawrence Seaway and Power Project*.

CHAPTER 5: LOST VILLAGES

1 Running east from Iroquois: Matilda Township (between Iroquois and Morrisburg, a distance of approximately 10 miles, a number of farms and residences); Williamsburg Township (extending roughly between Morrisburg and Aultsville for 7.5 miles, predominantly farming but also included a large number of cottages); Osnabruck Township (approximately 9.5 miles along the river, including the hamlets of Aultsville, Farran's Point, Dickinson's Landing, and Wales); Cornwall Township (approximately 11 miles along the river and completely surrounds the City of Cornwall – a large portion was amalgamated into the City of Cornwall during the St. Lawrence project – and also included the flooded hamlets of Mille Roches and Moulinette).
2 For a useful technical and logistical overview of the rehabilitation project, see an article by one of the HEPCO engineers: J.H. Jackson, "The St. Lawrence Power Project Rehabilitation: A Review of Major Features," *Engineering Journal* 43, 5 (February 1960): 67-73.
3 The inundation phase was covered, albeit often perfunctorily, in the book-length histories that appeared within a decade of the opening of the seaway. The accounts produced in the first decade or two after the seaway was opened, such as Mabee, Chevrier, and Willoughby, "effectively effaced the villagers and denied them any agency in the story," according to Rosemary O'Flaherty. Willoughby, *The St. Lawrence Waterway*; Mabee, *The Seaway Story*; Chevrier, *The St. Lawrence Seaway*; Rosemary O'Flaherty, "Community Legacies: 50th Anniversary Seaway Celebrations," paper presented at Rifts in the Rapids: The St. Lawrence Seaway Then and Now panel, annual conference of the Canadian Historical Association, Montreal, June 1, 2010.

Local histories include Clive and Frances Marin's history, *Stormont, Dundas, and Glengarry, 1945-1978* (Belleville, ON: Mike Publishing, 1982); Rosemary Rutley's *Voices from the Lost Villages* (Ingleside, ON: Old Crone, 1998); Anne-Marie Shields's *Lost Villages, Found Communities: A Pictorial History of the Lost Villages of the St. Lawrence Seaway* (Cornwall, ON: Astro Printing, 2004); Jean C. Jeacle's *To Make a House a Home: The Story of Ingleside, Ontario* (Township of Osnabruck, ON, 1975); E.W. Morgan's *"Up the Front": A Story of Morrisburg* (Toronto: Ryerson, 1964). The Lost Villages Historical Society is also an excellent resource for local history. Locally produced fiction includes Maggie Wheeler's four historical crime/mystery novels set in the Lost Villages area – *The Brother of Sleep, All*

Mortall Things, A Violent End, and *On a Darkling Plain* (Renfrew, ON: General Store Publishing House, 2004, 2006, 2007, 2009) – and Jennifer DeBruin's *A Walk with Mary* (Renfrew, ON: General Store Publishing House, 2012). Anne Michaels's *Winter Vault,* which was nominated for the Giller Prize, and Johanna Skibsrud's *The Sentimentalists* (Vancouver: Douglas and McIntyre, 2010), which won the Giller Prize, also deals with the Lost Villages. A play titled "A Seaway Story" (written and directed by Janet Irwin and based on interviews with former residents of the Lost Villages) played in Cornwall in 2008. Artist Louis Helbig has done aerial photography of the submerged remains of the inundated villages, which can be seen on his website and have been shown at various galleries and exhibits: http://www.louishelbig.com/sunkenvillagesst.html.

4 Ontario, RG 19-61-1, Municipal Affairs, Research Branch – Special Studies, St. Lawrence Seaway Study, box 21, file 14.1.5, Minutes of Meetings – St. Lawrence Seaway #1, "Department Takes Over Seaway Area Planning; Hydro 'Put in Its Place,'" *Globe and Mail,* January 7, 1954.

5 For example, Northern Canada Power flooded out part of the Mattagami First Nation, which lived along the eponymous river, in the 1920s, and compensation was still an ongoing issue in the 1950s and 1960s. Jean L. Manore, *Cross-Currents: Hydroelectricity and the Engineering of Northern Ontario* (Waterloo, ON: Wilfrid Laurier University Press, 1999), chaps. 2 and 5. On national parks and expropriation, see Alan MacEachern, *Natural Selections: National Parks in Atlantic Canada, 1935-1970* (Montreal: McGill-Queen's University Press, 2001).

6 HEPCO, SPP Series, C.E. Blee (Chief Engineer, Tennessee Valley Authority) to J.R. Montague, Director of Engineering, HEPCO), February 5, 1952. On the TVA and resettlement, see also Michael J. McDonald and John Muldowny, *TVA and the Dispossessed: The Resettlement of Population in the Norris Dam Area* (Knoxville: University of Tennessee Press, 1982).

7 HEPCO, SPP Series, Norman Moore (American Society of Civil Engineers) to Otto Holden, (n.d.) 1952.

8 Ontario, RG 19-61-1, Municipal Affairs, Research Branch – Special Studies, St. Lawrence Seaway Study, box 21, file 14.1.5, Minutes of Meetings – St. Lawrence Seaway #1, Memorandum to Bunnell, September 16, 1952.

9 Ontario, RG 19-61-1, Municipal Affairs, Research Branch – Special Studies, St. Lawrence Seaway Study, box 21, file 14.1.5, Minutes of Meetings – St. Lawrence Seaway #1, Memorandum of Meeting (by E.B. Easson), September 30, 1952.

10 At the start of construction, the *Globe and Mail* complained that there was a lack of governmental consultation with local residents, though it soon rescinded this complaint. Opinion, "The People Should Plan, Too," *Globe and Mail,* May 20, 1954.

11 HEPCO, SPP Series, Radio Report: "Saunders over CFRB and other Ontario Stations," June 8, 1954.

12 HEPCO, SPP Series, "Cornwall Editors Speaks to Kiwanis Club on Human Aspect of Plans for Seaway," *Owen Sound Sun Times,* September 3, 1954.

13 St. Lawrence University Archives, St. Lawrence Seaway Series, box 66, Power Authority of the State of New York, "Land Acquisition on the American Side of the St. Lawrence Seaway and Power Projects," July 18, 1955.

14 Ibid. Carol Sheriff's study of the Erie Canal, one of a number of works on that waterway, makes for an interesting comparison with the seaway, since Sheriff focuses on various conceptions of progress, expropriation, and views of property in the context of a waterway

constructed by the State of New York a century earlier. Carol Sheriff, *The Artificial River: The Erie Canal and the Paradox of Progress, 1817-1862* (New York: Hill and Wang, 1996).
15 PASNY, Minutes of Trustee Meetings, May 26, 1955.
16 PASNY, Minutes of Trustee Meetings, July 18, 1955.
17 HEPCO, SPP Series, Supplementary Report to James S. Duncan (Chairman, and HEPCO Commissioners): "The Acquisition of Lands and Related Matters for the St. Lawrence Power Project," January 2, 1957.
18 HEPCO, SPP Series, Memorandum for file: St. Lawrence Project – Dikes, July 14, 1953.
19 Iroquois: 332; Matilda Township: 161; Morrisburg: 327; Williamsburg Township: 215; Osnabruck Township: 565; Cornwall Township: 435. HEPCO, SPP Series, Memorandum to Carrick: Property Transactions – St. Lawrence Seaway, July 12, 1954.
20 PASNY, Minutes of Trustee Meetings, June 7, 1954.
21 Ontario, RG 34-3, container 27R, file: St. Lawrence Waterway, file: St. Lawrence Seaway, 1948-June 1954, Clerk of Iroquois to Minister: Town of Iroquois: General Land Use Plan and Text, August 13, 1954.
22 Ontario, RG 19-61-1, Municipal Affairs, Research Branch – Special Studies, St. Lawrence Seaway Study, box 21, file 14.1.5, Minutes of Meetings – St. Lawrence Seaway #1, Joint Submission to Ontario Hydro, August 12-13, 1954. [Note: Where original documents use "Ontario Hydro" that usage will be retained.]
23 Don O'Hearn, "Confuse Methods of Compensation," *Port Arthur News-Chronicle*, July 19, 1954.
24 HEPCO, SPP Series, copy of Editorial, "It's an Old Story," *Peterborough Examiner*, August 26, 1954.
25 HEPCO, SPP Series, Memorandum to Chairman: Rehabilitation – St. Lawrence River – Power Project, August 6, 1954.
26 Mabee, *The Seaway Story*, 207.
27 For a study of expropriation in Canada, see Eric C.E. Todd, *The Law of Expropriation and Compensation in Canada*, 2nd ed. (Toronto: Carswell, 1992).
28 Freeman, *The Politics of Power*, 105.
29 HEPCO, SPP Series, copy of "Cornwall Editors Speaks to Kiwanis Club on Human Aspect of Plans for Seaway," *Owen Sound Sun Times*, September 3, 1954.
30 HEPCO, SPP Series, copy of Reginald Hardy, "Many Fine Homes Soon to Disappear – St. Lawrence Families' Plight," *Ottawa Evening Citizen*, August 11, 1954.
31 Ontario, RG 34-3, container 27R, file: St. Lawrence Waterway, file: St. Lawrence Seaway, 1948-June 1954, Memorandum to Bunnell, Subject: Preliminary Survey, St. Lawrence Area (September 13, 14, 15, 16, 17, 1954), September 23, 1954.
32 The massive frequency standardization effort, which began in 1949 and took close to a decade to finish, involved converting from 25-cyle to 60-cyle electricity and saw HEPCO give out legions of new appliances in order to induce conversion and provide a market for electricity. See Freeman, *The Politics of Power*.
33 Ontario, RG 34-3, container 27R, file: St. Lawrence Waterway, file: St. Lawrence Seaway, 1948-June 1954, Warrender to Premier Frost re: Dr. Wells Coates, August 17, 1954.
34 Ibid.
35 HEPCO, SPP Series, HEPCO press conference, St. Lawrence – Rehabilitation of Iroquois and Matilda, October 13, 1954.
36 HEPCO, SPP Series, St. Lawrence Rehabilitation, October 7, 1954. In the end, 152 homes were moved to the new sites, and 86 new buildings were built. Richardson, Rooke, and McNevin, *Developing Water Resources*, 28.

37 HEPCO, SPP Series, St. Lawrence Rehabilitation: Meeting at Osnabruck, November 23, 1954.
38 HEPCO, SPP Series, St. Lawrence Rehabilitation: Meeting with Matilda Township, November 22, 1954.
39 HEPCO, SPP Series, St. Lawrence Rehabilitation: Meeting at Osnabruck, November 23, 1954.
40 Ontario, RG 19-61-1, Municipal Affairs, Research Branch – Special Studies, St. Lawrence Seaway Study, box 21, file 14.1.5, Minutes of Meetings – St. Lawrence Seaway #1, Memorandum of Meeting re: Iroquois, December 21, 1954.
41 In some cases, the council of a township was involved with the planning of the town, and there were obvious conflicts of interest. Ontario, RG 19-61-1, Municipal Affairs, Research Branch – Special Studies, St. Lawrence Seaway Study, box 21, file 14.1.5, Minutes of Meetings – St. Lawrence Seaway #1, Progress Report: "Current Situation in Seaway Valley," July 15, 1955; HEPCO, SPP Series, St. Lawrence Board of Review: Minutes of a Meeting in the Parliament Buildings, Toronto, February 2, 1955.
42 HEPCO, SPP Series, Memorandum to Lamport: House to House Survey – Village of Farran's Point, February 10, 1955.
43 These included Iroquois, Morrisburg, Aultsville, Farran's Point, Dickinson's Landing, Wales, and Moulinette. HEPCO, SPP Series, House to House Survey of Communities Affected by the St. Lawrence Power Development, February 25, 1955.
44 HEPCO, SPP Series, House to House Survey: Village of Wales, February 17, 1955.
45 Ontario, RG 19-61-1, Municipal Affairs, Research Branch – Special Studies, St. Lawrence Seaway Study, box 21, file 14.1.5, Minutes of Meetings – St. Lawrence Seaway #1, copy of Clark Davey, "Compensation Still Biggest Worry: Seaway Valley," *Globe and Mail*, April 26, 1955.
46 Members of council complained that there was undue pressure by agents on property owners; no breakdown on how property values were arrived at; apparent wide discrepancies in offers made to property owners who had properties of similar value; and, in many instances, the amount of 15 percent for forcible taking was not enough.
47 HEPCO, SPP Series, Meeting of the St. Lawrence Fruit Growers' Association with HEPCO, November 2, 1955.
48 HEPCO, SPP Series, General Comments re: Fact Finding Survey – Iroquois and Morrisburg, 23 December 1954.
49 J.H. Jackson, "Rehabilitation," 70-71.
50 HEPCO built observation stations at the Iroquois Control Dam and the Cornwall power dam. Visitor statistics by year: 1955: 28,700; 1956: 315,400; 1957: 219,500; 1958: 1,030,00. David Nelson Scharf, "The Effect of the St. Lawrence Seaway Project on Recreational Land Use in the International Rapids Section of the Seaway Valley" (MA thesis, Carleton University, 1970).
51 HEPCO, SPP Series, Draft: St. Lawrence Power Project Rehabilitation, Meeting held in Morrisburg, October 17, 1955.
52 The naming of the towns was a matter of debate. For example, see Marin and Marin, *Stormont, Dundas, and Glengarry,* 32-33.
53 According to Mabee, Morrisburg citizens had the most direct participation in the planning. Mabee, *The Seaway Story,* 215-16.
54 There were exceptions, such as Robin Cross, who opened a marina near Long Sault. According to his daughter, it was likely because of his personal connection with Harry

Notes to pages 158-62   269

    Hustler, as Cross gave Hustler's daughter rides to school. This led to some resentment in the community. Phone interview with Bonnie Clarke, 15 June 2011. HEPCO, SPP Series, Report of Meeting in Morrisburg (August 9, 1956), Outstanding Problems Related to the Rehabilitation Problem in the St. Lawrence Seaway Valley (Chairman: AEK Burnell, Consultant, Ontario Department of Planning and Development), August 31, 1956.

55 Challies had been one of the key HEPCO representatives initially dealing with the people and councils, which caused local controversy.
56 Dennis Dack, a speechwriter at the time for Saunders, speculates that Frost may have envied Saunders's popularity and influence with Prime Minister St. Laurent. Interview with Dennis Dack, Toronto, May 2, 2011.
57 The nature of liberalism, particularly as it relates to property, is a prominent theme in Canadian historiography in recent years, in large part because of Ian McKay's "liberal order framework" proposal in which Canada can be interpreted as a project of liberal rule. This theory, which McKay does not take beyond 1945, stresses the supremacy of the individual and posits that the state's ability to exert its hegemony rests on three fundamental liberal values: (1) equality before the law; (2) the enjoyment of certain civil liberties; and (3) the sacred right to private property, to which the first two are subordinated. See Ian McKay, "The Liberal Order Framework: A Prospectus for a Reconnaissance of Canadian History," *Canadian Historical Review* 81, 3 (2000): 617-45; Jean-Francois Constance and Michel Ducharme, eds., *Liberalism and Hegemony: Debating the Canadian Liberal Revolution* (Toronto: University of Toronto Press, 2009), 9. On earlier engagements with Canadian liberalism see also Fernande Roy, *Progrès, harmonie, liberté* (Montreal, Boréal, 1988); Tina Loo, *Making Law, Order, and Authority in British Columbia, 1821-1871* (Toronto: University of Toronto Press, 1994). In addition to the Castonguay and Kinsey chapter in *Liberalism and Hegemony*, other studies that explicitly connect liberalism and environmental history in Canada include James Murton, *Creating a Modern Countryside: Liberalism and Land Resettlement in British Columbia* (Vancouver: UBC Press, 2007); Stunden Bower, *Wet Prairie*.
58 HEPCO, SPP Series, Report on the Acquisition of Lands and Related Matters for the St. Lawrence Power Project (by Property Office), 1955-56.
59 Mabee, *The Seaway Story*, 210.
60 HEPCO, SPP Series, Moses to Kuykendall, July 26, 1955.
61 HEPCO, SPP Series, Meeting of the St. Lawrence Fruit Growers' Association with HEPCO, November 2, 1955.
62 Ibid.
63 HEPCO, SPP Series, Moses to McEwen, August 25, 1955.
64 HEPCO, SPP Series, J.J. Wingfelder (St. Lawrence Board of Review) to Secretary HEPCO, November 10, 1955.
65 HEPCO, SPP Series, Memorandum to Commission from Hustler re: W.H. Reddick, Part of Lot 7, Concession I, Township of Matilda, April 27, 1956.
66 Interview with Joan McEwan, Long Sault (Lost Villages Historical Society Museum), June 22, 2011.
67 Eighty-four units of row housing were built once moved homes were exhausted as options for renters.
68 Ontario, RG 19-61-1, Municipal Affairs, Research Branch – Special Studies, St. Lawrence Seaway Study, box 21, file 14.1.2, St. Lawrence Seaway Authority, Ontario–St. Lawrence Development Commission, Progress Report, 2, January 27, 1956. According to Harold A.

Wood, before the St. Lawrence project, there were 246 cottages on the portions of the riverfront unencumbered by the old canals, and only one cottage on the old canals. Harold A. Wood, "Recreational Land Use Planning in the St. Lawrence Seaway Area, Ontario," *Community Planning Review,* March 1955: 24.
69 Maple Grove was apparently given its own cemetery because residents had to be moved earlier than those from the rest of the Lost Villages, since this hamlet just west of Cornwall stood in the way of powerhouse and dike construction.
70 PASNY, Minutes of Trustee Meetings, July 13, 1955.
71 "Town Loses $4 Million Valuation," *Massena Observer,* August 8, 1955, 1.
72 For recollections revealing animosity toward the PASNY agents see Heather M. Cox et al., "Drowning Voices and Drowning Shoreline: A Riverside View of the Social and Ecological Impacts of the St. Lawrence Seaway and Power Project," *Rural History* (1999): 242-43.
73 PASNY, Minutes of Trustee Meetings, February 17, 1958.
74 PASNY, Minutes of Trustee Meetings, July 19, 1956.
75 PASNY, Minutes of Trustee Meetings, December 19, 1956.
76 Interviews with Grace McBath, Kathy Dupray, Donna Dunn, Alice Dumas, Nancy Badlam, and Helen Badlam, Waddington, December 6, 2012.
77 A Waddington milk processing plant close to the river was also lost. "Historic Waddington Island House Must Give Way to St. Lawrence Seaway Work," *Massena Observer,* August 8, 1955, 5; Interviews with Grace McBath, Kathy Dupray, Donna Dunn, Alice Dumas, Nancy Badlam, and Helen Badlam, Waddington, December 6, 2012.
78 Interview with Russell Strait, Norfolk, May 21, 2012.
79 Excavation of 845,000 cubic yards of material at an estimated cost of $760,500,00 near the Crysler Park Memorial. Ontario, RG 10-61-1, Municipal Affairs, Research Branch – Special Studies, St. Lawrence Seaway Study, 23, file: 14.1.14 #1: Foreshore Improvements, St. Lawrence Seaway, James Duncan (Chairman HEPCO) to Premier Frost: Shore Line Improvements, St. Lawrence Development, March 19, 1957; Ontario, RG 10-61-1, Municipal Affairs, Research Branch – Special Studies, St. Lawrence Seaway Study, 23, file: 14.1.14 #1: Foreshore Improvements, St. Lawrence Seaway, Report on Shore Grading on the St. Lawrence River from the Advisory Committee to the Parks Integration Board (by W.B. Greenwood, Chairman, Advisory Committee to the Parks Integration Board), May 7, 1957.
80 St. Lawrence University Archives, St. Lawrence Seaway Series, Mabee Series (D. Canadian Material), box 66, Power Authority of the State of New York, "Land Acquisition on the American Side of the St. Lawrence Seaway and Power Projects," July 18, 1955.
81 In August 1954, the Ontario Department of Game and Fisheries had, in conjunction with discussions on fishways, suggested a bird sanctuary. HEPCO, SPP Series, St. Lawrence Power Project, Discussion with Mr. J.D. Millar by Mr. G. Mitchell and Dr. Holden on July 30, 1954, in Dr. Holden's Office, August 10, 1954.
82 For more specific information on the development of the recreational areas connected to the St. Lawrence project consult Scharf, "Effect of the St. Lawrence Seaway Project"; Wood, "Recreational Land Use Planning"; Anthony Adamson, "Crysler's Park" *Canadian Architect* 3, 2 (February 1958): 41-42.
83 LAC, RG 25, 6779, file 1268-D-40, pt. 49, St. Lawrence Seaway and Power Project – General File (May 31-September 26, 1956), "Seaway Valley Residents Satisfied with Hydro's Efforts in Moving Homes to New Sites," *Globe and Mail,* July 11, 1956.
84 "Seaway Valley Residents Satisfied with Hydro's Efforts in Moving Homes to New Sites," *Globe and Mail,* July 11, 1956.

85 HEPCO, SPP Series, Report of Meeting in Morrisburg (August 9, 1956), Outstanding Problems Related to the Rehabilitation Problem in the St. Lawrence Seaway Valley (Chairman: AEK Burnell, Consultant, Ontario Department of Planning and Development), August 31, 1956.
86 HEPCO, SPP Series, Supplementary Report to James S. Duncan (Chairman, and HEPCO Commissioners): "The Acquisition of Lands and Related Matters for the St. Lawrence Power Project," January 2, 1957.
87 HEPCO, SPP Series, Ontario Hydro News Release no. 86: Open Morrisburg Shopping Centre and Move Last House in Hydro's St. Lawrence Rehabilitation Program, December 5, 1957.
88 *Cornwall Daily Standard Freeholder,* July 2, 1958.
89 Interview with Joan McEwan, Long Sault (Lost Villages Historical Society Museum), June 22, 2011; interview with Jane Craig, Long Sault (Lost Villages Historical Society Museum), June 22, 2011.
90 Ontario, RG 10-61-1, Municipal Affairs. Research Branch – Special Studies, St. Lawrence Seaway Study 23, file R-1-1-1, Memorandum to M.H. Sinclair (Area Studies Group), St. Lawrence Area Field Trip. From A.H. Holmes (Area Studies Group) re: St. Lawrence Area Field Trip, July 17, 1959.
91 Iroquois: 151; Matilda Township: 28; Morrisburg: 93; Williamsburg Township: 21; Ingleside: 108; Long Sault: 130.
92 HEPCO, SPP Series, Hustler to Duncan: St. Lawrence Power Project Lands, July 11, 1958.
93 Ibid.
94 J.H. Jackson, "Rehabilitation," 67. See also Alan MacEachern's account of expropriations for national parks in eastern Canada in roughly the same period, where those losing land displayed a similar deference to governmental power. MacEachern, *Natural Selections.*
95 University of Toronto, School of Social Work, *Roundtable on Man and Industry, Community Survey Report, St. Lawrence Impact Area,* vol. 4, pt. 2, Town of Iroquois, 40, 39, 41.
96 For an interesting comparison see Loo, "People in the Way," 170-72.
97 Many oral interviews have been done since the 1970s with former Lost Villagers, including those done by this author. These interviews tend to reflect a certain generation (e.g., the older generations from the 1950s are generally not available or able to share their memories). In addition, for various reasons, over the past few decades the same people have tended to be interviewed – this means there is a chance that certain perspectives seem more dominant, though interviews with this range of people does produce different and even contradictory responses to many of the same questions.
98 Mabee, *The Seaway Story,* 239.
99 Trailer parks sprouted up in Massena. For example, in 1957 10 percent of Massena lived in trailer courts. Ibid., 242.
100 Ibid., 238-40.
101 See Rutley, *Voices*; Shields, *Lost Villages, Found Communities*; and the interviews conducted by the author.
102 Parr, chap. 4, "Movement and Sound: A Walking Village Remade: Iroquois and the St. Lawrence Seaway," in *Sensing Changes.*
103 Ibid., 94. See also Loo, "Disturbing the Peace," 906-8.
104 Rutley, *Voices,* 100.

105 Ontario, RG 34-3, container 27R, file: St. Lawrence Waterway, file: St. Lawrence Seaway, 1948-June 1954, Memorandum to Bunnell, Subject: Preliminary Survey, St. Lawrence Area (September 13, 14, 15, 16, 17, 1954), September 23, 1954.
106 Writing as the St. Lawrence project was completed, Peter Stokes, who was critical of many elements of the rehabilitation, contends that "the improvement of the loop streets are unappreciated since the previous towns weren't large enough to appreciate the traffic hazards of the old grid system." Peter Stokes, "St. Lawrence, a Criticism," *Canadian Architect* 3, 2 (February 1958): 43-48. See also Sarah Bowser, "The Planner's Part," *Canadian Architect* 3, 2 (February 1958): 38-40.
107 In addition to first-hand accounts of the construction process, Claire Parham's oral history of the construction of the St. Lawrence project contains fascinating insights into the social and cultural aspects. Parham, *St. Lawrence Seaway,* 68.
108 Ibid., 71.
109 Interview with Jim Brownell, Long Sault (Lost Villages Historical Society Museum), May 16, 2011.
110 The Lost Villages had one-room schoolhouses, and the schools were combined into new structures in the new towns. Churches were also replaced, and in many cases congregations from the different municipalities were combined in the new towns. One church slated for inundation, the Holy Trinity Memorial Church, was dismantled by its congregants and reassembled out of the path of the flooding.
111 Interview with Bonnie Clarke, Long Sault (Lost Villages Historical Society Museum), June 22, 2011.
112 Interview with Lyall Manson, Long Sault (Lost Villages Historical Society Museum), June 22, 2011.
113 Joan McEwan was a teacher at Moulinette and then Ingleside. Interview with Joan McEwan, Long Sault (Lost Villages Historical Society Museum), June 22, 2011.
114 Interview with David Hill, Long Sault (Lost Villages Historical Society Museum), June 22, 2011.
115 An example is Morgan's 1964 memoir-history *"Up the Front."*
116 Elderly citizens dying soon after their removal and the inundation is a common theme in oral histories and interviews.
117 Interview with Vale Brownell, Long Sault (Lost Villages Historical Society Museum), June 22, 2011.
118 A CD by James Gordon (and the students of Rothwell-Osnabruck School) titled *Songs from the Lost Villages of the St. Lawrence* is available for purchase from the Lost Villages Historical Society, along with a songbook. Some of these songs, as well as poetry, can be accessed at the *Ottawa Citizen*'s online feature (created in 2008) commemorating the Lost Villages: http://www2.canada.com/ottawacitizen/.
119 Lost Villages Historical Society, "Lost Villages," http://www.lostvillages.ca/. As Rosemary O'Flaherty describes it, "The movement of people resulted in a hybrid spatialization that spawned the idea of the 'Lost Villages,' a historically specific idea that constructed an alternative identity for the former villagers. Identity became equated with lost place and the sacralization of that place confirmed its importance and, hence, strengthened their sense of attachment to the disappeared landscape ... The transmutation of landscape to riverscape in the Seaway Valley is an iconic narrative that tells a story about a place and its various representations at particular moments in time." O'Flaherty, "Community Legacies," 8.

120 IJC, Canadian Section, docket 68-8-1:2, St. Lawrence Power Application, Material Distributed at the Opening of the St. Lawrence Power Project, Luncheon Address by James S. Duncan, Chairman, Ontario Hydro, Official Opening St. Lawrence Power Project, September 5, 1958.

121 Iroquois named a crescent after Hustler, reflecting the level of respect for him in that community.

122 Nelles, *The Politics of Development*, 425, 465.

123 HEPCO, SPP Series, General Comments re: Fact Finding Survey – Iroquois and Morrisburg, December 23, 1954.

124 HEPCO, SPP Series, Letter from Duncan to Miss I.K. Farlinger, October 10, 1957.

125 Cox et al., "Drowning Voices," 242-43.

126 HEPCO allocated 29 percent of all its power to industry, while Quebec Hydro only allocated 13 percent of its Beauharnois power to industry. Mabee, *The Seaway Story*, 232-33. On Alcoa, see George David Smith, *From Monopoly to Competition: The Transformation of Alcoa, 1888-1986* (New York: Cambridge University Press, 1988).

127 Cox et al., "Drowning Voices," 241-42.

128 Mabee, *The Seaway Story*, 204.

129 Todd, *The Law of Expropriation*, 2.

130 Patrick McGreevy has made this contention regarding another border waterway, the Niagara River and Niagara Falls, and it is also applicable to the St. Lawrence River. Patrick McGreevy, *Imagining Niagara: The Meaning and Making of Niagara Falls* (Amherst: University of Massachusetts Press, 1994).

131 Interview with Russell Strait, Norfolk, May 21, 2012.

132 Eventually some additional private property along the water was allowed, such as east of Morrisburg.

133 Ontario, RG 10-61-1, Municipal Affairs, Research Branch – Special Studies, St. Lawrence Seaway Study 23, file: 14.1.14 #1: Foreshore Improvements, St. Lawrence Seaway, Memorandum to Challies, from M.A.R. Laird, May 29, 1956.

134 Interview with David Hill, Long Sault (Lost Villages Historical Society Museum), June 22, 2011.

135 On the issue of mobility and the St. Lawrence Seaway and Power Project, see Daniel Macfarlane, "Creating the Seaway: Mobility and a Modern Megaproject," in *Moving Natures: Environments and Mobility in Canadian History*, ed. Ben Bradley, Colin Coates, and Jay Young (Calgary: NiCHE-University of Calgary Press Environmental History Series, forthcoming).

136 HEPCO, SPP Series, Remarks of Robert Moses at the Canadian Club, Château Laurier, Ottawa, on "The International St. Lawrence," June 13, 1956.

Chapter 6: Flowing Forward

1 It consisted of 3.2 million cubic yards of concrete; 2 million tons of sand; 3.2 million tons of stone; 28,000 tons of structural steel; and 20,200 tons of gates, hoists, and cranes. IJC, Canadian Section, docket 68-8-1:2, St. Lawrence Power Application, Material Distributed at the Opening of the St. Lawrence Power Project, St. Lawrence Power Project – Information Booklet (1958).

2 IJC, Canadian Section, docket 67-2-5:6, Lake Ontario Levels Reference, Meetings, McNaughton, Burbridge, Cote 1953/01/16, Memorandum to General McNaughton re:

August 29, 1952, meeting, September 2, 1952. A number of claims were made by US Lake Ontario shore owners because of Gut Dam. They unsuccessfully tried to sue Canada, and requested that the US Foreign Claims Settlement Commission examine the claims. Finally, in 1968, Canada agreed to pay a token $350,000 as settlement for the alleged damage. Carl F. Goodman, "Canada–United States Settlement of Gut Dam Claims: Report of the Agent of the United States before the Lake Ontario Claims Tribunal," *International Legal Materials* 8, 1 (January 1969): 118-43.

3 These experts were discovering, largely for the first time, that levels in the Great Lakes have always fluctuated under the influence of natural forces including the major ones of precipitation and evaporation and also winds, barometric pressure, ice jams, glacial rebound, aquatic weed growth and, to some extent, tides. Long-term fluctuations occur over periods of consecutive years and have varied dramatically since water levels have been recorded for the Great Lakes. Continuous wet and cold years will cause water levels to rise. Conversely, consecutive warm and dry years will cause water levels to decline. The Great Lakes system experienced extremely low levels in the late 1920s, mid-1930s, again in the mid-1960s, and currently in the upper Great Lakes. Extremely high water levels were experienced in the 1870s, early 1950s, early 1970s, mid-1980s, and mid-1990s.

4 LAC, RG 25, 6352, file 1268-AD-40, pt. 1, St. Lawrence Project: Dredging at Cornwall Island (December 1, 1954, to March 25, 1955), Memorandum for the Minister – St. Lawrence Project, January 24, 1955.

5 The long-term average flow (1860-1954) was determined to be 240,000 cfs, which was about 4,000 cfs more than the average used for Method of Regulation 5. IJC, Canadian Section, docket 68-5-1, St. Lawrence Project, Miscellaneous Memoranda, March 1954 – Memorandum. Studies showed that the impact of the levels of Gut Dam had been exaggerated and was really about four and a half inches, which was approximately half of what had been believed by some. IJC, Canadian Section, 68-2-5:6-1: St. Lawrence Power Application, Minutes of IJC Meetings, July 1952 and April 1962, St. Lawrence Power Application: Modification of Order of Approval (Executive Session, Boston), April 9, 1954; "Effects on Lake Ontario Water Levels of the Gut Dam and Channel Changes in the Galop Rapids Reach of the St. Lawrence River, Main Report," Report to the International Joint Commission by the International Lake Ontario Board of Engineers, October 1958.

6 IJC, Canadian Section, 68-3-V2: St. Lawrence Power Application, Correspondence from January 1 to December 21, 1954, Henry to McNaughton re: 238-242, Controlled Single Stage Project, International Rapids Section, May 12, 1954.

7 IJC, Canadian Section, 68-3-V2: St. Lawrence Power Application, Correspondence from January 1 to December 21, 1954, Memorandum of Meeting, July 3, 1954.

8 There is a large body of literature on scientific uncertainty and policy making in general, with much of it concerning the Great Lakes environment, such as Terence Kehoe, "The Burden of Proof: Pollution Control and Scientific Uncertainty," chap. 5 in *Cleaning Up the Great Lakes: From Cooperation to Confrontation* (Dekalb: Northern Illinois University Press, 1997). On expert scientific control of nature in the Canadian context see also: Stephen Bocking, *Nature's Experts: Science, Politics, and the Environment* (New Brunswick, NJ: Rutgers University Press, 2004); Dean Bavington, *Managed Annihilation: An Unnatural History of the Newfoundland Cod Collapse* (Vancouver: UBC Press, 2010).

9 The regulation criteria outlined that the water level of Montreal Harbour would be no lower than would have occurred if the power project had not been built. Bryce, *Hydraulic Engineering History*, 94; LAC, RG 25, 6778, file 1268-D-40, pt. 43.2, St. Lawrence Seaway

and Power Project – General File, DEA Memorandum: Lake Ontario Levels, April 26, 1955.
10 IJC, Canadian Section, docket 68-2-5:1-9, St. Lawrence Power Application. Executive Session April 1957 and October 1957, IJC, St. Lawrence Power Development, Semi-Annual Meeting (Washington), April 9, 1957.
11 IJC, Canadian Section, St. Lawrence Power Application, Model Studies – I, Importance to Canada of the Construction of a Hydraulic Model for the Determination of the Effects of the Gut Dam and Channel Improvements in the Galop Rapids Section of the St. Lawrence River (McNaughton), December 9, 1953.
12 IJC Canadian Section, docket 68-8-6:3, St. Lawrence Power Application, FPC in the United States Court of Appeals 1953-54, St. Lawrence Power Application, Model Studies – I, Associate Committee of the National Research Council on St. Lawrence River Models, draft, October 15, 1953.
13 LAC, RG 25, 6778, file 1268-D-40, pt. 45, St. Lawrence Seaway and Power Project – General File, DEA Memorandum: St. Lawrence Seaway and Power Project: Visit to Ontario Hydro Models, July 5, 1955.
14 IJC, Canadian Section, Committee on St. Lawrence River Model Studies, Progress Memorandum 2, National Research Council, January 15, 1958.
15 Ibid.
16 Tina Loo and Meg Stanley make this point in the context of British Columbia dam building in the 1960s and 1970s, calling it "high modernist local knowledge." Tina Loo with Meg Stanley, "An Environmental History of Progress: Damming the Peace and Columbia Rivers," *Canadian Historical Review* 92, 3 (September 2011): 399-427. Evenden, *Fish versus Power*, also provides an excellent discussion of the expertise and authority of fisheries scientists in the context of 1950s dam building.
17 According to Passfield, the SLSA alone employed 120 Canadian engineers; HEPCO used 66 engineers in design, and 50 more to supervise construction of the powerhouse. Passfield, "Construction," 41. On the use of gauges, see Bryce, *Hydraulic Engineering History*, 75-81.
18 Loo with Stanley, "An Environmental History"; Parr, *Sensing Changes*.
19 IJC, Canadian Section, docket 68-2-5, Joint Board of Engineers, 1, Meeting 28 of St. Lawrence River Joint Board of Engineers, July 3, 1958.
20 In 2012, after almost fifteen years of study, the IJC announced a new method of regulation, titled Bv7, that allows for more natural fluctuation cycles and greater variability. An updated version of Bv7, named Plan 2014, then emerged which framers hope will better protect all interests through the use of trigger levels for adjusting Lake Ontario outflows during extreme water level fluctuations on the lake. As this book went to press, the IJC was still studying Plan 2014.
21 Six ice booms were subsequently installed to prevent such ice formations. IJC, Canadian Section, docket 68-3-V10, St. Lawrence Power Application, Correspondence re: Interim Measures Regulation, Memorandum of Telephone Conversation re: St. Lawrence, January 14, 1959; IJC, Canadian Section, docket 68-8-2:2, St. Lawrence Power Application, SLRJBE – Basic Documents, Brief to the St. Lawrence River Joint Board of Engineers on Ice-Boom Installation for St. Lawrence Power Project, June 10, 1959.
22 LAC, RG 25, 5026, file 1268-D-40, pt. 54, St. Lawrence Seaway Project – General File, January 8, 1960-February 27, 1962, Memorandum: Power Generation at Barnhart Island, October 13, 1961.

23 Method of regulation 1958-DD, which incorporates 1958-D, was eventually developed to deal with such fluctuating conditions. As noted above, this method is in the process of potentially being replaced.
24 Bryce, *Hydraulic Engineering History*, 108.
25 Dwight D. Eisenhower Library and Archives, Staff Files, Records of John S. Bragdon as Special Assistant to the President for Public Works Planning, 1949-61, Research Problem – General, Subject: Effect on Great Lakes and St. Lawrence River of an Increase of 1,000 Cubic Feet per Second in the Diversion at Chicago, January 29, 1957.
26 The channels connecting Lake Superior with Lake Huron and Lake Huron with Lake Erie had a controlling depth of only twenty-five feet downbound and twenty-one feet upbound. LAC, RG 25, 6777, file 1268-D-40, vol. 42.1, St. Lawrence Seaway Project – General File, DEA: Item for Weekly Division Notes, March 9, 1955.
27 Such as Toronto, Hamilton, Duluth-Superior, Cleveland, and Thunder Bay. See F.J. Bullock, *Ships and the Seaway* (Toronto: Dent, 1959).
28 LAC, RG 25, file 1268-AE-40, pt. 2.1, Tolls on the St. Lawrence Seaway, Memorandum of Agreement between the SLSDC and the SLSA respecting the St. Lawrence Seaway Tariff of Tolls, January 29, 1959.
29 Dwight D. Eisenhower Presidential Library and Archives, Central Files – President's Personal File, PPF, 1-F-120, box 59, file 1-F-123, Canada: St. Lawrence Seaway Celebration Left and Returned, June 26, 1959 (2), Formal Opening, Saint Lawrence Seaway (June 1959), July 7, 1958.
30 The walls were expected to cost about $7.5 million. LAC, RG 25, 5026, file 1268-D-40, vol. 53.2, St. Lawrence Seaway Project – General File, May 2, 1958-December 8, 1959, SLSA press release 206, August 21, 1959.
31 Nevertheless, 1,971 acres of land had been expropriated for $4 million by May 1966. See Jenish, *The St. Lawrence Seaway*, 39-40.
32 Canada finished its hydraulic work in December 1960, the US Army Corps of Engineers had finished its engineering work in November 1960, and all dredging was carried out by January 1961. Eisenhower Archives, White House Central Files, Official File, 1953-61, OF 3-JJ-2 Armed Forces Network, box 92, file: OF 3-JJ: Saint Lawrence Seaway Development Corporation (4); Castle to President, February 17, 1960.
33 Where this large number is derived from is unclear. Perhaps it included all service and ancillary jobs indirectly connected to construction.
34 LAC, RG 25, 6786, file 1415-40, pt. 4, FP 1, Text of Queen's Speech Delivered to the Opening of the St. Lawrence Seaway Ceremonies, June 26, 1959.
35 LAC, RG 25, 6786, file 1415-40, pt. 4, FP 1, Text of Remarks by President Eisenhower at St. Lawrence Opening Ceremonies, June 26, 1959.
36 LAC, RG 25, 5026, file 1268-D-40, vol. 54, St. Lawrence Seaway Project – General File, January 8, 1960-February 27, 1962, SLSA press release 223, 1960.
37 Traffic patterns according to the origin or destination of the voyages showed that 30.8 percent of the total movement was between two Canadian ports, one-third moved between Canada and US ports, and 36.6 percent consisted of foreign trade to and from Canada and the United States. LAC, RG 25, 5026, file 1268-D-40, pt. 54, St. Lawrence Seaway Project – General File, SLSA press release 235, April 18, 1961.
38 As of 1962, direct overseas trade accounted for 35 percent of total tonnage; 30 percent of total traffic moved between two Canadian ports; 37 percent moved between Canada and US ports, and 32 percent consisted of foreign trade to and from Canada and the United

States. Of the total cargo traffic, that of Canadian origin or destination accounted for 75 percent of the total. Canadian ships carried 56 percent of the cargo through the seaway in 1962, with 3.7 percent destined for the United States. LAC, RG 25, 5026, file 1268-D-40, pt. 54, St. Lawrence Seaway Project – General File, SLSA – Final Statistics for the 1962 Navigation Season, May 27, 1962.

39 LAC, RG 25, 5078, file 1268-AE-40, pt. 4, Memorandum from Rankin to Taylor, February 21, 1963. It is estimated that Canada spent $533 million in the seaway ($320 million in the Montreal-Lake Ontario section; $213 million in the Welland Canal section) between 1959 and 1966; during the same period, the United States spent $133 million, plus the cost of dredging and channel maintenance in the Great Lakes connecting channels. Richardson, Rooke, and McNevin, *Developing Water Resources*, 50.

40 For more on the history of the seaway in the post-1959 period, see Jenish, *St. Lawrence Seaway*; Stagg, *The Golden Dream*; Mabee *The Seaway Story*.

41 K.J. Rea, *The Prosperous Years: The Economic History of Ontario, 1939-1975* (Toronto: University of Toronto Press, 1985), 71.

42 Alexander, *Pandora's Locks*, xvii; Marc Levinson, *The Box: How the Shipping Container Made the World Smaller and the World Economy Bigger* (Princeton, NJ: Princeton University Press, 2006).

43 It was only in the twenty-first century that Canada formally renounced its all-Canadian seaway intentions.

44 For example, see Sussman, *Quebec and the St. Lawrence Seaway*, 30-32. Even before the completion of the seaway, the McGill-sponsored Montreal Research Council tentatively predicted that the waterway would have a beneficial economic impact on the city: Montreal Research Council, *The Impact of the St. Lawrence Seaway on the Montreal Area* (Montreal: Montreal Research Council, McGill University, 1958). See also Witol, *The St. Lawrence Seaway and Quebec*.

45 St. Lawrence Seaway Management Corporation, annual report 2011-12: *Delivering Economic Value,* http://grandslacs-voiemaritime.ca/.

46 Sussman, *The St. Lawrence Seaway*, 3.

47 HEPCO, 91.123, Department of Lands and Fisheries, Memorandum to Mr. F.A. MacDougall, Deputy Minister, re: Meeting – St. Lawrence and Ottawa Rivers Fish and Wildlife Studies, December 4, 1955; HEPCO, 91.123, Memorandum re: St. Lawrence Power Project: Conference with Dr. Harness and Dr. Clarke of Department of Lands and Forests on Proposed Fish and Wildlife Studies on Ottawa and St. Lawrence Rivers, December 7, 1955; HEPCO, 91.123, St. Lawrence Power Project, Discussion with Mr. J.D. Millar, Mr. G. Mitchell, and Dr. Holden on July 30, 1954, in Dr. Holden's Office, August 10, 1954.

48 HEPCO, 91.123, Department of Lands and Fisheries, Memorandum to Mr. F.A. MacDougall, Deputy Minister, re: Meeting – St. Lawrence and Ottawa Rivers Fish and Wildlife Studies, December 4, 1955; HEPCO, 91.123, Memorandum re: St. Lawrence Power Project: Conference with Dr. Harness and Dr. Clarke of Department of Lands and Forests on Proposed Fish and Wildlife Studies on Ottawa and St. Lawrence Rivers, December 7, 1955; HEPCO, 91.123, St. Lawrence Power Project, Discussion with Mr. J.D. Millar, Mr. G. Mitchell, and Dr. Holden on July 30, 1954, in Dr. Holden's Office, August 10, 1954.

49 "Remove Fish at Dewatered Dam," *Massena Observer*, August 8, 1955, 4.

50 In the 1990s, thirty-five professors and fifty students at the University of Ottawa conducted a $2.2 million study of the impact of the dam on fish habitat. "Hydro Dam Blamed for Tainting Fish," *Ottawa Citizen*, November 3, 1998.

51 Cox et al., "Drowning Voices," 249.
52 Great Lakes Fisheries Commission, "Fish Community Objectives for the St. Lawrence River," draft, December 14 (Ann Arbor, MI: Great Lakes Fisheries Commission, 2001), 6; J.R. Greeley and C.W. Greene, "Fishes of the Area with Annotated List," in New York State Conservation Department, *A Biological Survey of the St. Lawrence Watershed, Supplement to 20th Annual Report, 1930* (Albany: New York State Conservation Department, 1931).
53 Great Lakes Fisheries Commission, "Fish Community Objectives," 5.
54 Richard Carignan, "Positionner le Quebec dans l'Histoire Environnementale Mondiale/ Positioning Quebec in Global Environmental History," paper presented at the Rivers panel, Ecosystem Dynamics conference, Montreal, September 3, 2005.
55 Michèle Dagenais, *Montréal et l'eau: Une histoire environnementale* (Montreal: Boréal, 2011).
56 The authors of the study qualify that this lack of "floristic interest" might have come from the fact that no systematic survey had previously been done. W.G. Dore and J.M. Gillette, *Botanical Survey of the St. Lawrence Seaway Area in Ontario* (Ottawa: Canadian Department of Agriculture, 1955), 1.
57 Cox et al., Drowning Voices," 250.
58 PASNY, Minutes of Trustee Meetings, Appendix M: Memorandum from William Latham, Subject: St. Lawrence Waterfowl Management Project, July 19, 1956.
59 LAC, RG 25, 5026, file 1268-D-40, 54, St. Lawrence Seaway Project – General File, January 8, 1960-February 27, 1962, Memorandum to Deputy Minister re: Peaking Tests at St. Lawrence River Power Project, September 17, 1962.
60 IJC, Canadian Section, 68-3-V8 – St. Lawrence Power Application, Correspondence, 1958, letter from PASNY to Publisher of *Ogdensburg Advance News,* "Underground Leak at Power Pool Feared Threat to Seaway Hydro Energy," November 3, 1958.
61 Interview with Lyle Manson, Long Sault (Lost Villages Historical Society Museum), June 22, 2011.
62 Interview with Joan McEwan, Long Sault (Lost Villages Historical Society Museum), June 22, 2011.
63 Ian Bowering, *Cornwall: From Factory Town to Seaway City, 1900-1999,* vol. 1 (Cornwall, ON: Cornwall Standard Freeholder and Stormont, Dundas, and Glengarry Historical Society, 1999), 96.
64 Gregory G. Beck and Bruce Littlejohn, *Voices for the Watershed: Environmental Issues in the Great Lakes–St. Lawrence Drainage Basin* (Montreal: McGill-Queen's University Press, 2000).
65 Alexander, *Pandora's Locks,* xix.
66 On invasive species also see Willoughby, *The Joint Organizations*; Dorsey, *The Dawn of Conservation Diplomacy*; Margaret Beattie Bogue, *Fishing the Great Lakes: An Environmental History, 1783-1933* (Madison, WI: University of Wisconsin Press, 2000); Philip Scarpino, "Great Lakes Fisheries: International Response to Their Decline and the Lamprey/Alewife Invasion," in *The Political Economy of Water,* A History of Water 1, 2, ed. Terje Tvedt and Richard Coopey (London: I.B. Tauris, 2006).
67 Great Lakes Fisheries Commission, "Fish Community Objectives," 5, 16; S.P. Patch and W.D. Busch, "Fisheries in the St. Lawrence River – Past and Present: A Review of Historical Natural Resources Information and Habitat Changes in the International Section of the St. Lawrence River" (Cortland, NY: US Fish and Wildlife Service, 1984).
68 Nancy Langston, "What Happened to the Lake Trout? Land Use Change, Pollution, Sea Lampreys, and Climate Change in Lake Superior," draft paper presented at Border Flows:

A Workshop on Water Relations along the Canada-US Border, August 19, 2012, Kingston, Ontario.
69 White, *Long Sault Rapids*; HEPCO, 91.123, Memorandum for Chief Engineer re: Power Possibilities at Morrisburg, October 3, 1918.
70 James L. Wuebben, ed., "Winter Navigation on the Great Lakes: A Review of Environmental Studies," CRREL Report 95-10 (US Army Corps of Engineers: Cold Regions Research and Engineering Laboratory, May 1995).
71 IJC, Canadian Section, 68-3-1:2, St. Lawrence Power Application, Correspondence, C. McGrath, Joint Report by HEPCO and PASNY, "Assessment of Shoreline Erosion and Marshland Recession Downstream of the St. Lawrence Power Project," March 1983.
72 Hauptman, *The Iroquois Struggle*, 144; K. Martin, "Akwesasne Environments, 1999."
73 IJC, Canadian Section, 68-5-6, St. Lawrence Power Application, General Memorandum 1955, a Report to the International Joint Commission on Concerns of the St. Regis Band Regarding Impacts from the St. Lawrence Seaway and Power Development, March 1982; IJC, Canadian Section, 68-3-1:2, St. Lawrence Power Application, Correspondence, C. McGrath, Joint Report by HEPCO and PASNY, "Assessment of Potential Effects of Peaking/Ponding Operations at the St. Lawrence Power Project on Downstream Muskrat Populations," New York Power Authority and Ontario Hydro, March 1983.
74 IJC, Canadian Section, 68-3-1:2, St. Lawrence Power Application, Correspondence, C. McGrath, Joint Report by HEPCO and PASNY, "Assessment of Shoreline Erosion and Marshland Recession Downstream of the St. Lawrence Power Project," March 1983. However, an environmental advisor to the US section of the IJC labelled as "cavalier" the conclusions about the relationship between peaking/ponding and muskrats. This comes from a memorandum attached to the March 1983 report: Memorandum from Joel Fisher, Environmental Advisor, US Section, April 4, 1983. For more information on peaking and ponding, see Bryce, *Hydraulic Engineering History*, 102-6.
75 IJC, Canadian Section, 68-5-6, St. Lawrence Power Application, General Memorandum 1955, a Report to the International Joint Commission on Concerns of the St. Regis Band Regarding Impacts from the St. Lawrence Seaway and Power Development, March 1982.
76 The "Save the River" campaign gained prominence in the 1970s. Interestingly, the infamous Abbie Hoffman helped start this campaign while hiding from the law under the pseudonym "Barry Freed."
77 For a study on this relicensing process see Thomas Snider, *Power Dam Politics: Dealing with the Politics of Power at the Local Level; An Insider's Story* (Ashland, OH: Bookmasters, 2003).
78 The upper St. Lawrence falls under the purview of the 2008 Great Lakes–St. Lawrence River Sustainable Water Resources Agreement (an agreement including Ontario, Quebec, and the eight states bordering the Great Lakes, which builds on the 1968 Great Lakes Basin Compact and 1985 Great Lakes Charter). A settlement was reached in June 2008 in which Ontario Power Generation, the successor to HEPCO/Ontario Hydro, gave Akwesasne $45 million and several islands as compensation, though that award is still under appeal.

CONCLUSION: TO THE HEART OF THE CONTINENT

1 LAC, RG 25, file 1268-D-40, pt. 36.2, 6350, St. Lawrence Seaway Project: General File (August 26-September 30, 1954), Address by the Rt. Hon. Louis S. St-Laurent on the Celebration of the Tenth Anniversary of Le Syndicat des Journalistes de Langue Française de Montreal, October 23, 1954.

2 Ibid.
3 LAC, RG 25, file 1268-D-40, pt. 36.2, 6350, St. Lawrence Seaway Project: General File (August 26-September 30, 1954), letter from St. Laurent to Editor of the *Ottawa Journal*, October 27, 1954.
4 NARA II, RG 84, file 322.2, St. Lawrence Seaway (1952), US Embassy, Ottawa, Classified General Records, 1950-61, Perkins to Secretary, St. Lawrence Seaway and Power Project, November 1, 1952.
5 Memorandum by the Under Secretary of State (Webb) to the President, Subject: Visit of Prime Minister St. Laurent on September 28 to Discuss the St. Lawrence Seaway and Power Project, September 27, 1951, *FRUS, 1951*, vol. 2, Canada, 916.
6 Others who have already endeavoured to strip some of the lustre include Robert Bothwell, *Canada and the United States: The Politics of Partnership* (Toronto, University of Toronto Press, 1992), 48; Denis Stairs, "Realists at Work: Canadian Policy Makers and the Politics of Transition from Hot War to Cold War," in *Canada and the Early Cold War, 1943-1957*, ed. Greg Donaghy (Ottawa: Department of Foreign Affairs and International Trade, 1998); Norman Hillmer, "The Foreign Policy That Never Was, 1900-1950," in *Canada, 1900-1950: Un pays prend sa place/A Country Comes of Age*, ed. Serge Bernier and John MacFarlane (Ottawa: Organization for the History of Canada, 2003); Greg Donaghy, "Coming off the Gold Standard: Reassessing the 'Golden Age' of Canadian Diplomacy" (lecture, Johnson-Shoyama Graduate School of Public Policy, University of Saskatchewan, December 2009); Hector Mackenzie, "Golden Decade(s)? Reappraising Canada's International Relations in the 1940s and 1950s," *British Journal of Canadian Studies* 23, 2 (2010): 179-206.
7 In terms of Canadian-American relations, the transition from the St. Laurent to the Diefenbaker government was characterized more by continuity than change, at least until the John F. Kennedy administration took office in 1961.
8 St. Laurent tended to have good working relationships with most of his colleagues. See Thomson, *Louis St. Laurent*; Robert Bothwell and William Kilbourn, *C.D. Howe: A Biography* (Toronto: McClelland and Stewart, 1979); J.L. Granatstein, *A Man of Influence: Norman A. Robertson and Canadian Statecraft, 1929-68* (Toronto: Deneau, 1981), 250-53.
9 John Hilliker and Donald Barry, "The Pearson Years, 1946-1957," pt. 1 in *Coming of Age, 1946-1968*, vol. 2 of *Canada's Department of External Affairs* (Kingston: McGill-Queen's University Press, 1989). See also chapters on Hume Wrong, A.D.P. Heeney, Gerry Riddell, and other key members of DEA in *Architects and Innovators: Building the Department of Foreign Affairs and International Trade, 1909-2009*, Queen's Policy Studies Series, ed. Greg Donaghy and Kim Richard Nossal (Montreal and Kingston: McGill-Queen's University Press, 2009).
10 In addition to the already mentioned works by Tina Loo on the Columbia treaty and project see Richardson, Rooke, and McNevin, *Developing Water Resources*; Wilson, *People in the Way*; Swainson, *Conflict over the Columbia*; Richard White, *The Organic Machine*; Meg Stanley, *Voices from Two Rivers*; Barbara Cosens, ed., *The Columbia River Treaty Revisited: Transboundary River Governance in the Face of Uncertainty* (Corvallis: Oregon State University Press, 2012).
11 Loo, "People in the Way," 165.
12 LAC, RG 25, file 1268-D-40, pt. 13.1, St. Lawrence and Niagara River Treaty Proposals – General Correspondence (July 4-October 13, 1951), 6344, Memorandum, Meeting between the President and the Prime Minister of Canada on the St. Lawrence Project on September 28, 1951, September 28, 1951.

13 For example, the *Globe and Mail* conjectured that the abandonment of the all-Canadian seaway had a negative impact on the Liberal showing in the November 1954 by-elections that followed the 1954 seaway agreement. Opinion, George Bain, "Ottawa Letter," *Globe and Mail,* November 19, 1954.
14 It is striking that Ontario, the province that had pressed the hardest for the St. Lawrence project and the province that benefited the most from it in terms of employment during the project and from the resulting hydroelectricity, elected only twenty-one Liberals in the 1957 federal election, compared with sixty-one Progressive Conservatives. Although other matters surely were at play, such as the St. Laurent government's handling of the Suez crisis (which was the result of Egypt's attempt to nationalize the canal so that the proceeds could be used to fund the construction of the Aswan High Dam) and the use of closure over the pipeline, these issues were directly connected to perceptions of Canada's relationship to the United States.
15 Scott, *Seeing like a State,* 89.
16 Scott contends that in pre-modern and/or early modern Europe, the state used the forerunners of high modernist techniques in order to amass "descriptive" information (e.g., maps), whereas high modernist states are "prescriptive." Scott, *Seeing like a State,* 90.
17 Ibid., 88.
18 Tina Loo, "People in the Way," 165. See also Tina Loo, "High Modernism and the Nature of Canada" (lecture, York University, Toronto, March 5, 2012); Loo with Stanley, "Environmental History of Progress." Other scholars who have considered the application of high modernism to Canadian hydroelectric projects include Kenny and Secord, "Engineering Modernity"; Van Huizen, "Building a Green Dam."
19 Wynn, *Canada and Arctic North America,* 284. Standard works on Canadian state building and state formation include Hugh Aiken, "Defensive Expansionism: The State and Economic Growth in Canada," in *Approaches to Canadian Economy History,* ed. W.T. Easterbrook and M.H. Watkins (Toronto: McClelland and Stewart, 1967); Leo Panitch, ed., *The Canadian State: Political Economy and Political Power* (Toronto: University of Toronto Press, 1977); Resnick, *The Masks of Proteus;* Allan Greer and Ian Radforth, eds., *Colonial Leviathan: State Formation in Mid-Nineteenth Century Canada* (Toronto: University of Toronto Press, 1992); Bruce Curtis, *The Politics of Population: State Formation, Statistics, and the Census of Canada, 1840-1875* (Toronto: University of Toronto Press, 2001).
20 Richard P. Tucker, "Containing Communism by Impounding Rivers: American Strategic Interests and the Global Spread of High Dams in the Early Cold War," in *Environmental Histories of the Cold War,* ed. J.R. McNeill and Corinna R. Unger (Cambridge: Cambridge University Press, 2010), 139.
21 In addition to Tucker, "Containing Communism," and Josephson, *Industrialized Nature,* see Ann Danaiya Usher, *Dams as Aid: A Political Anatomy of Nordic Development Thinking* (New York: Routledge, 1997); Nicholas Cullather, "Damming Afghanistan: Modernization in a Buffer State," *Journal of American History* 89 (September 2002): 512-37; Nils Gilman, *Mandarins of the Future: Modernization Theory in Cold War America* (Baltimore: Johns Hopkins University Press, 2003); David Ekbladh, *The Great American Mission: Modernization and the Construction of an American World Order* (Princeton, NJ: Princeton University Press, 2010). The various contributions to the series *A History of Water* demonstrate that ambitious water control programs were indeed spread across the globe.
22 Alexander Missal, *Seaway to the Future: American Social Visions and the Construction of the Panama Canal* (Madison, WI: University of Wisconsin Press, 2008).

23 Scott notes in the introduction to *Seeing Like a State* that he had included the TVA as an American example of a high modernist project in an earlier draft of his book. This research was later published as James C. Scott, "High Modernist Social Engineering: The Case of the Tennessee Valley Authority," in *Experiencing the State,* ed. Lloyd I. Rudolph and John Kurt Jacobsen (Toronto: Oxford University Press, 2006).

24 Matthew Evenden indicates that, in addition to American inspirations, British influences shaped the patterns of Canadian hydro dam developments. Evenden, *Fish versus Power,* 268.

25 In total, Soviet dams flooded 2,600 villages and 165 cities, and almost 50,000 square miles, including over 19,000 square miles of agricultural land and an equivalent amount of forestland. See Josephson, chap. 1, "Pyramids of Concrete: Rivers, Dams, and the Ideological Roots of Brute Force Technology," in *Industrialized Nature.* On the St. Laurent government's policies toward the Soviet Union see Jamie Glazov, *Canadian Policy toward Khrushchev's Soviet Union* (Montreal: McGill-Queen's University Press, 2002).

26 The HEPCO chairman also boasted that "the free and independent workers of the USA and Canada" would accomplish their hydro development in less time than the Soviets. IJC, Canadian Section, docket 68-8-1:2, St. Lawrence Power Application, Material Distributed at the Opening of the St. Lawrence Power Project, Luncheon Address by James S. Duncan, Chairman, Ontario Hydro, Official Opening St. Lawrence Power Project, September 5, 1958.

27 On the Aswan High Dam, see Hussein M. Fahim, *Dams, People and Development: The Aswan High Dam Case* (New York: Pergamon Press, 1981); John Waterbury, *Hydropolitics of the Nile Valley* (Syracuse, NY: Syracuse University Press, 1979). For consideration of Aswan and an extended discussion of technology, environment, and modernity see Mitchell, *Rule of Experts.*

28 Cullather, "Damming Afghanistan," 521.

29 The classic exposition of this view is the 1967 essay by Lynn White Jr., "The Historical Roots of Our Ecological Crisis," *Science* (10 March 1967): 1203-7. For recent engagements with White's thesis see Willis Jenkins, *Ecologies of Grace: Environmental Ethics and Christian Theology* (New York: Oxford University Press, 2008); Willis Jenkins, "After Lynn White: Religious Ethics and Environmental Problems," *Journal of Religious Ethics* 37, 2 (2009): 283-309; Richard Bauckham, *The Bible and Ecology: Rediscovering the Community of Creation* (Waco, TX: Baylor University Press, 2010).

30 These included the Hoover/Boulder and Grand Coulee, as well as the recently completed Hungry Horse Dam in Montana. On the link between St. Lawrence construction and the Hungry Horse Dam, as well as background on the workforce, see Parham, chap. 3, "The Workforce," in *St. Lawrence Seaway.*

31 This engineering knowledge was further disseminated and "institutionalized" through professional conferences and journals, as Tina Loo and Meg Stanley, among others, have demonstrated. Loo with Stanley, "Environmental History of Progress," 419-21. For a more global view prominent around the turn of the twentieth century, see Jessica B. Teisch, *Engineering Nature: Water, Development and the Global Spread of American Environmental Expertise* (Chapel Hill: University of North Carolina Press, 2011), as well as Worster, "Water in the Age of Imperialism," in Tvedt and Oestigaard, *The World of Water.*

32 On bulk diversion proposals see J.C. Day and Frank Quinn, *Water Diversion and Export: Learning from Canadian Experience,* University of Waterloo, Department of Geography

Publication Series 36 (Waterloo, ON: University of Waterloo, 1992); Frederic Lasserre, "Water Diversion from Canada to the United States: An Old Idea Born Again?" *International Journal* 62, 1 (Winter 2006/2007): 81-92; Frank Quinn, "Water Diversion, Export and Canada-US Relations: A Brief History," research paper (Toronto: Program on Water Issues, Munk Centre for International Studies, University of Toronto, August 2007); Benjamin Forest and Patrick Forest, "Engineering the North American Waterscape: The High Modernist Mapping of Continental Water Transfer Projects," *Political Geography* (March 2012): 167-83.

33 Desbiens, "Producing North and South"; Desbiens, *Power from the North*. On the James Bay hydro projects, see also McCutcheon, *Electric Rivers*; Carlson, *Home Is the Hunter*; Stéphane Savard, "Retour sur un projet du siècle: Hydro-Québec comme vecteur des représentations symboliques et identitaires du Québec, 1944 à 2005" (PhD diss., Université Laval, 2010).

34 Such postwar high modernist projects are imbricated in discussions on gender and masculinity, a topic that deserves further exploration. For an example in the Canadian context see Christopher Dummitt, *The Manly Modern: Masculinity in Postwar Canada* (Vancouver: UBC Press, 2007).

35 Nelles, *The Politics of Development*, xxvii.

36 Loo, "People in the Way," 169-71; Loo with Stanley, "An Environmental History."

37 Cox et al., "Drowning Voices," 251.

38 Thomas P. Hughes, "The Evolution of Large Technological Systems," in *The Social Construction of Technological Systems: New Directions in the Sociology and History of Technology*, ed. Deborah G. Douglas, Wiebe E. Bijker, and Thomas P. Hughes (Cambridge, MA: MIT Press, 1987); William Cronon, *Nature's Metropolis: Chicago and the Great West* (New York: W.W. Norton, 1991); Donald Worster, *The Wealth of Nature: Environmental History and the Ecological Imagination* (New York: Oxford University Press, 1994); White, *The Organic Machine*.

39 Liza Piper, *The Industrial Transformation of Subarctic Canada* (Vancouver: UBC Press, 2009), 10; Sara B. Pritchard and Thomas Zeller, "The Nature of Industrialization," in Cutcliffe and Reuss, *The Illusory Boundary*, 70. See also Pritchard, *Confluence*.

40 Scott identifies three elements: improvement, bureaucratic management, and aesthetic dimension. Scott, *Seeing like a State*, 224; Scott, "High Modernist Social Engineering," 19.

41 Scott, *Seeing like a State*, 101-2.

42 Jess Gilbert, "Low Modernism and the Agrarian New Deal: A Different Kind of State," in *Fighting for the Farm: Rural America Transformed*, ed. Jane Adams (Philadelphia: University of Pennsylvania Press, 2003).

43 One could also speculate that North American political institutions in the 1950s were not adequately representative or democratic.

44 Loo with Stanley, "An Environmental History." James Murton suggests that studies of state attempts to organize society and the environment need to take into account the "state-idea," which is the "historically specific discursive and ideological formations" that develop in conjunction with the state structure. Matthew Farish and Whitney Lackenbauer argue in their study of Canadian Arctic planning that the high modernist concept needs to be more precisely attuned to the complexities of history and geography. Paul Josephson identifies national and local technology-shaping factors that can help account for societal complexities and historically specific conditions: national culture, level of economic development,

climate, technological sophistication, degree of centralization of decision making and production, and the ideological importance of various artifacts. Paul R. Josephson, *Industrialized Nature: Brute Force Technology and the Transformation of the Natural World* (Washington, DC: Island Press/Shearwater, 2002), 10; James Murton, "Creating Order: The Liberals, the Landowners, and the Draining of Sumas Lake, British Columbia," *Environmental History* 13, 1 (January 2008): 96, 104; Matthew Farish and P. Whitney Lackenbauer, "High Modernism in the Arctic: Planning Frobisher Bay and Inuvik," *Journal of Historical Geography* 35, 3 (January 2009): 519.

45 Andrew Biro provides the term "hydrological nationalism," but since St. Lawrence nationalism is tied up in the waters of the St. Lawrence as well as the manipulation of these waters, it is both hydrological and hydraulic, since the former is generally accepted as referring to the water itself and the latter to the ways it is manipulated and modified, such as canal and hydroelectric works. Andrew Biro, "Half-Empty or Half-Full? Water Politics and the Canadian National Imaginary," in *Eau Canada: The Future of Canada's Water*, ed. Karen Bakker (Vancouver: UBC Press, 2007), 323.

46 Carolyn Johns, "Introduction," in Sproule-Jones, Johns, and Heinmiller, eds., *Canadian Water Politics*, 4. See also the essay on Canadian water and identity on the Environment Canada website: "Water and the Canadian Identity," http://www.ec.gc.ca/.

47 Jean L. Manore, "Rivers as Text: From Pre-Modern to Post-Modern Understandings of Development, Technology and the Environment in Canada and Abroad," in *The World of Water*, A History of Water 1, 3, ed. Terje Tvedt and Eva Jakobsson (London: I.B. Tauris, 2006), 229.

48 Armstrong, Evenden, and Nelles, *The River Returns*.

49 Biro, "Half-Empty or Half-Full?" in Bakker, *Eau Canada*.

50 For a more in-depth discussion see Marco Adria, *Technology and Nationalism*; R. Douglas Francis, *The Technological Imperative in Canada: An Intellectual History* (Vancouver: UBC Press, 2009); Grant, *Lament for a Nation*; Grant, *Technology and Empire*.

51 See W.L. Morton, "The Relevance of Canadian Historians," chap. 4 in *The Canadian Identity* (Madison, WI: University of Wisconsin Press, 1962). Donald Worster in his chapter on Canadian-American differences cites Marilyn Dubasak, Margaret Atwood, and Northrop Frye in support of this hostility argument. Worster argues that these factors include cultural differences (e.g., fusion between freedom/liberty and wilderness in American thinking), the greater Canadian reliance on resource extraction industries, a relatively greater abundance of wilderness, and lack of federal control over land in Canada. Donald Worster, "Wild, Tame, and Free: Comparing Canadian and US Views of Nature," in Findlay and Coates, *Parallel Destinies*, 257-60. On the links between the environment and identity in Canada and the United States see also Marx, *Machine in the Garden;* Harris, "The Myth of the Land in Canadian Nationalism"; Perry Miller, *Nature's Nation* (Cambridge, MA: Belknap Press, 1967); Marcia B. Kline, *Beyond the Land Itself: Views of Nature in Canada and the United States* (Cambridge, MA: Harvard University Press, 1970); Carl Berger, "The Truth North Strong and Free," in Russell, *Nationalism in Canada*; George Altmeyer, "Three Ideas of Nature in Canada, 1893-1914," *Journal of Canadian Studies* 11(1976): 21-36; Ramsay Cook, *Canada, Quebec and the Uses of Nationalism* (Toronto: McClelland and Stewart, 1986); Marilyn Dubasak, *Wilderness Preservation: A Cross-Cultural Comparison of Canada and the United States* (New York: Garland, 1990); Nye, *American Technological Sublime*; Eric Kaufmann, "Naturalizing the Nation: The Rise of Naturalistic Nationalism in the

United States and Canada," *Comparative Studies in Society and History* 40 (1998): 666-95; Claire Elizabeth Campbell, *Shaped by the West Wind: Nature and History in Georgian Bay* (Vancouver: UBC Press, 2005); Alan MacEachern, "A Little Essay on Big: Towards a History of Canada's Size," in *Big Country, Big Issues: Canada's Environment, Culture, and History,* Rachel Carson Perspectives 4 (2011).

52 Creighton, *The Empire of the St. Lawrence,* 383.

Bibliography

ARCHIVAL SOURCES

Canada
Government of Ontario
Hydro-Electric Power Commission of Ontario (now Ontario Power Generation)
Library and Archives Canada
MG 3 A 6 (C.P. Wright fonds)
MG 26 J (William Lyon Mackenzie King fonds)
MG 26 K (R.B. Bennett fonds)
MG 26 L (Louis St. Laurent fonds)
MG 26 N (Lester Pearson fonds)
MG 27 B 20 (C.D. Howe fonds)
MG 30 D 33 (O.D. Skelton fonds)
MG 30 E 101 (Hume Wrong fonds)
MG 30 E 133 (Andrew G.L. McNaughton fonds)
MG 30 E 163 (Norman Robertson fonds)
MG 31 E 59 (Robert B. Bryce fonds)
MG 32 B 34 (J.W. Pickersgill fonds)
RG 2 (Privy Council Office)
RG 12 (Department of Transport)
RG 24 (Department of Defence)
RG 25 (Department of External Affairs)
RG 32 (Department of Indians Affairs)
RG 51 (International Joint Commission)
RG 52 (St. Lawrence Seaway Authority)
Lost Villages Historical Society
Stormont, Dundas, and Glengarry Historical Society

United States
Dwight D. Eisenhower Presidential Library
Franklin D. Roosevelt Presidential Library
Harry S. Truman Presidential Library
National Archives and Records Administration II
RG 59 (Department of State)
RG 84 (Department of State – Canada)
RG 128 (Federal Power Commission)
Power Authority of the State of New York
St. Lawrence University Archives – St. Lawrence Seaway Collection

International Joint Commission
International Joint Commission – Canadian Section

Interviews
Helen Badlam, resident of Waddington (December 6, 2012)
Nancy Badlam, resident of Waddington (December 6, 2012)
Jim Brownell, former resident of Lost Villages (May 16, 2011)
Vale Brownell, former resident of Lost Villages (June 22, 2011)
Bonnie Clarke, former resident of Lost Villages (June 15, 2011)
Jane Craig, former resident of Lost Villages (June 22, 2011)
Dennis Dack, former HEPCO employee (May 2, 2011)
Alice Dumas, resident of Waddington (December 6, 2012)
Donna Dunn, resident of Waddington (December 6, 2012)
Kathy Dupray, resident of Waddington (December 6, 2012)
Tim Gault, Lost Villages Historical Society (April 21, 2011)
David Hill, former resident of Lost Villages (June 22, 2011)
Lyall Manson, former resident of Lost Villages (June 22, 2011)
Grace McBath, resident of Waddington (December 6, 2012)
Joan McEwan, former resident of Lost Villages (June 22, 2011)
Russell B. Strait, resident of Waddington (May 21, 2012)
Noella Whorrall, former resident of Lost Villages (June 22, 2011)

PRINTED PRIMARY SOURCES

Baxter, R.R. *Documents on the St. Lawrence Seaway*. New York: Frederick A. Praeger, 1960.
Berle, Beatrice Bishop, and Travis Beal Jacobs, eds. *Navigating the Rapids, 1918-1971: From the Papers of Adolf A. Berle*. New York: Harcourt Brace Jovanovich, 1973.
Conference of Canadian Engineers. *St. Lawrence Waterway Project: Report of Conference of Canadian Engineers on the International Rapids Section; December 30, 1929*. Ottawa: Queen's Printer, 1930.
Cooper, Hugh L., and Co. *Report to the International Joint Commission on Navigation and Power in the St. Lawrence River*. New York: Hugh L. Cooper and Co., 1921.
Ferrell, Robert H., ed. *Off the Record: The Private Papers of Harry S. Truman*. New York: Harper and Row, 1980.

Government of Canada. House of Commons. *Correspondence and Documents Relating to St. Lawrence Deep Waterway Treaty 1932, Niagara Convention 1929, and Ogoki River and Kenogami River (Long Lake) Projects and Export of Electrical Power.* Ottawa: J.O. Patenaude, 1938.

–. *Correspondence and Documents Relating to the Great Lakes–St. Lawrence Basin Development, 1938-41.* Ottawa: Edmond Cloutier, 1941.

–. *Debates.* Ottawa: Queen's Printer, 1921-59.

Government of Canada. Department of External Affairs. *Documents on Canadian External Relations.* Ottawa: Queen's Printer, various years.

Government of the United States. *Confidential Files of the Eisenhower White House, 1953-61.* LexisNexis microfilm. Abilene, KS: Eisenhower Library.

–. *Diaries of Dwight D. Eisenhower, 1953-1961.* LexisNexis microfilm. Abilene, KS: Eisenhower Library.

–. *Foreign Relations of the United States.* Washington, DC: US Government Printing Office, various years.

–. *The Papers of John Foster Dulles and of Christian A. Herter, 1953-1961.* LexisNexis microfilm. Abilene, KS: Eisenhower Library.

–. *The Personal Papers of John Foster Dulles.* Pts. 1-5. Wilmington, DE: Scholarly Resources, 1994.

–. *President Franklin D. Roosevelt's Office Files, 1933-45.* Edited by William E. Leuchtenburg. Washington: University Publications of America, 1990.

–. *President Harry S. Truman's Office Files, 1945-53.* Edited by William E. Leuchtenburg. Washington: University Publications of America, 1989.

–. Department of Commerce (N.R. Danielian). *The St. Lawrence Survey.* 7 vols. Washington, DC: US Government Printing Office, 1941.

–. Saint Lawrence Seaway Development Corporation, Office of Information. "Great Lakes–St. Lawrence Seaway: Collection of Pamphlets." Washington, DC: US Government Printing Office, 1955.

–. US Congress. *St. Lawrence Seaway Manual: A Compilation of Documents on the Great Lakes Seaway Project and Correlated Power Development.* US Senate Document 165, 1955. Washington: US Government Printing Office, 1955.

–. US Congressional Records. 72nd-86th sessions.

–. US Senate. *St. Lawrence Waterway: Report of the United States and Canadian Government Engineers on the Improvement of the St. Lawrence River from Montreal to Lake Ontario made to the International Joint Commission (Wooten-Bowden Report).* Supplementary to Senate Document 114, 67th Congress. Washington, DC: Government Printing Office, 1922.

Kates (J.), and Associates. *St. Lawrence Seaway Traffic Studies.* Toronto: Kates (J.), and Associates, 1966.

Keyser, C. Frank. *The St. Lawrence Seaway Project: A Brief Historical Background.* Library of Congress: Legislative Reference Service. Public Affairs bulletin 58. Washington, DC: US Government Printing Office, 1947.

Nixon, Edgar B., ed. *Franklin D. Roosevelt and Foreign Affairs.* Cambridge, MA: Belknap Press, 1933-1945.

Wilson, Norman D. *The Rehabilitation of the St. Lawrence Communities: A Report on the Factors in the Rehabilitation of the St. Lawrence Communities Partly or Wholly Inundated in the Development for Power and Navigation of the International Rapids Section of the St. Lawrence River.* Ottawa: Canadian Advisory Committee on Reconstruction, 1943.

Bibliography

Secondary Sources

Acheson, Dean. *Present at the Creation: My Years in the State Department.* New York: W.W. Norton, 1969.

Adamson, Anthony. "Crysler's Park." *Canadian Architect* 3, 2 (February 1958): 41-42.

Adria, Marco. *Technology and Nationalism.* Montreal: McGill-Queen's University Press, 2010.

Alexander, Jeff. *Pandora's Locks: The Opening of the Great Lakes–St. Lawrence Seaway.* East Lansing, MI: Michigan State University Press, 2009.

Alfred, Gerald R. (Taiaiake). *Heeding the Voices of Our Ancestors: Kahnawake Mohawk Politics and the Rise of Native Nationalism.* Toronto: Oxford University Press, 1995.

Altmeyer, George. "Three Ideas of Nature in Canada, 1893-1914." *Journal of Canadian Studies* 11 (1976): 21-36.

Andrew, Arthur. *The Rise and Fall of a Middle Power: Canadian Diplomacy from King to Mulroney.* Toronto: James Lorimer, 1993.

Angus, Ian H. *A Border Within: National Identity, Cultural Plurality, and Wilderness.* Montreal: McGill-Queen's University Press, 1997.

Annin, Peter. *The Great Lakes Water Wars.* Washington, DC: Island Press, 2006.

Armstrong, Christopher. *The Politics of Federalism: Ontario's Relations with the Federal Government, 1867-1942.* Toronto: University of Toronto Press, 1981.

Armstrong, Christopher, Matthew Evenden, and H.V. Nelles. *The River Returns: An Environmental History of the Bow.* Montreal: McGill-Queen's University Press, 2009.

Armstrong, Christopher, and H.V. Nelles. *Monopoly's Moment: The Organization and Regulation of Canadian Utilities, 1830-1930.* Philadelphia: Temple University Press, 1986.

–. *Wilderness and Waterpower: How Banff National Park Became a Hydro-Electric Storage Reservoir.* Calgary: University of Calgary Press, 2013.

Aronsen, Lawrence. *American National Security and Economic Relations with Canada, 1945-1954.* Westport, CT: Praeger, 1997.

Ashworth, William. *The Late, Great Lakes: An Environmental History.* Detroit: Wayne State University Press, 1997.

Atwood, Margaret. *Survival: A Thematic Guide to Canadian Literature.* Toronto: McClelland and Stewart, 2004.

Austin, Alvyn. "The Lost Villages." *Canadian Magazine,* March 26, 1977, 24-26.

Azzi, Stephen. *Walter Gordon and the Rise of Canadian Nationalism.* Montreal: McGill-Queen's University Press, 1999.

Ball, Norman R. *"Mind, Heart, and Vision": Professional Engineering in Canada, 1887 to 1987.* Ottawa: National Museum of Science and Technology and the Engineering Centennial Board, 1987.

Bavington, Dean. *Managed Annihilation: An Unnatural History of the Newfoundland Cod Collapse.* Vancouver: UBC Press, 2010.

Beck, Mary Celeste. "An Historical Evaluation of the St. Lawrence Seaway Controversy, 1950-1953." PhD diss., St. John's University, 1954.

Beck, Gregory G., and Bruce Littlejohn. *Voices for the Watershed: Environmental Issues in the Great Lakes–St. Lawrence Drainage Basin.* Montreal: McGill-Queen's University Press, 2000.

Becker, William H. *From the Atlantic to the Great Lakes: A History of the US Army Corps of Engineers and the St. Lawrence Seaway.* Washington, DC: US Army Corps of Engineers, 1984.

Behiels, Michael, and Reginald Stuart, eds. *Transnationalism: Canada–United States History into the Twenty-First Century*. Montreal: McGill-Queen's University Press, 2010.
Beigie, Carl E., and Alfred O. Hero Jr., eds. *Natural Resources in US-Canadian Relations*. 3 vols. Boulder, CO: Westview Press, 1980.
Beisner, Robert L. *Dean Acheson: A Life in the Cold War*. New York: Oxford University Press, 2006.
Bélanger, Damien-Claude. *Prejudice and Pride: Canadian Intellectuals Confront the United States, 1895-1945*. Toronto: University of Toronto Press, 2011.
Bercuson, David J. *True Patriot: The Life of Brooke Claxton, 1898-1960*. Toronto: University of Toronto Press, 1993.
Berger, Carl. *The Writing of Canadian History: Aspects of English-Canadian Historical Writing since 1900*. 2nd ed. Toronto: University of Toronto Press, 1986.
Beston, Henry. *The St. Lawrence*. Toronto: Rinehart, 1945.
Biggs, David. *Quagmire: Nation-Building and Nature in the Mekong Delta*. Seattle: University of Washington Press, 2010.
Billington, David P., and Donald C. Jackson. *Big Dams of the New Deal Era: A Confluence of Engineering and Politics*. Norman: University of Oklahoma Press, 2006.
Binnema, Ted. "'Most Fruitful Results': Transborder Approaches to Canadian-American Environmental History." In *A Companion to American Environmental History*, edited by Douglas Cazaux Sackman, 615-34. Chichester: Wiley-Blackwell, 2010.
Blanchard, Raoul. *Le Canada francais: Province de Quebec*. Montreal: Fayard, 1960.
Blatter, Joachim, and Helen M. Ingram. *Reflections on Water: New Approaches to Transboundary Conflicts and Cooperation*. Cambridge, MA: MIT Press, 2001.
Bliss, Michael. *Northern Enterprise: Five Centuries of Canadian Business*. Toronto: McClelland and Stewart, 1987.
Bloomfield, L.M., and Gerald F. Fitzgerald. *Boundary Waters Problems of Canada and the United States: The International Joint Commission, 1912-1958*. Toronto: Carswell, 1958.
Bocking, Stephen. *Nature's Experts: Science, Politics, and the Environment*. New Brunswick, NJ: Rutgers University Press, 2004.
Bogue, Margaret Beattie. *Fishing the Great Lakes: An Environmental History, 1783-1933*. Madison, WI: University of Wisconsin Press, 2000.
Bothwell, Robert. *Alliance and Illusion: Canada and the World, 1945-1984*. Vancouver: UBC Press, 2007.
–. *The Big Chill: Canada and the Cold War*. Toronto: Irwin, 1998.
–. *Canada and the United States: The Politics of Partnership*. Toronto: University of Toronto Press, 1992.
Bothwell, Robert, Ian Drummond, and John English. *Canada since 1945: Power, Politics, and Provincialism*. Rev. ed. Toronto: University of Toronto Press, 1989.
Bothwell, Robert, and Norman Hillmer, eds. *The In-Between Time: Canadian External Policy in the 1930s*. Vancouver: C. Clark, 1975.
Bothwell, Robert, and William Kilbourn. *C.D. Howe: A Biography*. Toronto: McClelland and Stewart, 1979.
Botts, Lee, and Paul Muldoon. *Evolution of the Great Lakes Water Quality Agreements*. East Lansing, MI: Michigan State University Press, 2005.
Bow, Brian J. *The Politics of Linkage: Power, Interdependence, and Ideas in Canada-US Relations*. Vancouver: UBC Press, 2009.

Bowering, Ian. *Cornwall: From Factory Town to Seaway City, 1900-1999*. Vol. 1. Cornwall, ON: Cornwall Standard Freeholder and Stormont, Dundas, and Glengarry Historical Society, 1999.

Bowlus, W. Bruce. *Iron Ore Transport on the Great Lakes: The Development of a Delivery System to Feed American Industry*. London: McFarland, 2011.

Bowser, Sarah. "The Planner's Part." *Canadian Architect* 3, 2 (February 1958): 38-40.

Briar, John H. *Taming of the Sault: A Story of the St. Lawrence Power Project: Heart of the Seaway*. Watertown, NY: Hungerford-Holbrook, 1960.

Brooks, Karl Boyd, ed. *The Environmental Legacy of Harry S. Truman*. Kirksville, MS: Truman State University Press, 2009.

–. *Public Power, Private Dams: The Hells Canyon Controversy*. Seattle: University of Washington Press, 2006.

Brown, D.D. *The St. Lawrence Seaway Tolls and the Quebec-Labrador Iron Ore Industry*. Mineral Policy Sector, internal report. MRI 81/10. Ottawa: Energy, Mines and Resources Canada, 1981.

Browne, George W. *The St. Lawrence River: Historical, Legendary, Picturesque*. New York: Putnam, 1905.

Bryce, J.B. *A Hydraulic Engineering History of the St. Lawrence Power Project with Special Reference to Regulation of Water Levels and Flows*. Toronto: Ontario Hydro, 1982.

Bukowczyk, John J., Nora Faires, David R. Smith, and Randy William Widdis. *Permeable Border: The Great Lakes Basin as Transnational Region, 1650-1990*. Pittsburgh: University of Pittsburgh Press, 2005.

Bullock, Frederick J. *Ships and the Seaway*. New York: Dent, 1959.

Burton, Thomas L. *Natural Resource Policy in Canada: Issues and Perspectives*. Toronto: McClelland and Stewart, 1972.

Cain, Louis P. "Unfouling the Public's Nest: Chicago's Sanitary Diversion of Lake Michigan Water." *Technology and Culture* 15, 14 (October 1974): 594-613.

Campbell, Claire Elizabeth. *Shaped by the West Wind: Nature and History in Georgian Bay*. Vancouver: UBC Press, 2005.

Campbell, Scott M. "Backwater Calculations for the St. Lawrence Seaway with the First Computer in Canada." *Canadian Journal of Civil Engineering* 36 (2009): 1164-69.

Camu, Pierre. *Le Saint-Laurent et les Grands Lacs au temps de la vapeur, 1850-1950*. Montreal: Hurtubise HMH, 2005.

–. *Problèmes des Transports dans la Région du Bas Saint-Laurent*. Montreal: Conseil d'Orientation Économique du Bas Saint-Laurent, 1960.

Careless, J.M.S. *Canada: A Story of Challenge*. Toronto: Macmillan, 1963.

Caro, Robert. *The Power Broker: Robert Moses and the Fall of New York*. New York: Vintage, 1975.

Carroll, John E. *Environmental Diplomacy: An Examination and a Prospective of Canadian-US Transboundary Environmental Relations*. Ann Arbor: University of Michigan Press, 1983.

Castonguay, Stéphane. "The Production of Flood as Natural Catastrophe: Extreme Events and the Construction of Vulnerability in the Drainage Basin of the St. Francis River, Mid-Nineteenth to Mid-Twentieth Century." *Environmental History* 12, 4 (October 2007): 820-44.

Castonguay, Stéphane, and Michèle Dagenais, eds. *Metropolitan Natures: Environmental Histories of Montreal*. Pittsburgh: University of Pittsburgh Press, 2011.

Castonguay, Stéphane, and Matthew Evenden, eds. *Urban Rivers: Remaking Rivers, Cities, and Space in Europe and North America*. Pittsburgh: University of Pittsburgh Press, 2012.

Chacko, Chirakaikaran Joseph. *The International Joint Commission between the United States of America and the Dominion of Canada*. New York: AMS Press, 1968.

Chapnick, Adam. *Canada's Voice: The Public Life of John Wendell Holmes*. Vancouver: UBC Press, 2009.

Charland, Maurice. "Technological Nationalism." *Canadian Journal of Political and Social Theory* 10, 1 (1986): 196-220.

Chevrier, Lionel. *The St. Lawrence Seaway*. Toronto: Macmillan, 1959.

Cioc, Mark. *The Rhine: An Eco-Biography, 1815-2000*. Seattle: University of Washington Press, 2002.

Cohen, Andrew. *While Canada Slept: How We Lost Our Place in the World*. Toronto: McClelland and Stewart, 2004.

Cohen, Warren I. *Empire without Tears: America's Foreign Relations, 1921-1933*. Philadelphia: Temple University Press, 1987.

Cole, Wayne S. *Roosevelt and the Isolationists, 1932-45*. Lincoln: University of Nebraska Press, 1983.

Comstock, R.S. "The St. Lawrence Seaway and Power Project: A Case Study in Presidential Leadership." PhD diss., Ohio State University, 1956.

Constant, Jean-François, and Michel Ducharme. *Liberalism and Hegemony: Debating the Canadian Liberal Revolution*. Toronto: University of Toronto Press, 2009.

Cook, Ramsay. *Canada, Quebec and the Uses of Nationalism*. Toronto: McClelland and Stewart, 1986.

Cosens, Barbara, ed. *The Columbia River Treaty Revisited: Transboundary River Governance in the Face of Uncertainty*. Corvallis: Oregon State University Press, 2012.

Cox, Heather M., Brendan G. DeMelle, Glenn R. Harris, Christopher P. Lee, and Laura K. Montondo. "Drowning Voices and Drowning Shoreline: A Riverside View of the Social and Ecological Impacts of the St. Lawrence Seaway and Power Project." *Rural History* 10, 2 (1999): 235-57.

Creighton, Donald. *Canada's First Century, 1867-1967*. Toronto: St. Martin's Press, 1970.

–. *The Commercial Empire of the St. Lawrence, 1760-1850*. Toronto: Ryerson Press, 1937.

–. "The Decline and Fall of the Empire of the St. Lawrence." Canadian Historical Association, *Historical Papers*, 1969: 14-25.

–. *Dominion of the North: A History of Canada*. Toronto: Houghton Mifflin, 1944.

–. *The Empire of the St. Lawrence: A Study in Commerce and Politics*. Toronto: Macmillan, 1956.

–. *The Forked Road: Canada, 1939-1957*. Toronto: McClelland and Stewart, 1976.

Cuff, R.D., and J.L. Granatstein. *American Dollars – Canadian Prosperity: Canadian-American Economic Relations, 1945-1950*. Toronto: Samuel Stevens, 1978.

–. *Ties That Bind: Canadian-American Relations in Wartime, from the Great War to the Cold War*. Toronto: Samuel Stevens, 1977.

Curtis, Bruce. *The Politics of Population: State Formation, Statistics, and the Census of Canada, 1840-1875*. Toronto: University of Toronto Press, 2001.

Curtis, Kenneth M., and John E. Carroll. *Canadian-American Relations: The Promise and the Challenge*. Toronto: Lexington, 1983.

Dagenais, Michèle. "Montreal and Its Waters: An Entangled History." In *Big Country, Big Issues: Canada's Environment, Culture, and History*. Rachel Carson Center Perspectives 4 (2011): 44-59.
–. *Montréal et l'eau: Une histoire environnementale*. Montreal: Boréal, 2011.
Dales, John. *Hydroelectricity and Industrial Development: Quebec, 1898-1940*. Cambridge, MA: Harvard University Press, 1957.
Dallek, Robert. *Franklin D. Roosevelt and American Foreign Policy, 1932-1945*. New York: Oxford University Press, 1979.
Damms, Richard V. *The Eisenhower Presidency, 1953-1961*. London: Longman, 2002.
Daniels, Ronald J., ed. *Ontario Hydro at the Millennium: Has Monopoly's Moment Passed?* Montreal: McGill-Queen's University Press, 1996.
Day, J.C., and Frank Quinn. *Water Diversion and Export: Learning from Canadian Experience*. University of Waterloo, Department of Geography Publication Series 36. Waterloo, ON: University of Waterloo, 1992.
Dealey, J.Q. "The Chicago Drainage Canal and St. Lawrence Development." *American Journal of International Law* 23, 2 (April 1929): 307-28.
DeBruin, Jennifer. *A Walk with Mary*. Renfrew, ON: General Store Publishing House, 2012.
Dempsey, Dave. *On the Brink: The Great Lakes in the 21st Century*. East Lansing, MI: Michigan State University Press, 2004.
Desbiens, Caroline. *Power from the North: Territory, Identity, and the Culture of Hydroelectricity in Quebec*. Vancouver: UBC Press, 2013.
–. "Producing North and South: A Political Geography of Hydro Development in Quebec." *Canadian Geographer* 48, 2 (2004): 101-18.
Desjardins, Pauline. "Navigation and Waterpower: Adaptation and Technology on Canadian Canals." *IA: The Journal of the Society for Industrial Archaeology* 29, 1 (NA 2003): 21-48.
Dinwoodie, D.H. "The Politics of International Pollution Control: The Trail Smelter Case." *International Journal* 27, 2 (Spring 1972): 219-35.
Divine, Robert A. *Eisenhower and the Cold War*. New York: Oxford University Press, 1981.
Doern, G. Bruce. *Green Diplomacy: How Environmental Policy Decisions Are Made*. Toronto: C.D. Howe Institute, 1993.
Doheny-Farina, Stephen. *The Grid and the Village: Losing Electricity, Finding Community, Surviving Disaster*. New Haven, CT: Yale University Press, 2001.
Dohler, G.C. *Current Survey: St. Lawrence River, Montreal-Quebec, 1960*. Ottawa: Canadian Hydrographic Service, Survey and Mapping Branch, Department of Mines and Technical Surveys, 1961.
Donaghy, Greg, ed. *Canada and the Early Cold War, 1943-1957*. Ottawa: Department of Foreign Affairs and International Trade, 1998.
–. "Coming off the Gold Standard: Reassessing the 'Golden Age' of Canadian Diplomacy." Lecture, Johnson-Shoyama Graduate School of Public Policy, University of Saskatchewan, December 2009.
–. *Tolerant Allies: Canada and the United States, 1963-1968*. Montreal: McGill-Queen's University Press, 2002.
Donaghy, Greg, and Michael K. Carroll, eds. *In the National Interest: Canadian Foreign Policy and the Department of Foreign Affairs and International Trade, 1909-2009*. Calgary: University of Calgary Press, 2011.

Donaghy, Greg, and Kim Richard Nossal, eds. *Architects and Innovators: Building the Department of Foreign Affairs and International Trade, 1909-2009*. Queen's Policy Studies Series. Montreal: McGill-Queen's University Press, 2009.

Donaghy, Greg, and Stephane Roussel. *Escott Reid: Diplomat and Scholar*. Montreal: McGill-Queen's University Press, 2004.

Donovan, Robert J. *Tumultuous Years: The Presidency of Harry S. Truman*. New York: Norton, 1982.

Dorsey, Kurkpatrick. *The Dawn of Conservation Diplomacy: US-Canadian Wildlife Protection Treaties in the Progressive Era*. Seattle: University of Washington Press, 1998.

–. "Dealing with the Dinosaur (and Its Swamp): Putting the Environment in Diplomatic History." *Diplomatic History* 29, 4 (September 2005): 573-87.

Drescher, Nuala. *Engineers for the Public Good: A History of the Buffalo District US Army Corps of Engineers*. Washington, DC: U.S Army Corps of Engineers, 1982.

Dubasak, Marilyn. *Wilderness Preservation: A Cross-Cultural Comparison of Canada and the United States*. New York: Garland, 1990.

Dubinsky, Karen. *The Second Greatest Disappointment: Honeymooning and Tourism at Niagara Falls*. Toronto: Between the Lines, 1999.

Dummitt, Christopher. *The Manly Modern: Masculinity in Postwar Canada*. Vancouver: UBC Press, 2007.

Eayrs, James George. *The Art of the Possible: Government and Foreign Policy in Canada*. Toronto: University of Toronto Press, 1961.

–. *In Defence of Canada*. Toronto: University of Toronto Press, 1964.

Ekbladh, David. *The Great American Mission: Modernization and the Construction of an American World Order*. Princeton, NJ: Princeton University Press, 2010.

Emerson, Norman. "Before the Flood." *Ontario History*, Winter 1958: 47-50.

–. "Farewell to Ault Park." *Ontario History*, Winter 1958: 6-7.

English, John. *The Life of Lester Pearson*. 2 vols. Toronto: Lester and Orpen Dennys, 1989 and 1992.

Environment Canada. *St. Lawrence River: Area of Concern (Canadian Section), Status of Beneficial Use Impairments*. Ontario: Ontario Ministry of the Environment, 2011.

–. "Water and the Canadian Identity." n.d. http://www.ec.gc.ca/.

Environmental Protection Agency. *State of the Great Lakes: What Are the Major Pressures Impacting the St. Lawrence River?* EPA 905-F-06-913. New York: Environmental Protection Agency, 2004.

Evenden, Matthew. *Fish versus Power: An Environmental History of the Fraser River*. New York: Cambridge University Press, 2004.

–. "Mobilizing Rivers: Hydro-Electricity, the State, and World War II in Canada." *Annals of the Association of American Geographers* 99, 5 (2009): 845-55.

Evenden, Matthew, and Graeme Wynn. "Fifty-Four, Forty, or Fight? Writing within and across Boundaries in North American Environmental History." In *Nature's End: History and the Environment*, edited by Sverker Sörlin and Paul Warde, 215-46. New York: Palgrave Macmillan, 2009.

Fahim, Hussein M. *Dams, People and Development: The Aswan High Dam Case*. New York: Pergamon Press, 1981.

Farish, Matthew, and P. Whitney Lackenbauer. "High Modernism in the Arctic: Planning Frobisher Bay and Inuvik." *Journal of Historical Geography* 35, 3 (January 2009): 517-44.

Ferrell, Robert H. *Harry S. Truman and the Cold War Revisionists*. Columbia: University of Missouri Press, 2006.
Findlay, John M., and Kenneth S. Coates. *Parallel Destinies: Canadian-American Relations West of the Rockies*. Seattle: University of Washington Press, 2002.
Fleming, Keith Robson. *Power at Cost: Ontario Hydro and Rural Electrification, 1911-1958*. Montreal: McGill-Queen's University Press, 1992.
Forest, Benjamin, and Patrick Forest. "Engineering the North American Waterscape: The High Modernist Mapping of Continental Water Transfer Projects." *Political Geography* (March 2012): 167-83.
Forkey, Neil S. "'Thinking like a River': The Making of Hugh MacLennan's Environmental Consciousness." *Journal of Canadian Studies* 41, 2 (Spring 2007): 42-64.
Fox, Annette Baker, Alfred O. Hero Jr., and Joseph S. Nye Jr. *Canada and the United States: Transnational and Transgovernmental Relations*. New York: Columbia University Press, 1976.
Franck, Alain. *Naviguer sur le fleuve au temps passé 1860-1960*. Quebec City: Publications Québec, 2000.
Fraser, Steven Wayne. "Seaway Valley in Ontario: A Study in the Relocation of Settlement." BA honours thesis, Carleton University, 1971.
Freeman, Neil B. *The Politics of Power: Ontario Hydro and Its Government, 1906-1995*. Toronto: University of Toronto Press, 1996.
From Dream to Reality: Ontario Hydro's Story of the Construction of the St. Lawrence Seaway and Power Project. DVD. Cornwall, ON: KAV Productions, 2009.
Froschauer, Karl. *White Gold: Hydroelectric Power in Canada*. Vancouver: UBC Press, 1999.
Gaddis, John Lewis. *The Cold War: A New History*. New York: Penguin, 2005.
–. *Strategies of Containment: A Critical Appraisal*. Rev. ed. New York: Oxford University Press, 2005.
Gagné, Jean. *À la découverte du Saint-Laurent*. Montreal: Éditions de l'Homme, 2005.
Gardner, Lloyd C. *Architects of Illusions: Men and Ideas in American Foreign Policy, 1941-1949*. Chicago: Franklin Watts, 1970.
Ghobashy, Omar Z. *The Caughnawaga Indians and the St. Lawrence Seaway*. New York: Devin-Adair, 1961.
Gibson, Frederick W., and Jonathan G. Rossie, eds. *The Road to Ogdensburg: The Queen's/St. Lawrence Conferences on Canadian-American Affairs, 1935-1941*. East Lansing, MI: Michigan State University, 1993.
Glazebrook, G.P. *A History of Transportation in Canada*. Toronto: Ryerson Press, 1938.
Glazov, Jamie. *Canadian Policy toward Khrushchev's Soviet Union*. Montreal: McGill-Queen's University Press, 2002.
Gogo, Jean L., ed. *Lights on the St. Lawrence: An Anthology*. Toronto: Ryerson Press, 1958.
Goldsmith, Edward, and Nicholas Hildyard. *The Social and Environmental Effects of Large Dams*. San Francisco: Sierra Club Books, 1984.
Good, Mabel Tinkiss. *Chevrier: Politician, Statesman, Diplomat and Entrepreneur of the St. Lawrence Seaway*. Montreal: Stanké, 1987.
Goodman, Carl F. "Canada-United States Settlement of Gut Dam Claims: Report of the Agent of the United States before the Lake Ontario Claims Tribunal." *International Legal Materials* 8, 1 (January 1969): 118-43.
Gossage, Peter. *Water in Canadian History: An Overview*. Research paper. Ottawa: Government of Canada, Inquiry on Federal Water Policy, 1985.

Graebner, Norman. *America as a World Power: A Realist Appraisal from Wilson to Reagan.* Wilmington, DE: Scholarly Resources, 1984.

Granatstein, J.L. *Canadian Foreign Policy since 1945: Middle Power or Satellite?* 3rd ed. Toronto: Copp Clark, 1973.

–. *How Britain's Weakness Forced Canada into the Arms of the United States.* Toronto: University of Toronto Press, 1989.

–. *A Man of Influence: Norman A. Robertson and Canadian Statecraft, 1929-68.* Ottawa: Deneau, 1981.

–. *Yankee, Go Home? Canadians and Anti-Americanism.* Toronto: HarperCollins, 1997.

Grant, George. *Lament for a Nation: The Defeat of Canadian Nationalism.* Toronto: McClelland and Stewart, 1965.

–. *Technology and Empire: Perspectives on North America.* Toronto: House of Anansi, 1969.

Grant, R.E., and Associates. *Fish Habitat Changes – Thousand Islands, Middle Corridor, and Lake St. Lawrence.* St. Lawrence River Discussion Paper. Brockville, ON: Ontario Ministry of Natural Resources and New York State Department of Environmental Conservation, n.d.

Great Lakes Fisheries Commission. "Fish Community Objectives for the St. Lawrence River." Draft, December 14. Ann Arbor, MI: Great Lakes Fisheries Commission, 2001.

Greeley, J.R., and C.W. Greene. "Fishes of the Area with Annotated List." In *A Biological Survey of the St. Lawrence Watershed, Supplement to 20th Annual Report, 1930.* Albany, NY: New York Conservation Department, 1931.

Greenstein, Fred I. *The Hidden-Hand Presidency: Eisenhower as Leader.* New York: Basic Books, 1991.

Hamby, Alonzo L. *Harry S. Truman and the Fair Deal.* Lexington, MA: D.C. Heath, 1974.

Hamilton, Janice. *The St. Lawrence River: History, Highway and Habitat.* Westmount, QC: Redlader, 2006.

Harrison, W.E.C. *Canada in World Affairs, 1949-1950.* Toronto: Oxford University Press, 1957.

Hauptman, Laurence M. *The Iroquois Struggle for Survival: World War II to Red Power.* Syracuse, NY: Syracuse University Press, 1986.

Heeney, A.D.P. *The Things That Are Caesar's: Memoirs of a Canadian Public Servant.* Toronto: University of Toronto Press, 1972.

Heintzman, Ralph. "Political Space and Economic Space: Quebec and the Empire of the St. Lawrence." *Journal of Canadian Studies* 29, 2 (1994): 19-63.

Henderson, Henry L., and David B. Woolner, eds. *FDR and the Environment.* New York: Palgrave Macmillan, 2009.

Hilliker, John. *Canada's Department of External Affairs.* Vol. 1, *The Early Years, 1909-1946.* Montreal: McGill-Queen's University Press, 1990.

Hilliker, John, and Donald Barry. *Canada's Department of External Affairs.* Vol. 2, *Coming of Age, 1946-1968.* Montreal: McGill-Queen's University Press, 1995.

Hillmer, Norman. "The Foreign Policy That Never Was, 1900-1950." In *Canada, 1900-1950: Un pays prend sa place/A Country Comes of Age,* edited by Serge Bernier and John MacFarlane, 141-53. Ottawa: Organization for the History of Canada, 2003.

–. "O.D. Skelton and the North American Mind." *International Journal* 60, 1 (Winter 2004/2005): 93-110.

–. *Partners Nevertheless: Canadian-American Relations in the Twentieth Century.* Toronto: Copp Clark Pitman, 1989.
–. "Reflections on the Unequal Border." *International Journal* 60, 2 (Spring 2005): 331-40.
Hillmer, Norman, and Adam Chapnick, eds. *Canadas of the Mind: The Making and Unmaking of Canadian Nationalisms in the Twentieth Century.* Montreal: McGill-Queen's University Press, 2007.
Hillmer, Norman, and Ian M. Drummond. *Negotiating Freer Trade: The United Kingdom, Canada, and the Trade Agreements of 1938.* Waterloo, ON: Wilfrid Laurier Press, 1989.
Hillmer, Norman, and J.L. Granatstein. *Empire to Umpire: Canada and the World into the Twenty-First Century.* 2nd ed. Toronto: Nelson College Indigenous, 2007.
–. *For Better or for Worse: Canada and the United States into the Twenty-First Century.* Toronto: Nelson Thomson, 2007.
Hillmer, Norman, and Garth Stevenson, eds. *A Foremost Nation: Canadian Foreign Policy and a Changing World.* Toronto: McClelland and Stewart, 1977.
Hills, Theo L. *The St. Lawrence Seaway.* London: Methuen, 1959.
Hixson, Walter L. *The Myth of American Diplomacy: National Identity and US Foreign Policy.* New Haven, CT: Yale University Press, 2008.
Hogan, Michael J. *America in the World: The Historiography of US Foreign Relations since 1941.* Cambridge: Cambridge University Press, 1996.
–. *A Cross of Iron: Harry S. Truman and the Origins of the National Security State.* New York: Cambridge University Press, 1998.
Holmes, John W. *The Better Part of Valour: Essays on Canadian Diplomacy.* Toronto: McClelland and Stewart, 1970.
–. *Canada: A Middle-Aged Power.* Toronto: McClelland and Stewart, 1976.
–. *Life with Uncle: The Canadian-American Relationship.* Toronto: University of Toronto Press, 1981.
–. *The Shaping of Peace: Canada and the Search for World Order 1943-1957.* Toronto: University of Toronto Press, 1979.
Hornsby, Stephen J., and John G. Reid. *New England and the Maritime Provinces: Connections and Comparisons.* Montreal: McGill-Queen's University Press, 2005.
Howard, Jane Mary. "Some Economic Aspects of the St. Lawrence Project." MA thesis, Catholic University of America, 1949.
Howe, Stanley Russell. "C.D. Howe and the Americans, 1940-1957." PhD diss., University of Maine, 1977.
Hunt, Michael H. *Ideology and US Foreign Policy.* New Haven, CT: Yale University Press, 1987.
HEPCO. *Power at Niagara.* Toronto: Government of Ontario, 1970.
Immerman, Richard H. *John Foster Dulles: Piety, Pragmatism, and Power in US Foreign Policy.* Wilmington, DE: Scholarly Resources, 1999.
Jackson, J.H. "The St. Lawrence Power Project Rehabilitation: A Review of Major Features." *Engineering Journal* 43, 5 (February 1960): 67-73.
Jackson, John N. *The Welland Canals and Their Communities: Engineering, Industrial, and Urban Transformation.* Toronto: University of Toronto Press, 1997.
Jackson, John N., John Burtniak, and Gregory P. Stein. *The Mighty Niagara: One River, Two Frontiers.* Amherst, NY: Prometheus, 2003.

Jakobsson, Eva. "Industrialization of Rivers: A Water System Approach to Hydropower Development." *Knowledge, Technology, and Policy* 14, 4 (Winter 2002): 41-56.

Jasen, Patricia Jane. *Wild Things: Nature, Culture, and Tourism in Ontario, 1790-1914.* Toronto: University of Toronto Press, 1995.

Jeacle, Jean C. *To Make a House a Home: The Story of Ingleside, Ontario.* Osnabruck, ON: Township of Osnabruck, 1975.

Jenish, D'arcy. *The St. Lawrence Seaway: Fifty Years and Counting.* Manotick, ON: Penumbra Press, 2009.

Jenkins, Phil. *River Song: Sailing the History of the St. Lawrence.* Toronto: Viking, 2001.

Jermyn, Chris. "Some St. Lawrence Seaway Communities, 1959-1969." *Canadian Geographical Journal,* November 1969: 154-63.

Jobbitt, Steve. "Re-Civilizing the Land: Conservation and Postwar Reconstruction in Ontario, 1939-1961." MA thesis, Lakehead University, 2001.

Jockel, Joseph. *Canada in NORAD, 1957-2007: A History.* Montreal: McGill-Queen's University Press, 2007.

Jones-Imhotep, Edward. "Nature, Technology, and Nation." *Journal of Canadian Studies* 38 (2004): 5-36.

Josephson, Paul R. *Industrialized Nature: Brute Force Technology and the Transformation of the Natural World.* Washington, DC: Island Press/Shearwater, 2002.

Judson, Clara Ingram. *St. Lawrence Seaway.* New York: Follett, 1959.

Kahnawà:ke Revisited: The St. Lawrence Seaway. DVD. Directed by Kakwirano:ron Cook. Kakari:io Pictures and Millennium Productions, 1999.

Kaufman, Burton Ira. *Trade and Aid: Eisenhower's Foreign Economic Policy, 1953-1961.* Baltimore: Johns Hopkins University Press, 1982.

Kaufmann, Eric. "Naturalizing the Nation: The Rise of Naturalistic Nationalism in the United States and Canada." *Comparative Studies in Society and History* 40 (1998): 666-95.

Keating, Thomas F. *Canada and World Order: The Multilateralist Tradition in Canadian Foreign Policy.* 2nd ed. Toronto: Oxford University Press, 2002.

Keesbury, Forrest. "The Role of Dwight D. Eisenhower in the Development of the St. Lawrence Seaway." MA thesis, Bowling Green State University, 1965.

Kehoe, Terence. *Cleaning Up the Great Lakes: From Cooperation to Confrontation.* Dekalb: Northern Illinois University Press, 1997.

Kenny, James L., and Andrew G. Secord. "Engineering Modernity: Hydroelectric Development in New Brunswick, 1945-1970." *Acadiensis* 39, 1 (Winter/Spring 2010): 3-26.

Kirkey, Stephanie Ann. "From the Friendly City to the Seaway City: The Impacts of Deindustrialization and the St. Lawrence Seaway and Power Project on the Seaway Valley." MA thesis, Queen's University, 1997.

Klingensmith, Daniel. *"One Valley and a Thousand": Dams, Nationalism, and Development.* New York: Oxford University Press, 2007.

Konrad, Victor, and Heather N. Nicol. *Beyond Walls: Reinventing the Canada-United States Borderlands.* London: Ashgate, 2008.

Kottman, Richard. "The Diplomatic Relations of the United States and Canada, 1927-1941." PhD diss., Vanderbilt University, 1958.

–. "Herbert Hoover and the St. Lawrence Treaty of 1932." *New York History* 56 (July 1975): 314-40.

–. *Reciprocity and the North Atlantic Triangle, 1932-1938*. Ithaca, NY: Cornell University Press, 1968.

Kreinen, Mordechai E. "Trade Creation and Diversion by the St. Lawrence Seaway." *Review of Economics and Statistics* 43, 3 (August 1961): 295-97.

Lackenbauer, P. Whitney, and Matthew Farish. "The Cold War on Canadian Soil: Militarizing a Northern Environment." *Environmental History* 12, 4 (October 2007): 921-50.

Lafrenière, Normand. *Canal Building on the St. Lawrence: Two Centuries of Work, 1779-1959*. Ottawa: Parks Canada, 1983.

Lasserre, Frederic. "Water Diversion from Canada to the United States: An Old Idea Born Again?" *International Journal* 62, 1 (Winter 2006/2007): 81-92.

Lasserre, Jean-Claude. *Le Saint-Laurent, grande porte de l'Amérique*. LaSalle, QC: Hurtubise, 1980.

Le Prestre, Philippe G., and Peter J. Stoett. *Bilateral Ecopolitics: Continuity and Change in Canadian-American Environmental Relations*. London: Ashgate, 2006.

Legget, Robert Ferguson. *Rideau Waterway*. 2nd ed. Toronto: University of Toronto Press, 1986.

–. *Seaway*. Toronto: Clarke, Irwin, 1979.

Lennox, Patrick. *At Home and Abroad: The Canada-US Relationship and Canada's Place in the World*. Vancouver: UBC Press, 2009.

LesStrang, Jacques. *Seaway: The Untold Story of North America's Fourth Seacoast*. Seattle: Superior, 1976.

Levinson, Marc. *The Box: How the Shipping Container Made the World Smaller and the World Economy Bigger*. Princeton, NJ: Princeton University Press, 2006.

Linton, Jamie. *What Is Water? The History of a Modern Abstraction*. Vancouver: UBC Press, 2009.

Loo, Tina. "Disturbing the Peace: Environmental Change and the Scales of Justice on a Northern River." *Environmental History* 12, 4 (October 2007): 895-919.

–. "High Modernism and the Nature of Canada." Lecture, York University, Toronto, March 5, 2012.

–. "People in the Way: Modernity, Environment, and Society on the Arrow Lakes." *BC Studies* 142/143 (Summer/Autumn 2004): 161-96.

Loo, Tina, with Meg Stanley. "An Environmental History of Progress: Damming the Peace and Columbia Rivers." *Canadian Historical Review* 92, 3 (September 2011): 399-427.

The Lost Village of Aultsville. DVD. Ingleside, ON: Lost Villages Historical Society, 1988.

The Lost Villages: Then and Now. DVD. Ingleside, ON: Lost Villages Historical Society and Gaby's Video Productions, 1988.

The Lost Village of Wales. DVD. Ingleside, ON: Lost Villages Historical Society, 1988.

Mabee, Carleton. *The Seaway Story*. New York: Macmillan, 1961.

MacEachern, Alan. "A Little Essay on Big: Towards a History of Canada's Size." In *Big Country, Big Issues: Canada's Environment, Culture, and History*. Rachel Carson Center Perspectives 4 (2011): 6-15.

–. *Natural Selections: National Parks in Atlantic Canada, 1935-1970*. Montreal: McGill-Queen's University Press, 2001.

Macfarlane, Daniel. "'Caught between Two Fires': St. Lawrence Seaway and Power Project, Canadian-American Relations, and Linkage." *International Journal* 67, 2 (Summer 2012): 465-82.

–. "'A Completely Man-Made and Artificial Cataract': The Transnational Manipulation of Niagara Falls." *Environmental History*, Vol. 18, No. 4 (October 2013): 759-84.
–. "Creating a Cataract: The Transnational Manipulation of Niagara Falls to the 1950s." In *Urban Explorations: Environmental Histories of the Toronto Region*, edited by Colin Coates, Stephen Bocking, Ken Cruikshank, and Anders Sandberg, 251-67. Hamilton, ON: L.R. Wilson Institute for Canadian Studies-McMaster University, 2013.
–. "Rapid Changes: Canada and the St. Lawrence Seaway and Power Project." Toronto: Program on Water Issues (POWI), Munk School of Global Affairs, University of Toronto, September 2011. http://powi.ca/.
–. "To the Heart of the Continent: Canada and the Negotiation of the St. Lawrence Seaway and Power Project, 1921-1954." PhD diss., University of Ottawa, 2010.
Mackenzie, Hector. "Golden Decade(s)? Reappraising Canada's International Relations in the 1940s and 1950s." *British Journal of Canadian Studies* 23, 2 (2010): 207-32.
–. "Myths of the Golden Age of Canadian Diplomacy." Lecture, University of British Columbia, October 20, 2006.
MacLennan, Hugh. *Two Solitudes*. Toronto: Collins, 1945.
Maddox, Robert J. *From War to Cold War: The Education of Harry S. Truman*. Boulder, CO: Westview Press, 1988.
Mahant, E.E., and Graeme S. Mount. *An Introduction to Canadian-American Relations*. Toronto: Methuen, 1984.
–. *Invisible and Inaudible in Washington: American Policies toward Canada*. Vancouver: UBC Press, 1999.
Mahood, Harry R. "The St. Lawrence Seaway Bill of 1954: A Case Study of Pressure Groups in Conflict." *Southwestern Social Science Quarterly* 53 (September 1966): 141-49.
Malkus, Alida. *Blue-Water Boundary: Epic Highway of the Great Lakes and the St. Lawrence*. New York: Hastings House, 1960.
Maloney, Sean. *Canada and UN Peacekeeping: Cold War by Other Means, 1945-1970*. St. Catharines, ON: Vanwell, 2002.
–. "Why Keep the Myth Alive?" *Canadian Military Journal*, Spring 2007: 100-2.
Manore, Jean L. *Cross-Currents: Hydroelectricity and the Engineering of Northern Ontario*. Waterloo, ON: Wilfrid Laurier University Press, 1999.
Marcus, Maeva. *Truman and the Steel Seizure Case: The Limits of Presidential Power*. New York: Columbia University Press, 1977.
Marin, Clive, and Frances Marin. *Stormont, Dundas, and Glengarry, 1945-1978*. Belleville, ON: Mika, 1982.
Marine, Gene. *America the Raped: The Engineering Mentality and the Devastation of a Continent*. New York: Simon and Schuster, 1969.
Marks, Frederick W. *Power and Peace: The Diplomacy of John Foster Dulles*. Westport, CT: Praeger, 1993.
–. *Wind over Sand: The Diplomacy of Franklin Roosevelt*. 2nd ed. Athens: University of Georgia Press, 1988.
Marr, William I., and Donald G. Paterson, *Canada: An Economic History*. Toronto: Macmillan, 1980.
Martin, Kallen. "Akwesasne Environments, 1999: Relicensing a Seaway after a Legacy of Destruction." *Native Americas* 16, 1 (1999): 24-27.
Martin, Lawrence. *The Presidents and the Prime Ministers: Washington and Ottawa Face to Face; The Myth of Bilateral Bliss, 1867-1982*. Toronto: Doubleday, 1982.

Martin, Thibault, and Steven M. Hoffman, eds. *Power Struggles: Hydro-Electric Development and First Nations in Manitoba and Quebec*. Winnipeg: University of Manitoba Press, 2009.

Martin-Nielsen, Janet. "South over the Wires: Hydroelectricity Exports from Canada, 1900-1925." *Water History* 1 (2009): 109-29.

Massell, David. *Amassing Power: J.B. Duke and the Saguenay River, 1897-1927*. Montreal: McGill-Queen's University Press, 2000.

–. *Quebec Hydropolitics: The Peribonka Concessions of the Second World War*. Montreal: McGill-Queen's University Press, 2011.

–. "A Question of Power: A Brief History of Hydroelectricity in Quebec." In *Quebec Questions: Quebec Studies for the Twenty-First Century*, edited by Stephan Gervais, Christopher Kirkey, and Jarrett Rudy, 338-56. Oxford: Oxford University Press, 2011.

Massolin, Philip A. *Canadian Intellectuals, the Tory Tradition and the Challenge of Modernity, 1939-1970*. Toronto: University of Toronto Press, 2001.

Masters, Donald C. *Canada in World Affairs, 1953-1955*. Toronto: Oxford University Press, 1959.

Matte, Gilles, and Gilles Pellerin. *Carnets du St-Laurent*. Montreal: Heures Blues, 1999.

Mauch, Christof, and Thomas Zeller, eds. *Rivers in History: Perspectives on Waterways in Europe and North America*. Pittsburgh: University of Pittsburgh Press, 2008.

Maxwell, Norman James. "The Development of the St. Lawrence Waterway: A Factor in Canadian-American Relations." MA thesis, McMaster University, 1960.

Mayar, Harold M. "Great Lakes–Overseas: An Expanding Trade Route." *Economic Geography* 30, 2 (April 1954): 117-43.

McConville, Daniel J. "Seaway to Nowhere." *Invention and Technology*, Fall 1995: 34-44.

McCoy, Donald R. *The Presidency of Harry S. Truman*. Lawrence: University of Kansas Press, 1984.

McCully, Patrick. *Silenced Rivers: The Ecology and Politics of Large Dams*. London: Zed, 2001.

McDougall, Robert. "The All-Red Dream: Technological Nationalism and the Trans-Canada Telephone System." In *Canadas of the Mind: The Making and Unmaking of Canadian Nationalisms in the Twentieth Century*, edited by Norman Hillmer and Adam Chapnick, 46-62. Montreal: McGill-Queen's University Press, 2007.

McFarlane, H.M. "Backwater Computations for the St. Lawrence Seaway and Power Project, Part A: Hydraulic Engineering Aspects of Computations." *Engineering Journal* 43, 2 (1960): 55-66.

McKay, Paul. *Electric Empire: The Inside Story of Ontario Hydro*. Toronto: Between the Lines, 1983.

McKenty, Neil. *Mitch Hepburn*. Toronto: McClelland and Stewart, 1967.

McKinsey, Elizabeth R. *Niagara Falls: Icon of the American Sublime*. Cambridge: Cambridge University Press, 1985.

McNeill, J.R., and Corinna R. Unger. *Environmental Histories of the Cold War*. Cambridge: Cambridge University Press, 2010.

Melakopides, Costas. *Pragmatic Idealism: Canadian Foreign Policy, 1945-1995*. Montreal: McGill-Queen's University Press, 1998.

Melanson, Richard A., and David Mayers, eds. *Reevaluating Eisenhower: American Foreign Policy in the 1950s*. Urbana: University of Illinois Press, 1987.

Merchant, Livingston T., ed. *Neighbors Taken for Granted: Canada and the United States*. New York: Praeger, 1966.

Meren, David. *With Friends like These: Entangled Nationalisms and the Canada-Quebec-France Triangle, 1944-1970*. Vancouver: UBC Press, 2012.
Michaels, Anne. *The Winter Vault*. Toronto: McClelland and Stewart, 2009.
Missal, Alexander. *Seaway to the Future: American Social Visions and the Construction of the Panama Canal*. Madison, WI: University of Wisconsin Press, 2008.
Mitchell, Timothy. *Rule of Experts: Egypt, Techno-Politics, Modernity*. Los Angeles: University of California Press, 2002.
Molle, François, Peter P. Mollinga, and Philipp Wester. "Hydraulic Bureaucracies: Flows of Water, Flows of Power." *Water Alternatives* 2, 3 (October 2009): 328-49.
Montreal Research Council. *The Impact of the St. Lawrence Seaway on the Montreal Area*. Montreal: Montreal Research Council, McGill University, 1958.
Morgan, E.W. *"Up the Front": A Story of Morrisburg*. Toronto: Ryerson, 1964.
Motiuk, Laurence, and Canadian Institute of International Affairs. *A Reading Guide to Canada in World Affairs, 1945-1971*. Toronto: Canadian Institute of International Affairs, 1972.
Mouat, Jeremy. *The Business of Power: Hydro-Electricity in South Eastern British Columbia, 1897-1997*. Victoria: Sono Nis, 1997.
Moulton, Harold Glenn. *The St. Lawrence Navigation and Power Project*. Washington, DC: Brookings Institution, 1929.
Muirhead, Bruce. *Dancing around the Elephant: Creating a Prosperous Canada in an Era of American Dominance, 1957-1973*. Toronto: University of Toronto Press, 2007.
Mullin, Alex. "How Will the Seaway Plan Affect Your Fishing and Hunting?" *Forest and Outdoors* 49 (July 3, 1953): 6-7.
Murton, James. *Creating a Modern Countryside: Liberalism and Land Resettlement in British Columbia*. Vancouver: UBC Press, 2007.
–. "Creating Order: The Liberals, the Landowners, and the Draining of Sumas Lake, British Columbia." *Environmental History* 13, 1 (January 2008): 92-124.
Neatby, H. Blair. *William Lyon Mackenzie King, 1924-1932*. Vol. 2, *The Lonely Heights*. Toronto: University of Toronto Press, 1963.
–. *William Lyon Mackenzie King, 1932-1939*. Vol. 3, *The Prism of Unity*. Toronto: University of Toronto Press, 1976.
Nelles, H.V. *The Politics of Development: Forests, Mines, and Hydro-Electric Power in Ontario, 1849-1941*. 2nd ed. Montreal: McGill-Queen's University Press, 2005.
Nelson, Gordon, with Susan Fournier, Christopher Lemieux, Eric Tucs, Natalie Korobaylo, Costas Farassogiou, and Lucy Sportza. *The Great River: A Heritage Landscape Guide to the Upper St. Lawrence Corridor*. Environments Publication. Waterloo, ON: University of Waterloo, 2005.
Netherton, Alexander. "The Political Economy of Canadian Hydro-Electricity: Between Old 'Provincial Hydros' and Neoliberal Regional Energy Regimes." *Canadian Political Science Review* 1, 1 (2007): 107-24.
New, W.H. "The Great River Theory: Reading MacLennan and Mulgan." *Essays on Canadian Writing* 56 (Fall 1995): 162-82.
–. *Land Sliding: Imagining Space, Presence, and Power in Canadian Writing*. Toronto: University of Toronto Press, 1997.
Newbigin, Marion. *Canada, the Great River, the Lands and the Men*. Toronto: Harcourt, Brace, 1986.

Newman, Peter Charles. *Renegade in Power: The Diefenbaker Years*. Toronto: McClelland and Stewart, 1973.
Newton, Jim. *Eisenhower: The White House Years*. New York: Doubleday, 2011.
Nicholson, Patrick. *Vision and Indecision*. Don Mills, ON: Longman, 1968.
Nossal, Kim Richard. *The Politics of Canadian Foreign Policy*. 3rd ed. Scarborough, ON: Prentice Hall, 1997.
Nye, David E. *American Technological Sublime*. Cambridge, MA: MIT Press, 1996.
–. *Electrifying America: Social Meanings of a New Technology, 1880-1940*. Cambridge, MA: MIT Press, 1992.
Offner, Arnold A. *Another Such Victory: President Truman and the Cold War, 1945-1953*. Stanford, CA: Stanford University Press, 2002.
"The Ontario Historical Society and the Preservation of the Historic Values of the St. Lawrence Seaway Area." *Ontario History*, Spring 1956: 81-85.
Opp, James, and John C. Walsh, eds. *Placing Memory and Remembering Place in Canada*. Vancouver: UBC Press, 2010.
Osborne, Brian S., and Donald Swainson. *The Sault Ste. Marie Canal: A Chapter in the History of Great Lakes Transport*. Ottawa: National Historic Parks and Sites Branch, Parks Canada, 1986.
Otter, A.A. den. *The Philosophy of Railways: The Transcontinental Railway Idea in British North America*. Toronto: University of Toronto Press, 1997.
Ouellet, Marie-Claude. *Le Saint-Laurent: Un Fleuve à découvrir*. Montreal: Éditions de l'Homme, 1999.
Page, Donald M., and Canadian Institute of International Affairs. *A Bibliography of Works on Canadian Foreign Relations, 1945-1970*. Toronto: Canadian Institute of International Affairs, 1973.
Panitch, Leo, ed. *The Canadian State: Political Economy and Political Power*. Toronto: University of Toronto Press, 1977.
Parham, Claire Puccia. *From Great Wilderness to Seaway Towns: A Comparative History of Cornwall, Ontario, and Massena, New York, 1784-2001*. Albany: State University of New York Press, 2004.
–. "The St. Lawrence Seaway: A Bi-National Political Marathon." *New York History* 85, 4 (Summer 2004): 359-85.
–. *The St. Lawrence Seaway and Power Project: An Oral History of the Greatest Construction Show on Earth*. Syracuse, NY: Syracuse University Press, 2009.
Parr, Joy. *Sensing Changes: Technologies, Environments, and the Everyday, 1953-2003*. Vancouver: UBC Press, 2009.
Passfield, Robert W. "Construction of the St. Lawrence Seaway." *Canal History and Technology Proceedings* 22 (2003): 1-55.
Patch, S.P., and W.D. Busch. "Fisheries in The St. Lawrence River – Past and Present: A Review of Historical Natural Resources Information and Habitat Changes in the International Section of the St. Lawrence River." Cortland, NY: US Fish and Wildlife Service, 1984.
Pattison, Christopher. "The St. Lawrence Seaway Question, 1950-1954: The Canadian Perspective." MA thesis, Carleton University, 1994.
Pearse, Peter H., Francois Bertrand, James W. MacLaren, and Government of Canada, Inquiry on Federal Water Policy. *Currents of Change: Final Report, Inquiry on Federal Water Policy*. Ottawa: Government of Canada, Inquiry on Federal Water Policy, 1985.

Peet, S.E. "Long Lake Diversion: An Environmental Evaluation." MA thesis, University of Waterloo, 1978.
Pelletier, Louis-Raphael. "Revolutionizing Landscapes: Hydroelectricity and the Heavy Industrialization of Society and Environment in the Comté de Beauharnois, 1927-1948." PhD Diss., Carleton University, 2005.
Pennanen, Gary. "Battle of the Titans: Mitchell Hepburn, Mackenzie King, Franklin Roosevelt, and the St. Lawrence Seaway." *Ontario History* 89, 1 (March 1997): 1-21.
Perras, Galen Roger. *Franklin Roosevelt and the Origins of the Canadian-American Security Alliance, 1933-1945: Necessary, but Not Necessary Enough*. Westport, CT: Praeger, 1998.
Phillips, Stephanie K. "The Kahnawake Mohawks and the St. Lawrence Seaway." MA thesis, McGill University, 2000.
Pickersgill, J.W. *The Mackenzie King Record*. Toronto: University of Toronto Press, 1960.
–. *My Years with Louis St. Laurent: A Political Memoir*. Toronto: University of Toronto Press, 1975.
Piper, Liza. *The Industrial Transformation of Subarctic Canada*. Vancouver: UBC Press, 2009.
Porter, Dale H. *The Thames Embankment: Environment, Technology, and Society in Victorian England*. Akron, OH: University of Akron Press, 1997.
Porter, Theodore M. *Trust in Numbers: The Pursuit of Objectivity in Science and Public Life*. Princeton, NJ: Princeton University Press, 1995.
Preston, Richard Arthur. *The Defence of the Undefended Border: Planning for War in North America, 1867-1939*. Montreal: McGill-Queen's University Press, 1977.
Pritchard, Sara B. *Confluence: The Nature of Technology and the Remaking of the Rhône*. Cambridge, MA: Harvard University Press, 2011.
Quinn, Frank. "Water Diversion, Export and Canada-US Relations: A Brief History." Research paper. Toronto: Program on Water Issues, Munk Centre for International Studies, University of Toronto, August 2007.
Rae, K.J. *The Prosperous Years: The Economic History of Ontario, 1939-1975*. Toronto: University of Toronto Press, 1985.
Reid, Escott. *Radical Mandarin: The Memoirs of Escott Reid*. Toronto: University of Toronto Press, 1989.
Resnick, Philip. *The Masks of Proteus: Canadian Reflections on the State*. Toronto: University of Toronto Press, 1990.
Reuss, Martin, and Stephen Cutcliffe, eds. *The Illusory Boundary: Environment and Technology in History*. Charlottesville: University of Virginia Press, 2010.
Richardson, Elmo. *Dams, Parks and Politics: Resource Development and Preservation in the Truman-Eisenhower Era*. Lexington: University Press of Kentucky, 1973.
Richardson, Ronald, Walter G. Rooke, and George McNevin. *Developing Water Resources: The St. Lawrence Seaway and the Columbia/Peace Power Projects*. Toronto: Ryerson Press, 1969.
Ritchie, Charles. *Diplomatic Passport: More Undiplomatic Diaries, 1946-1962*. Toronto: Macmillan, 1981.
A River Lost: The Story of the St. Lawrence Seaway. DVD. Produced by David Jones. Morrisburg, ON: David Jones Productions in association with KAV Productions and the Lost Villages Historical Society, 2006.
Robinson, H. Basil. *Diefenbaker's World: A Populist in Foreign Affairs*. Toronto: University of Toronto Press, 1989.
Roemer, Angelika. *The St. Lawrence Seaway, Its Ports and Its Hinterland*. Tübingen, Germany: Tübinger geographisce Studien, 1981.

Bibliography

Rolin, F. "The St. Lawrence Seaway and Power Project: A Study of Pressure Groups at Work." MA thesis, University of California, Berkeley, 1955.

Russell, Peter, ed. *Nationalism in Canada.* Toronto: McGraw-Hill, 1966.

Rutley, Rosemary. *Voices from the Lost Villages.* Ingleside, ON: Old Crone, 1998.

Savard, Stéphane. "Retour sur un projet du siècle: Hydro-Québec comme vecteur des représentations symboliques et identitaires du Québec, 1944 à 2005." PhD diss., Université Laval, 2010.

Saywell, John T. *Just Call Me Mitch: The Life of Mitchell F. Hepburn.* Toronto: University of Toronto Press, 1991.

Schama, Simon. *Landscape and Memory.* Toronto: Random House, 1995.

Scharf, David Nelson. "The Effect of the St. Lawrence Seaway Project on Recreational Land Use in the International Rapids Section of the Seaway Valley." MA thesis, Carleton University, 1970.

Scott, James C. "High Modernist Social Engineering: The Case of the Tennessee Valley Authority." In *Experiencing the State,* edited by Lloyd I. Rudolph and John Kurt Jacobsen, 3-52. Toronto: Oxford University Press, 2006.

–. *Seeing like a State: How Certain Schemes to Improve the Human Condition Have Failed.* New Haven, CT: Yale University Press, 1998.

Shelvey, Bruce H. "Skagit Scenes: Landscape Formation in the Pacific Northwest." PhD diss., Arizona State University, 1999.

Sheriff, Carol. *The Artificial River: The Erie Canal and the Paradox of Progress, 1817-1862.* New York: Hill and Wang, 1996.

Shields, Anne-Marie. *Lost Villages, Found Communities: A Pictorial History of the Lost Villages of the St. Lawrence Seaway.* Cornwall, ON: Astro Printing, 2004.

Skibsrud, Johanna. *The Sentimentalists.* Vancouver: Douglas and McIntyre, 2010.

Smith, Denis. *Diplomacy of Fear: Canada and the Cold War, 1941-1948.* Toronto: University of Toronto Press, 1988.

–. *Rogue Tory: The Life and Legend of John G. Diefenbaker.* Macfarlane Walter and Ross, 1997.

Smith, George David. *From Monopoly to Competition: The Transformation of Alcoa, 1888-1986.* New York: Cambridge University Press, 1988.

Smith, Jean Edward. *Eisenhower in War and Peace.* New York: Random House, 2011.

Snider, Thomas J. *Power Dam Politics: Dealing with the Politics of Power at the Local Level; An Insider's Story.* Ashland, OH: Bookmasters, 2003.

Spencer, Richard M. "Winter Navigation on the Great Lakes and St. Lawrence Seaway: A Study in Congressional Decision Making." MA thesis, Cornell University, 1992.

Spencer, Robert. *Canada in World Affairs: From UN to NATO, 1946-1949.* Toronto: Oxford University Press, 1959.

Spencer, Robert, John Kirton, and Kim Richard Nossal, eds. *The International Joint Commission Seventy Years On.* Toronto: Centre for International Studies, University of Toronto, 1981.

Spieler, Cliff, and Tom Hewitt. *Niagara Power: From Joncaire to Moses.* Lewiston, NY: Niagara Power, 1959.

Sproule-Jones, Mark, Carolyn Johns, and B. Timothy Heinmiller, eds. *Canadian Water Politics: Conflicts and Institutions.* Montreal: McGill-Queen's University Press, 2008.

"The St. Lawrence Seaway: Gateway to the World." Canadian Broadcasting Corporation Digital Archives. http://archives.cbc.ca/.

Stacey, C.P. *Canada and the Age of Conflict: A History of Canadian External Policies.* 2 vols. Toronto: Macmillan, 1977 and 1981.

–. *A Very Double Life: The Private World of Mackenzie King*. Toronto: Macmillan, 1977.
Stagg, Ronald. *The Golden Dream: A History of the St. Lawrence Seaway*. Toronto: Dundurn, 2010.
Stanley, Meg. *Voices from Two Rivers: Harnessing the Power of the Peace and Columbia*. Vancouver: Douglas and McIntyre, 2011.
Steinberg, Theodore. *Nature Incorporated: Industrialization and the Waters of New England*. Cambridge: Cambridge University Press, 1991.
–. "'That World's Fair Feeling': Control of Water in Twentieth Century America." *Technology and Culture* 34 (April 1993): 401-9.
Stephens, George Washington. *The St. Lawrence Waterway Project: The Story of the St. Lawrence River as an International Highway for Water-Borne Commerce*. Montreal: L. Carrier, 1930.
Stewart, Gordon T. *The American Response to Canada since 1776*. East Lansing, MI: Michigan State University Press, 1992.
Stokes, Peter. "St. Lawrence, a Criticism." *Canadian Architect* 3, 2 (February 1958): 43-48.
–. *A Village Arising: The Story of the Building of Upper Canada Village, 1957-1961 and After*. Port Hope, ON: ATS-PJS, 2011.
Story, Donald, and R. Bruce Shepard, eds. *The Diefenbaker Legacy: Canadian Politics, Law, and Society since 1957*. Regina: Canadian Plains Research Centre, 1998.
Stradling, David. *The Nature of New York: An Environmental History of the Empire State*. Ithaca, NY: Cornell University Press, 2010.
Stuart, Reginald C. *Dispersed Relations: Americans and Canadians in Upper North America*. Washington, DC: Woodrow Wilson Center Press and Johns Hopkins University Press, 2007.
Stunden Bower, Shannon. *Wet Prairie: People, Land, and Water in Agricultural Manitoba*. Vancouver: UBC Press, 2011.
Stursberg, Peter. *Diefenbaker: Leadership Gained, 1956-62*. Toronto: University of Toronto Press, 1975.
Styran, Roberta M., and Robert R. Taylor. *The Great Swivel Link: Canada's Welland Canal*. Toronto: Champlain Society, 2001.
–. *This Great National Object: Building the Nineteenth-Century Welland Canals*. Montreal: McGill-Queen's University Press, 2012.
Submerged. DVD. Directed by John Earle and Frank Burelle. Cornwall, ON: Fishrizzo Productions, 2007.
Sunken Villages. Louis Helbig. http://www.louishelbig.com/.
Sussman, Gennifer. *Quebec and the St. Lawrence Seaway*. Montreal: C.D. Howe Institute, 1979.
–. *The St. Lawrence Seaway: History and Analysis of a Joint Water Highway*. Montreal: C.D. Howe Research Institute, 1978.
Swainson, Neil A. *Conflict over the Columbia: The Canadian Background to an Historic Treaty*. Montreal: McGill-Queen's University Press, 1979.
Swanson, Roger Frank. *Intergovernmental Perspectives on the Canada-US Relationship*. New York: New York University Press, 1978.
Swettenham, John. *McNaughton*. 3 vols. Toronto: Ryerson Press, 1968-69.
Swift, Jamie, and Keith Stewart. *Hydro: The Decline and Fall of Ontario's Electric Empire*. Toronto: Between the Lines, 2004.

Swyngedouw, Erik. "Modernity and Hybridity: Nature, *Regeneracionismo,* and the Production of the Spanish Waterscape, 1890-1930." *Annals of the Association of American Geographers* 89, 3 (September 1999): 443-65.
Tallamy, B.D., and T.M. Sedweek. *The St. Lawrence Seaway Project.* Buffalo, NY: Niagara Frontier Planning Board, 1940.
Tansill, Charles Callan. *Canadian-American Relations, 1875-1911.* Gloucester, MA: Smith, 1964.
Taylor, Graham D., and Peter A. Baskerville. *A Concise History of Business in Canada.* Toronto: Oxford University Press, 1994.
Teigrob, Robert. *Warming Up to the Cold War: Canada and the United States' Coalition of the Willing, from Hiroshima to Korea.* Toronto: University of Toronto Press, 2009.
Thomas, Evan. *Ike's Bluff: President Eisenhower's Secret Battle to Save the World.* New York: Little, Brown, 2012.
Thomas, Lowell. *Story of the St. Lawrence Seaway.* Buffalo, NY: H. Stewart, 1958.
Thompson, Dwayne T. "The St. Lawrence Project: A Case Study in American Politics." PhD diss., George Peabody School of Education, 1957.
Thompson, John Herd, and Stephen J. Randall. *Canada and the United States: Ambivalent Allies.* 4th ed. Montreal: McGill-Queen's University Press, 2008.
Thomson, Dale C. *Louis St. Laurent: Canadian.* Toronto: Macmillan, 1967.
Todd, Eric C.E. *The Law of Expropriation and Compensation in Canada.* 2nd ed. Toronto: Carswell, 1992.
Toye, William. *The St. Lawrence.* Toronto: Oxford University Press, 1959.
Treasures of the Lost Villages. DVD. Directed by John Earle and Frank Burelle. Cornwall, ON: Fishrizzo Productions, 2008.
Trofimenkoff, Susan Mann. *The Dream of Nation: A Social and Intellectual History of Quebec.* Toronto: Macmillan, 1982.
Truman, Harry S. *Memoirs.* Vol. 1, *Year of Decisions.* Garden City, NY: Doubleday, 1955.
–. *Memoirs.* Vol. 2, *Years of Trial and Hope.* Garden City, NY: Doubleday, 1956.
Tucker, Richard P., and Edmund Russell, eds. *Natural Enemy, Natural Ally: Toward an Environmental History of Warfare.* Corvallis: Oregon State University Press, 2004.
Turpin, Trevor. *Dam.* London: Reaktion, 2008.
Tvedt, Terje, and Richard Coopey, eds. *The Political Economy of Water.* A History of Water 1, 2. London: I.B. Tauris, 2006.
–, eds. *Rivers and Society: From Early Civilizations to Modern Times.* A History of Water 2, 2. London: I.B. Tauris, 2010.
Tvedt, Terje, and Eva Jakobsson, eds. *Water Control and River Biographies.* A History of Water 1, 1. London: I.B. Tauris, 2006.
Tvedt, Terje, and T. Oestigaard, eds. *Ideas of Water from Ancient Societies to the Modern World.* A History of Water 2, 1. London: I.B. Tauris, 2009.
–. *The World of Water.* A History of Water 1, 3. London: I.B. Tauris, 2006.
Van der Aa, Hans. *Gateway to the World: A Picture Story of the St. Lawrence Seaway.* Montreal: Chomedy, 1959.
Van der Vleuten, Erik, and Cornelius Disco. "Water Wizards: Reshaping Wet Nature and Society." *History and Technology* 20, 3 (October 2009): 328-49.
Van Huizen, Philip. "Building a Green Dam: Environmental Modernism and the Canadian-American Libby Dam Project." *Pacific Historical Review* 79, 3 (August 2010): 418-53.

Vigod, Bernard L. *Quebec before Duplessis: The Political Career of Louis-Alexandre Taschereau*. Montreal: McGill-Queen's University Press, 1986.
Wagner, Eric. "The Peaceable Kingdom? The National Myth of Canadian Peacekeeping and the Cold War." *Canadian Military Journal*, Winter 2006/2007: 45-54.
Wagner, James Richard. "Partnership: American Foreign Policy toward Canada, 1953-57." PhD diss., University of Colorado, 1966.
Waldrum, James. *As Long as the Rivers Run: Hydroelectric Development and Native Communities*. Winnipeg: University of Manitoba Press, 1993.
Warkentin, John, ed. *So Vast and Various: Interpreting Canada's Regions in the Nineteenth and Twentieth Centuries*. Montreal: McGill-Queen's University Press, 2011.
Warshaw, Shirley Anne, ed. *Reexamining the Eisenhower Presidency*. Westport, CT: Greenwood Press, 1993.
Waterbury, John. *Hydropolitics of the Nile Valley*. Syracuse, NY: Syracuse University Press, 1979.
Waterway to the Heartland: Saga of Long Sault. DVD. Ottawa: Carleton Productions, 1999.
Webster, David. *Fire and the Full Moon: Canada and Indonesia in a Decolonizing World*. Vancouver: UBC Press, 2010.
Weller, Phil. *Fresh Water Seas: Saving the Great Lakes*. Toronto: Between the Lines, 1990.
The Welland Canals: Past and Present. DVD. St. Catharines, ON: St. Catharines Museum, 2008.
Wheeler, Maggie. *All Mortall Things*. Renfrew, ON: General Store Publishing House, 2006.
–. *The Brother of Sleep*. Renfrew, ON: General Store Publishing House, 2004.
–. *On a Darkling Plain*. Renfrew, ON: General Store Publishing House, 2009.
–. *A Violent End*. Renfrew, ON: General Store Publishing House, 2007.
Whitaker, Reginald, and Steve Hewitt. *Canada and the Cold War*. Toronto: James Lorimer, 2003.
Whitaker, Reginald, and Gary Marcuse. *Cold War Canada: The Making of a National Insecurity State, 1945-1957*. Toronto: University of Toronto Press, 1994.
White, Arthur V. *Long Sault Rapids, St. Lawrence River: An Enquiry into the Constitutional and Other Aspects of the Project to Develop Power Therefrom*. Ottawa: Mortimer, 1913.
White, Randall. *Ontario, 1610-1985: A Political and Economic History*. Toronto: Dundurn Press, 1996.
White, Richard. *The Organic Machine: The Remaking of the Columbia River*. New York: Hill and Wang, 1995.
Williams, William Appleman. *The Tragedy of American Diplomacy*. Rev. and enl. ed. New York: Dell, 1962.
Willoughby, William R. *The St. Lawrence Waterway: A Study in Politics and Diplomacy*. Madison, WI: University of Wisconsin Press, 1961.
Wilson, James W. *People in the Way: The Human Aspect of the Columbia River Project*. Toronto: University of Toronto Press, 1973.
Wirth, John D. *Smelter Smoke in North America: The Politics of Transborder Pollution*. Lawrence: University of Kansas Press, 2000.
Witol, Gregory, ed. *The St. Lawrence Seaway and Quebec*. Nepean, ON: Naval Officers' Association of Canada, 1997.
Wittfogel, Karl. *Oriental Despotism: A Comparative Study of Total Power*. New Haven, CT: Yale University Press, 1957.

Woo, Ming-Ko. "Water in Canada, Water for Canada." *Canadian Geographer* 45, 1 (2001): 85-92.
Wood, Harold A. "Recreational Land Use Planning in the St. Lawrence Seaway Area, Ontario." *Community Planning Review,* March 1955: 23-30.
–. "The St. Lawrence Seaway and Urban Geography, Cornwall-Cardinal, Ontario." *Geographical Review* 45, 4 (1955): 509-30.
Wood, Donald F. "The St. Lawrence Seaway: Some Considerations of Its Impact." *Land Economics* 34, 1 (February 1958): 61-73.
Worster, Donald. *Rivers of Empire: Water, Aridity, and the Growth of the American West.* New York: Pantheon, 1985.
–. "Wild, Tame, and Free: Comparing Canadian and US Views of Nature." In *Parallel Destinies: Canadian-American Relations West of the Rockies,* edited by John M. Findlay and Kenneth S. Coates, 246-73. Montreal: McGill-Queen's University Press, 2002.
Wright, Conrad Payling. *The St. Lawrence Deep Waterway: A Canadian Appraisal.* Toronto: Macmillan, 1935.
Wynn, Graeme. *Canada and Arctic North America: An Environmental History.* Santa Barbara: ABC-CLIO, 2006.

Index

Note: SLSPP stands for St. Lawrence Seaway and Power Project; "(f)" after a page number indicates a figure; "(m)" after a page number indicates a map; "(t)" after a page number indicates a table

Acheson, Dean, 69, 81
Adams Island, 181
advances (technological, scientific, engineering) on SLSPP, 135-37. *See also* construction of SLSPP; expertise
Advisory Committee on Reconstruction, Canada, 45
Aguasabon River, 45
aide-mémoire of August 12, 1954, 102-3
Akwesasne. *See* St. Regis (Akwesasne) First Nation and SLSPP
Alaskan boundary dispute, impact on Canadian-British-American relations, 23
Albany River (Ontario), diversion of, 32. *See also* Ogoki and Long Lac (Lake) diversions
Albany River (New York), 176
Alcan (Alaska) Highway, xi
Alcoa. *See* Aluminum Company of America
alewife, 205
Alexander, Jeff, 204-5
all-Canadian seaway, xxiv, 9, 14, 34, 36, 46, 50, 57-67, 70, 76, 80, 86, 89, 93-108, 207-18; as bluff, 13, 53, 56, 63, 65, 72-73; Canadian Cabinet division over, 94-99; Canadian decision for, 60-63, 69-72, 74-75; as economic and security threat to US, 9, 74, 76, 87, 93; future development of, 101-8, 119-20, 121, 130, 198, 209; as gamble, 95, 214-15; parallel lock at Cornwall, 98, 101, 119; renouncing of, 277n43; threats to by US, 10, 52, 80-85, 88-90, 97-101, 208, 214, 261n18. *See also* Federal Power Commission (FPC), US; United States, alternate deep waterway routes
Aluminum Company of America (Alcoa), 24; application for dam in St. Lawrence, 24, 31; diversion canal at Massena, 24, 31, 190
aluminum industry, 49, 63-64
American Superpower Company, 31
Anderson, Robert, 99, 101-2, 119
anti-Americanism in Canada, xx, 10, 64, 213
aquaculture, ecological impact of SLSPP on, 200-6. *See also* environmental impact of SLSPP

Index   311

Army Corps of Engineers (US), 27, 35, 37-38, 42, 113-15, 133, 141, 187, 189, 191, 230. *See also* engineering plans for SLSPP: evolution of; engineers and SLSPP; expertise
Association of American Railroads, 27, 82. *See also* railroads
Aswan High Dam, xvi, 221, 281n14
Atlantic Canada, opinion of SLSPP in, 28, 41, 56. *See also* public opinion and SLSPP, in Canada
Atlantic Ocean, 5, 21, 22, 35
Atwood, Margaret, 229
Ault Park, xv, 173
Aultsville, xv, 137, 139, 173, 223(f)
automobile industry, 63, 84

Barnhart Island, 24, 25, 31, 37, 42, 126, 130, 162, 179
Beauharnois Canal, 33, 44, 117, 129-30, 244n36
Beauharnois Locks, 129
Beauharnois Power Dam, 33, 44, 179, 244n36
Beck, Sir Adam, 29
Bennett, R.B., government of, 13, 34-40, 54
Berton, Pierre, xiii
Bertrand H. Snell Lock, 114, 133, 196(f), 262n28
bioregion, St. Lawrence as, xxi, 16, 227. *See also* borders, in environmental history
blueprints, 132. *See also* engineering plans for SLSPP; engineers and SLSPP; expertise
Bonneville Dam, 220
Borden, Robert, government of, and SLSPP, 26-27
border, bisecting power dam, 7
borderlands, approaches to, 10-11, 236n10
borders, in environmental history, xxi-xxii. *See also* bioregion, St. Lawrence as
Bouchette, Joseph, xvii-xviii
Boulder Dam, 220, 282n30
Boundary Waters Treaty (1909), 9, 23, 120
Brebner, John Bartlet, xix, 235n8

bridge, types, 263n48
Britannia (royal yacht), 193
Brockville, Ontario, 24
Brownell, Herbert, 92
Brownell, Jim, 172
Brownell, Vale, 173
brute-force technology, xiv, 16
buildings, movement of due to seaway, xv, 45. *See also* relocation, connected to SLSPP
Bureau of Reclamation (US) dams, 22

Caldwell Linen Mills, 151. *See also* Iroquois (town of)
Canada, relationship to Britain, xix, xx, 11, 48, 218, 230
Canada, relationship to US. *See* Canada-US relationship
Canada: The Great River, the Lands and the Men (Newbigin), xviii-xix
Canada Cotton Mills, 144
Canada-US relations. *See* Canada-US relationship
Canada-US relationship: American approach to Canada, 10, 88, 212-14; asymmetry in, 213-14; continuity in, 88, 280n7; formal vs informal relationship, 10-11, 106; impact of SLSPP on, 13, 18, 65-67, 74, 76, 88-89, 92-96, 99, 105-7, 121, 208, 212, 230; nature of, xx, 9-11, 48-49, 63-66, 76, 87-88, 106-7, 212-15, 235n7; routine vs punctuated issues, 213. *See also* foreign policy, Canada; joint seaway; linkage, in Canada-US relations
Canadian Commission of Conservation, 25, 242n11
Canadian Deep Waterways and Power Association, 26
Canadian National Railway. *See* railroads
Canadian Pacific Railway. *See* railroads
canals: compensation for 14-foot, 74, 98-105; maintenance of 14-foot canals during SLSPP construction, 119; as nation building, 16, 21, 23, 79; pre-SLSPP, 14, 21, 22, 22(m), 26(f), 43, 63, 79, 102, 176, 195, 242n2
Cardinal, Ontario, 132

Carnegie Endowment for International Peace: series on Canadian-American relations, xix-xx, 235n8
Carr, Emily, 229
Cartier, Jacques, xvii-xviii
Casgrain, J.P., 30
Castle, Lewis, 116-18
Caughnawaga. *See* Kahnawake First Nation and SLSPP
Cavell, Janice, 15
cemeteries, affected by SLSPP, 46, 140, 162, 167, 264n57, 270n69. *See also* Lost Villages and SLSPP; rehabilitation, connected to SLSPP; relocation, connected to SLSPP
ceremonies, for SLSPP, 100(f), 104(f), 115, 180(f), 192-94, 194(f), 195(f), 208, 261n10
certiori, Canadian petition for American use of, 257n52. *See also* Federal Power Commission (FPC), US: and PASNY power licence
Challies, George H., 158, 269n55
Chevrier, Lionel, 11, 60-62, 64-65, 71-73, 88, 94, 101, 116, 125-28, 215
Chicago diversion, 32, 35-36, 38-39, 44, 189-90, 245n50, 246n58
churches, affected by SLSPP, 156, 163, 165, 167-68, 170-71, 173-74, 272n11. *See also* Lost Villages and SLSPP; rehabilitation, connected to SLSPP; relocation, connected to SLSPP
Citizens' Joint Action Committee, 67, 68(f)
claims, by construction agencies, 137, 264n65
Clark, Bonnie, 172
Claxton, Brooke, 56, 215
Coates, Wells, plans for Iroquois, 150-52. *See also* Iroquois (town of)
cofferdams, 131, 132(f), 167
Cold War, xvi, 8, 10, 17, 49, 63, 77
Coles Creek Marina, 162
Columbia River Treaty, and power developments, 7, 215, 220
combined single-stage dam. *See* engineering plans for SLSPP: evolution of

Commercial Empire of the St. Lawrence (Creighton), 13
common works dredging controversy around Cornwall Island, 99-105, 119-21, 126-27, 205, 262n26
computer, first used in Canada, 137
Conference of Canadian Engineers, 246n54. *See also* engineers and SLSPP
Congress (US), and SLSPP, 22, 24-29, 38-39, 49-52, 59, 66, 70-73, 75, 80-81, 84, 86-87, 91, 93, 95, 118-19, 190, 208, 214-15
connecting channels, in Great Lakes basin, 33, 37, 190, 276n26
consent, manufacturing of. *See* high modernism; high modernism, negotiated
constitution, Canada, and jurisdiction over water resources, 31, 35, 40-41, 43, 62-63. *See also* hydroelectricity: federal vs provincial development of
constitution, US, and jurisdiction over water resources, 70-71. *See also* hydroelectricity: federal vs provincial development of
construction of SLSPP: accelerated schedule of, 112-13, 135, 137; difficulties, 119, 133; equipment, 111, 130-31, 131(f), 137, 264n51; flooding, amount of land, 7, 111, 128, 140; number of people moved by, 7; raising of power pool, 135, 167; rights to water and power from, 104; subsurface conditions, 122, 132-33, 137, 188; technical aspects and measurements of, xiv, 111, 130, 133, 273n1; unfinished work, 192, 276n32; in winter, 133, 134(f), 136-37
container shipping, 197-98
continental defence. *See* national security, and SLSPP
continentalist view of Canadian-American relationship, xix-xx, 235n8. *See also* critical nationalist view of Canadian-American relationship; Laurentian thesis
Cook, Ramsay, 78
Cornwall, Ontario, 7, 24, 29, 31, 62, 122, 144, 156

Cornwall Canal, 21, 63, 124
Cornwall Island, 119-22
Cornwall Township, 124, 156
Côte Ste-Catherine Lock, 124, 128
cottages, relocation of, 140, 162-63, 269n68. *See also* Lost Villages and SLSPP; rehabilitation, connected to SLSPP; relocation, connected to SLSPP
Cox, Gordon, 81-82
Creighton, Donald, xix, 3, 13
critical nationalist view of Canadian-American relationship, xx. *See also* continentalist view of Canadian-American relationship; Laurentian thesis
critical path methods, SLSPP as forerunner of, 137
Crysler's Farm, Battle of: flooding of, 46, 165, 222
Crysler Island Power Dam, proposed, 37
Crysler Park, 164

Danielian, N.R., 50, 81-82, 249n18
Dawn of Conservation Diplomacy: US-Canadian Wildlife Protection Treaties in the Progressive Era (Dorsey). *See* Dorsey, Kurkpatrick
de Luccia, E. Robert, 65
deaths and injuries on the job, 135. *See also* labour and working conditions on SLSPP
DeCew Falls, hydroelectric station, 55
deep waterway. *See* St. Lawrence Seaway and Power Project (SLSPP)
deference to authority, and SLSPP, 127-28, 159-60, 168, 176, 224, 271n94
democracies, and environmental impact, xxii, 178, 227
Department of External Affairs (Canada), 48, 70, 81-82, 94, 105-6, 209, 213
Department of Planning and Development (Ontario), 141, 146, 149, 153, 176. *See also* Lost Villages and SLSPP; Ontario, Government of, involvement in relocation and rehabilitation of dislocated communities; rehabilitation, connected to SLSPP; relocation, connected to SLSPP

Department of State (US), 10, 32, 36, 39, 53, 66-67, 71, 80, 214
Department of the Interior (US), 53, 58-59, 83
Department of Transport (Canada), 115, 123
Depression, 35, 62
Detroit River, 5, 190
Dewey, Thomas, 52, 54, 59
Diabo, Louis, 125
d'Iberville (shipbreaker), 192
Dickinson's Landing, xv, 139, 153
Diefenbaker, John G., 125, 193
dikes, 45, 119, 126, 130, 132, 136, 144-45, 204
division of costs agreements: between HEPCO and PASNY, 117-18, 146, 190; between Ontario and federal government (1932), 37; between Ontario and federal government (1941), 43-44; between Ontario and federal government (1951), 69-71, 75; between Ontario and Quebec (1943), 43; between PASNY and US Army Corps of Engineers (1932), 38; between PASNY and US Army Corps of Engineers (1941), 44. *See also* St. Lawrence Seaway and Power Project
Dondero Bill. *See* Congress (US), and SLSPP
Dorsey, Kurkpatrick, xxi-xxii
Drew, George, 52
dual-stage dam. *See* engineering plans for SLSPP: evolution of
Dulles, John Foster, 92, 94, 256n40
Duplessis, Maurice, 40-41, 63, 70, 116-17
Dwight D. Eisenhower Lock, 115, 130, 133, 138, 193, 197(f), 262n28

earth tilt, phenomenon of, 181. *See also* engineering plans for SLSPP; expertise
earthquake (1944) near Cornwall and Massena, 181
ecological impact of SLSPP. *See* environmental impact of SLSPP
economic growth, in North America during early Cold War period, 62-64

eel (American): impact of SLSPP on, 200, 206; ladders, 200

Eisenhower, Dwight D., administration of, 76, 83-84, 87-90, 114, 120, 160, 190, 214

electricity: export contracts from Quebec to Ontario for, 32; growth of distribution systems, 23, 175; impact of SLSPP on prices and distribution, 175-76, 273n126; as progress, 23-24, 221; shortages, 28, 31, 32, 45. *See also* hydroelectricity

Elizabeth II, Queen of Great Britain, 180(f), 193-94

empire, of the St. Lawrence, 13-14, 75, 230. *See also* Laurentian thesis

Empire of the St. Lawrence (Creighton), 13

energy resources, early Cold War Canadian development of, 64

engineering plans for SLSPP: evolution of, 24-25, 28, 31-32, 36-37, 41-42, 46, 87, 112, 123-24, 128-29, 133, 135, 180-89, 218-19, 282n23; and water levels, 181-89, 225, 274n3, 274n5, 275n20, 276n23. *See also* engineers and SLSPP; expertise

engineers and SLSPP, 141, 180-89, 225, 275n17; from foreign countries viewing SLSPP, 130; mistakes, 117-18, 137-38, 170, 181-89, 207, 225; pressures on, 183-84; public image of, 183-84, 225; uncertainty and rationalization, 183-84, 187. *See also* expertise

environment, attitudes toward. *See* expertise; nationalism, Canadian

environmental diplomacy, xxi-xxii, 9, 23, 77, 212, 215-16

Environmental Histories of the Cold War (edited volume), xxi

environmental impact of SLSPP, xvi, 126, 130, 167, 172, 178, 180, 200-6, 224-26; consideration and studies by involved states of, 200, 202-4, 278n56; creation of different nearshore aquatic habitat, 202; flora and flauna, 202-6; on flow regimes and fluctuations, 202-6; on ice formation, 202, 205; and International Joint Commission, 205-7; on muskrat, 206; on pollution, 202, 206; removal and alteration of islands, 129, 202, 206; and rise of environmental movement, 206; on scouring and silting processes, 202, 206; on shoreline erosion, 204, 206; on St. Regis (Akwesasne) First Nation, 205-7; and Upper Canada Migratory Bird Sanctuary, 204; on water conditions, 201; and Wilson Hill conservation project, 204. *See also* eel (American); fish, impact of SLSPP on; invasive species

envirotech, approaches of, 16

envirotechnical nationalism. *See* nationalism, Canadian

Erie Canal, 21, 266n14

exchanges of notes between Canada and US regarding SLSPP: (1952) June 30, 73, 86, 102; (1952) November 4, 83; (1953) November, 92; (1954) August 17, 102-6

executive agreement. *See* Great Lakes–St. Lawrence Basin Agreement (1941)

expertise: and bureaucrat as hero, 76; as Cold War weapon, 17; and engineer as hero, 77-80; engineering and scientific, 7, 25, 180-89, 218-30; limits of and faults with, 25, 180-89, 228; technocratic, 17, 76; transnational spread of, 221-22, 282n31. *See also* high modernism; technology

Farran's Point, xv, 139, 153, 173

Federal Power Commission (FPC), US, 31, 53, 58; and PASNY power licence, 50, 53-55, 59-60, 65, 72-74, 76, 81-84, 89-93, 95, 97, 143, 175, 208, 214, 256n40

federalism, impact of SLSPP on, 218

Ferguson, Howard, 32, 34-35

fire code, revision of Canadian, 137

First Nations, 3, 49, 164; and colonial assimilation in SLSPP, 127-28. *See also* Kahnawake First Nation and SLSPP; St. Regis (Akwesasne) First Nation and SLSPP

First World War, 10, 27-28, 33

fish, impact of SLSPP on, 200-6, 227, 277n50. *See also* environmental impact of SLSPP

Index

fishways, decision not to utilize, 200, 270n81
foreign policy, Canada: cultural sources of, 11, 236n11; domestic sources of, 217; environmental and technological sources of, 11; functionalism, 54; golden age, 48, 54, 212; as middle power, 48; as self-interested and pragmatic, 212, 215; support for US Cold War aims, 48. *See also* Canada-US relationship
FPC. *See* Federal Power Commission (FPC), US
Francis, R. Douglas, 79. *See also* technological imperative
Franklin D. Roosevelt power house. *See* Robert Moses–Robert H. Saunders Power Dam
frazil ice. *See* ice, control of formation of
frequency standardization, 142, 150, 267n32
the Front. *See* Lost Villages and SLSPP
Frost, Leslie, 64, 70, 145-46, 148, 159
Frye, Northop, 229

Galop Island, 181
General Motors, 175, 206
Georgian Bay Ship Canal, 26
Gilbert, Jess, 227
glacial till, 137
Godbout, Adélard, 42
Gordon Report, 64
government, systems of: Canada vs United States, 12, 28, 57, 249n29
grain and SLSPP, 61(t), 76, 199
Grand Coulee Dam, 7, 282n30
Grand Trunk Railway. *See* railroads
Grant, George, 78
Grasse River Lock. *See* Bertrand H. Snell Lock
Great Lakes, 4, 22-23, 37
Great Lakes Fishery Commission, 205
Great Lakes–St. Lawrence Association, 84
Great Lakes–St. Lawrence Basin Agreement (1941), 40-47, 49, 53, 62, 65, 69, 82-84, 107, 128, 190, 209
Great Lakes–St. Lawrence Basin Commission, 43

Great Lakes–St. Lawrence waterway, 4-5, 4(m), 6, 6(m), 22, 28, 60
Great Lakes Waterway Treaty (1932), 34-40, 46, 54, 59, 107, 209
groundbreaking events. *See* ceremonies, for SLSPP
Group of Seven, 229
Gut Dam, 82, 181, 274n2

Harris, Cole, xvii
Hartshorne House Movers Co., 152, 155, 155(f), 169. *See also* relocation, connected to SLSPP: and house moving
Hearn, R.L, 65
Heeney, A.D.P., 96, 214
Helbig, Louis, 8(f)
Henry, George, 35
Henry, R.A.C., 77, 94, 96, 183
Hepburn, Mitchell, 39-41
HEPCO. *See* Hydro-Electric Power Commission of Ontario (HEPCO)
Herridge, William, 35-37
Hickerson, John D., 39
high modernism, xiv-xvi, xxiv, 16-17, 112, 165, 174, 178, 218-19, 281n16, 281n18, 282n23, 283n34, 283n40, 283nn43-44. *See also* high modernism, negotiated
high modernism, negotiated, 226-29
Highway 2 (Ontario), 140, 156, 164, 169, 223(f). *See also* highways and roads displaced by SLSPP
Highway 37 (New York), 126. *See also* highways and roads displaced by SLSPP
Highway 401 (Ontario), 170. *See also* highways and roads displaced by SLSPP
highways and roads displaced by SLSPP, 126, 140, 156. *See also* rehabilitation, connected to SLSPP; relocation, connected to SLSPP
Hills, Theo, 11, 28
Holden, Otto, 183
Hollinger-Hanna Group, 63
Hoover, Herbert, 34-38
Hoover Dam. *See* Boulder Dam
hostility thesis, of Canadian identity and environment, 229-30, 284n51. *See also* nationalism, Canadian

house moving. *See* relocation, connected to SLSPP
Howe, C.D., xxiii, 13, 62, 64-65, 76, 80, 89, 91, 94, 101, 119, 215, 248n14, 256n44, 259n85
Hudson River, 22, 33, 176
Hungry Horse Dam, 282n30
Hustler, Harry, 154, 159, 174, 268n54, 273n121
Hutchinson, Bruce, 47
hybrid waterscape, SLSPP as, xxii, 111, 226. *See also* waterscape
hybridity, and ecosystems, 111, 225-26
hydraulic bureaucracies, 17, 11
hydraulic nationalism. *See* hydro nationalism; nationalism, Canadian; water
hydro nationalism, 229, 284n45. *See also* nationalism, Canadian; water
Hydro-Electric Power Commission of Ontario (HEPCO), 24, 29-31, 37, 41, 52-53, 55, 112-13, 115-18, 121, 135, 139-78, 182(f), 183, 187, 190, 224; perceptions of as fair and responsive, 149-50, 153-54, 159-60, 168-69, 171-78, 224. *See also* construction of SLSPP; Lost Villages and SLSPP; rehabilitation, connected to SLSPP; relocation, connected to SLSPP
hydroelectricity: in Canadian history and historiography, 16; dam designs and types, 245n40, 246n54; facilities in Canada, xiv, 24, 49, 64, 216-17, 222; facilities in North America, 24, 221; federal vs provincial development of, 30, 32, 34, 37, 40-42; federal vs state development of, 27, 36-37, 53, 59, 74, 87, 114, 141; path dependencies, 49, 220; private vs public development of, 25, 27, 34-37, 42, 87, 141; on St. Lawrence River before SLSPP, 24. *See also* electricity
hydropolitics, 9, 213

ice, control of formation of, 26, 29, 137, 187, 189, 202, 205, 228, 243n12. *See also* engineering plans for SLSPP: evolution of; engineers and SLSPP; expertise

Illinois diversion. *See* Chicago diversion
imperialism. *See* St. Lawrence Seaway and Power Project (SLSPP): as ecological imperialism; St. Lawrence Seaway and Power Project (SLSPP): as social imperialism; United States, imperialism toward Canada
industrial capitalism, 17, 78, 159, 178, 183, 222-24, 225
Ingleside, Ontario, 140, 156, 157(f). *See also* model towns: design of
Innis, Harold, xiii, xix, 3, 78
integration, Canadian-American, xix, 10-11, 48-49, 63, 76, 87-89, 91-92, 106; and SLSPP, xvi, 8, 199, 212-13
interdepartmental committee on SLSPP (Canada), 51, 57, 59, 62, 69, 76, 94, 96-99
International Deep Waterways Commission, 22-24
International Deep Waterways Convention, 22
International Joint Commission (IJC), 10, 23, 26-27, 29, 31, 46, 50, 79-81, 113, 120, 184-89, 204, 208, 217, 225; Canada-US application on Lake Ontario levels to, 74, 181-89; Canada-US application on St. Lawrence Power Project to, 50, 52, 54-55, 62, 70-76, 80-82, 86, 101, 182
International Lake Ontario Board of Engineers, 113, 115, 181-89. *See also* International Joint Commission (IJC)
International Massena Board of Control, 244n28
International Rapids section (IRS), 7, 24, 25, 37, 114-15, 115(f), 129, 209, 226
International St. Lawrence River Board of Control, 113, 115-16, 189. *See also* International Joint Commission (IJC)
International Waterways Commission, 24
Inundation Day, 166-67, 179-80, 188
invasive species, 204-6, 225, 230, 232n7
iron ore, 9, 49, 51-52, 59-60, 61(t), 63, 76, 78, 93, 198-200, 214, 222
Iron Ore Company of Canada, 63
Iroquois (town of), 7, 45, 124, 132-33, 139, 146, 151(f), 154, 160(f), 161, 170, 264n57;

Index

evolution of relocation plans, 150-52, 151(f), 161
Iroquois Contractors, 116
Iroquois Control Dam, 12(f), 42, 117-18, 130, 132, 179, 187, 210(f)
Iroquois Lock, 12(f), 129; Canadian decision for, 101-8, 118-19, 209, 210(f), 214, 259n85, 259n87, 262n19
IRS. *See* International Rapids section (IRS)
Itaipu Dam, 221

Jacques Cartier Bridge, 130, 263n49
Joint Chiefs of Staff (US), 90, 92
joint seaway: and Canada keeping door open to US involvement, 62, 72, 81, 84-85, 89, 96-97; Canadian acquiescence to, 76, 97-108; Canadian rationalizing of, 105-6; as defeat for Canada, 209; US specific proposal for, 86, 97-99. *See also* Canada-US relationship
Josephson, Paul, 16

Kahnawake First Nation and SLSPP, 123-28, 127(f), 132
Kemano-Kitimat project, 64. *See also* aluminum industry
Kennan, George, xxii
Kenogami River, 45
Kilbourn, William, 13
King, William Lyon Mackenzie, 13, 29-34, 40-46, 48, 51, 53-54, 80
Kingston, Ontario, 7
knowledge, local, 17, 170, 188, 226. *See also* expertise
Korean War, 51, 65-66, 74
Kuibyshev hydroelectric project, 221
Kuykendall, Jerome, 90

labour and working conditions on SLSPP, 116-17, 133, 134(f), 169, 264n64
Lachine Canal, 21, 25
Lachine Power Dam, plans for, 53, 116, 123, 261n15
Lachine Rapids, 125, 128
Lachine section, 7, 128
Lake Erie, 21, 37, 205
Lake Nipigon, 45

Lake Ontario, xviii, 7, 29, 82, 129, 205; levels of, 181-82; shore property owners, 181-89
Lake Ontario Land Owners and Beach Protection Association, 182. *See also* Lake Ontario: shore property owners
Lake St. Clair, 38, 190
Lake St. Francis, 7, 129
Lake St. Lawrence, 7, 111, 119, 139-40, 140(m), 167, 179, 181, 223(f)
Lake St. Louis, 128
Lake Superior, 32, 39, 41, 45
lamprey, 205
land acquisition and expropriations for SLSPP, 122-28, 141-63, 276n31; changes to expropriation laws in Canada, 148-49, 176, 267n27; complaints, appeals, and resistance regarding, 124-26, 143, 145, 147(f), 149, 159-61, 165, 174-77, 224, 268n46; and determination of property value and compensation, 141, 147-56, 165; and development of "surplus" lands, 162, 177, 164; and differences between approaches of Canada and US (and Ontario and New York), xv, 125, 140, 142-44, 160-63, 175-77; precedents for, 141, 271n94; and property transaction statistics, 126, 142, 145, 167-68, 267n36. *See also* Lost Villages and SLSPP; relocation, connected to SLSPP
landscape, 17, 126, 132, 152, 167, 172, 229, 238n25. *See also* waterscape
Laprairie basin, 123
Laurentian. *See* St. Lawrence River
Laurentian thesis, xix-xx, 3, 13-15, 211, 219, 239n27; as anti-American, 14; as geographically determinist, 14
Laurier, Sir Wilfrid Laurier, government of, 25-26
legal (immigration, waiver, and duty) issues stemming from transborder construction, 117
legibility. *See* high modernism
liberalism and property, conceptions of, 159, 161, 176, 269n57
Lindsay, Guy, 59, 77, 215

linkage, in Canada-US relations, 11, 32, 34, 38-39, 60, 66, 84, 88-89, 91-92, 96, 99, 101, 106, 213-14, 236n11. *See also* Canada-US relationship
Lippmann, Walter, 94
local, knowledge. *See* knowledge, local
Long Lac (Lake), diversion. *See* Ogoki and Long Lac (Lake) diversions
Long Sault (town of), Ontario, xv, 140, 156, 158(f), 166(f), 171(f). *See also* model towns, design of
Long Sault Canal. *See* Wiley-Dondero Ship Canal
Long Sault Control Dam, 42, 179, 201(f), 216(f)
Long Sault Development Corporation, 24-25
Long Sault Island, 130
Long Sault Park, 46
Long Sault Parkway, 162
Long Sault Rapids, 25, 26(f), 46, 130, 167, 172, 201
Longueuil, Quebec, 124
Loo, Tina, 219, 227
Lost Villages and SLSPP: 7, 127, 137, 139-78, 140(m); as agricultural and milling area, 140, 164; commemoration of, xvi, 140, 207, 265n2; and coping mechanisms, 159, 173; differences in generational impact, 161, 173, 272n116; early plans for rehabilitation of, 45-46; expectations of prosperity, 152, 168; foundations of visible, 8(f), 173-74; levels of prosperity, 169, 198-200; and lower-income groups, 161-62, 269n67; meetings with residents and representatives of, 146-53, 174; and negative impacts on residents, 169-78; nostalgia and intangibles, 149, 172-74; plans to improve residents and living standards, 142, 146; representations of, 140, 265n2; spatial distribution of communities, 265n1; surveys of residents of, 149, 153-54, 161, 174-75, 224. *See also* land acquisition and expropriations for SLSPP; rehabilitation, connected to SLSPP; relocation, connected to SLSPP

Lost Villages Historical Society, xv, 173
Louisville, New York, 149
Louisville Landing, New York, 149
Lower, A.R.M., xix
Loyalists. *See* United Empire Loyalists

Mabee, Carleton, 11, 71, 148, 156
MacEachern, Alan, 77, 271n94
macro theories of Canadian history. *See* Laurentian thesis; metropolitan-hinterland thesis; staples thesis
Manson, Lyall, 172
Maple Grove, 139-40, 162, 270n69
marine clay, 122, 132, 137, 151. *See also* construction of SLSPP: subsurface conditions
Massena, New York, 7, 24, 140, 162, 199
Massena Power Canal. *See* Aluminum Company of America
Massey Commission, 64
Matilda Township, 133, 150-51, 161
McEwan, Joan, 172
McKillop, A.B., xix
McLennan, Hugh, 3
McLuhan, Marshall, 78
McNaughton, A.G.L., 13, 36, 76, 79-80, 94, 96, 99, 104, 183, 215, 217
McWhorter, Roger, 82, 217
megaprojects, era of, 49, 77, 222, 224, 230
Mercier Bridge, 130
metatheories of Canadian history. *See* Laurentian thesis; metropolitan-hinterland thesis; staples thesis
method of regulation, of St. Lawrence River, 181-89, 274n5, 275n20, 276n23. *See also* engineering plans for SLSPP
metropolitan-hinterland thesis, 15, 218, 238n25
Michaels, Anne, xvi, xxiv-xxv
Midwest (US), opinion of SLSPP in, 29, 34
Mille Roches, Ontario, xv, 139, 144, 145(f), 156, 161
Mississippi River, 27, 34
Mitchell, Timothy, 241n43
mobility, 130, 178. *See also* transportation
model towns: design of, 139, 146, 154, 156, 167-71, 222, 272n106; identity and

Index 319

socialization, 170. *See also* Ingleside, Ontario; Long Sault (town of), Ontario; rehabilitation, connected to SLSPP; relocation, connected to SLSPP

models (hydraulic), use of, 117, 133, 137, 182(f), 183-89, 207, 224, 228; problems with, 187-88. *See also* engineers and SLSPP; expertise; progress

modern amenities and SLSPP, xv, 78, 124-25, 142, 148-50, 168-72, 222. *See also* Lost Villages and SLSPP; rehabilitation, connected to SLSPP; relocation, connected to SLSPP

modernization, xiv, xv, 8, 14, 29, 78, 79, 139, 142, 169, 172, 221-22. *See also* high modernism; progress; technology

Montreal, 7, 29, 30, 36, 116, 123, 128, 130, 203(f); and construction of SLSPP at, 116-17, 123-30; economic impact of SLSPP on, 198-99, 277n44; impact of SLSPP downstream on, 186, 274n9; opinion of SLSPP in, 28, 30 (*see also* public opinion and SLSPP, in Canada); port and harbour facilities at, 26, 130

Morrisburg, Ontario, 45, 139, 156, 170, 176; proposed power dam near, 29-30

Morton, W.L., 229

Moses, Robert, xxii, 112, 121-22, 123(f), 135, 142-43, 151-52, 160-62, 175-76, 178, 224, 227

Moses power house. *See* Robert Moses–Robert H. Saunders Power Dam

Moulinette, Ontario, 139, 144(f), 156

nation building, xiii, 16, 74, 76, 215, 220, 281n19. *See also* nationalism, Canadian; progress; technology; transportation

National Advisory Committee, 31

National Coal Association, 82

national imagination. *See* nationalism, Canadian

national interest: Canada and the, 11, 106, 212, 217; United States and the, 11, 92, 212. *See also* Canada-US relationship

national security, and SLSPP, 52, 60, 66, 76, 84, 90-91, 93, 211, 214

National Security Council (US), 90, 92

National Security Resources Board (US), 59-60

National St. Lawrence Association, 50

National St. Lawrence Project Conference, 27

nationalism, Canadian: and environment, 12-18, 64, 78-80, 211, 220, 229-30, 284n51; after Second World War, 49, 64; and SLSPP, 62-64, 68, 71-80, 92-93, 95-96, 101-5, 207, 209, 211, 220, 229; and St. Lawrence River, 14-16, 23, 30, 32, 76-80, 89, 211, 220, 229; and technology, 12-18, 64, 78-80, 211, 220, 229; and water, 229

natural resources: statist conservation, 25, 37, 58-59, 183; statist mobilization, 49, 51, 64, 211

"natural" state of St. Lawrence River. *See* engineering plans for SLSPP; engineers and SLSPP; expertise

nature: attitudes toward, 29, 139, 211-12, 224, 229; differing Canadian and American views of, 229-30, 284n51; resiliency and adaptability of, 202-4, 225-26. *See also* environmental impact of SLSPP

negotiated high modernism. *See* high modernism, negotiated

negotiated state, Canada as, 229

Nelles, H.V., 16, 174, 224

New Deal, approach to water resources, 34, 37, 58-59, 227. *See also* Tennessee Valley Authority (TVA), dam projects

New York Central Railroad, 121-22

New York Power Authority. *See* Power Authority of the State of New York (PASNY)

New York State, and SLSPP, 7, 27, 36-38, 44, 49, 52-59, 70, 76, 82-83, 76, 87, 90-91, 107, 112, 114, 124-26, 139, 141-43, 161, 163-64, 175-77, 199, 204, 206, 214. *See also* Lost Villages and SLSPP; rehabilitation, connected to SLSPP; relocation, connected to SLSPP

Newbigin, Marion: influence on Laurentian thesis, xviii-xix, 239n27

Niagara Mohawk, 175

Niagara region, and power developments, 24, 32-34, 38-41, 44-45, 47, 55-56,

64, 116, 122, 141-42, 220-22, 244n36, 247n83
North American school of Canada-US relations, 10, 235n8
North Atlantic Triangle (Brebner), xx, 235n8
North Atlantic triangle, concept of, 11, 23, 48
North Country, 176-77. *See also* St. Lawrence County, New York
nostalgia, impact of, xv, 173
Nye, David, 16

obsolescence, SLSPP and technological, xxiv, 198
Ogden Island, 163
Ogden Mansion, 163
Ogdensburg, New York, 130
Ogilby, John, xvii
Ogoki and Long Lac (Lake) diversions, 39-41, 44-45, 104, 190. *See also* Albany River (Ontario), diversion of
oil, and postwar economic growth, 49
oil sands and pipelines, parallels between SLSPP and, xxiii-xxiv
Ontario, economic, manufacturing, and industrial growth, 32, 52, 63, 65, 175
Ontario, Government of, involvement in relocation and rehabilitation of dislocated communities, 139, 141-53. *See also* Lost Villages and SLSPP; rehabilitation, connected to SLSPP; relocation, connected to SLSPP
Ontario, opinion of SLSPP in, 28, 63, 65. *See also* public opinion and SLSPP, in Canada
Ontario Historical Society, 164
Ontario Hydro. *See* Hydro-Electric Power Commission of Ontario (HEPCO)
Ontario Municipal Board of Review, 161, 165, 168
Ontario–St. Lawrence Development Commission, 150, 162, 164. *See also* Ontario, Government of, involvement in relocation and rehabilitation of dislocated communities; rehabilitation, connected to SLSPP
opening of SLSPP: fiftieth anniversary of, xv, 12. *See* ceremonies, for SLSPP

opposition groups, against SLSPP, 27-29, 38-42, 44, 50, 82-83, 118-20, 257n50. *See also* Congress (US), and SLSPP; special interests, ability to block legislation in US
oral interviews, and SLSPP, 170, 172, 271n97
orientation, of Canada: east-west vs north-south, xviii, 13, 79, 211; size and spatial reality, 79
Ottawa River, 7, 26, 45, 116, 141

Paley Commission, 51. *See also* United States, imperialism toward Canada
Panama Canal, 220
Pandora's Locks: The Opening the Great Lakes–St. Lawrence Seaway (Alexander), 204-5
Parham, Claire Puccia, 137-38, 171
parkland, 121-22, 130, 142, 162-65. *See also* recreational developments and SLSPP
Parliament (Canada), and SLSPP, 26, 38-39, 71, 75
Parr, Joy, 170
PASNY. *See* Power Authority of the State of New York (PASNY)
Pearson, Lester B., 13, 60, 65-66, 70, 73, 81, 88, 91, 94, 101, 102, 105, 215, 256n40, 259n87
Permanent Joint Board on Defence, 79-80, 90, 215
Pioneer Memorial, 162. *See also* cemeteries, affected by SLSPP; Upper Canada Village
Piper, Liza, 225
Point Rockaway Lock, 42, 118-19. *See also* Iroquois Lock
Pollys Gut bridge, 122, 130
power. *See* electricity; hydroelectricity
Power Authority of the State of New York (PASNY), 36-38, 53, 59, 73, 92, 112-13, 115-16, 121-23, 135, 139-43, 146, 161-64, 169, 175-76, 189-90, 224. *See also* construction of SLSPP; rehabilitation, connected to SLSPP; relocation, connected to SLSPP
power priority plan. *See* St. Lawrence Seaway and Power Project: separate Ontario/New York power project

Index

Prairie provinces, opinion of SLSPP in, 28, 67-68. *See also* public opinion and SLSPP, in Canada
predictions, for traffic on SLSPP, 42-43, 50-51, 61(t), 195-97, 250n42. *See* traffic on SLSPP
Prescott, Ontario, 24, 130
Pritchard, Sara B., 226
productive disagreement, 97, 101
progress: idea of and faith in, xiv, xvi, 3, 164-65, 172, 181-83, 218-30; SLSPP as, xv, 29, 62, 76-77, 142, 149, 159-60, 168, 172, 174-78, 207, 217-30
Project 2000. *See* Federal Power Commission (FPC), US: and PASNY power licence
proposals, for SLSPP, 22, 28-30
provincial diplomacy, by Ontario, 56
Provincial Paper Mill, 144
public opinion and SLSPP, in Canada, 56, 61-62, 67-72, 76, 91, 92, 95, 101-6, 209, 211

Quebec, impact of SLSPP on, 198-99
Quebec, opinion of SLSPP in, 28, 30, 63
Quebec City, opinion of SLSPP, 28

radar defences, across the north, 49, 92
railways. *See* railroads
railroads, xiii, xviii, 25, 28, 35, 63, 79; part of opposition groups against SLSPP, 27, 32; relocation because of SLSPP, 140, 156
razing, of infrastructure before flooding, 167, 202. *See also* relocation, connected to SLSPP: and house burning
recreational developments and SLSPP, 162-65, 168, 201(f). *See also* parkland
rehabilitation, connected to SLSPP: of area flooded by power project, 155-78; evolution of plans for, 139-55; largest in Canadian history, 7, 111, 139; scenic quality and beauty, 142, 145, 164, 177. *See also* Lost Villages and SLSPP; model towns
Rehabilitation of the St. Lawrence Communities (Wilson), 45-46, 141
relicencing, of St. Lawrence Power Project, 207

relocation, connected to SLSPP: evolution of plans for, 139-65; and house burning, 167, 172, 174; and house moving, 151-56, 155(f), 159, 161, 167-68; new towns, xv, 78; in Ontario from seaway, flooding from power project, xv, 4, 111; in Quebec from seaway, 139. *See also* land acquisition and expropriations for SLSPP; Lost Villages and SLSPP; model towns
replacement value. *See* land acquisition and expropriations for SLSPP: and determination of property value and compensation
reservoir, Lake St. Lawrence as, 167, 180-81, 188
Reynolds Metals, 175, 206
Rideau Canal, xviii
rights of vessels, in foreign waters, 99-105
river profile, establishment of. *See* engineering plans for SLSPP
Robert Moses–Robert H. Saunders Power Dam, 7, 111, 121-22, 129(f), 130, 133, 135, 136(f), 179, 184(f), 185(f), 195(f), 207, 216(f), 222
Robertson, Norman, 215
Robinson Bay Lock. *See* Dwight D. Eisenhower Lock
Roosevelt, Franklin D., and SLSPP: as governor of New York, 36-38; as president, 38-42, 48, 141
Roundtable on Man and Industry (1956), 168-69
Rowell-Sirois Commission. *See* constitution, Canada, and jurisdiction over water resources
Royal Commission on Dominion-Provincial Relations. *See* constitution, Canada, and jurisdiction over water resources
Royal Commission on Radio Broadcasting, xiii
Rutley, Rosemary, 170

Saint-Hubert, Quebec, 124
Saint-Lambert, Quebec, 124
Saint-Lambert Lock, 128, 130, 194(f), 203(f)

Saint Lawrence Seaway Development
  Corporation (SLSDC), 113-14, 116-17,
  120, 122-23, 130, 191-92, 195, 198, 200
salmon, Pacific, fisheries treaty, 34-36
Santa Cruz, Ontario, 139
Sault Ste. Marie Canals, 21, 45, 63
Saunders, Robert, 55-56, 55(f), 64-65, 142,
  150-53, 159, 248n14
Saunders power dam. *See* Robert Moses–
  Robert H. Saunders Power Dam
schools, affected by SLSPP, 122, 165, 170-
  72, 272n110. *See also* Lost Villages and
  SLSPP; rehabilitation, connected to
  SLSPP; relocation, connected to SLSPP
science. *See* technology; engineering plans
  for SLSPP; progress
seasonality, 205
Second World War, 41, 44, 46-48, 50, 52
self-liquidation, of SLSPP. *See* tolls, and
  self-liquidation, for SLSPP
Sentimentalists (Skibsrud), xvi
Scott, F.G., 30
Scott, James. C., 16, 219, 227, 282n23
Seaway International Bridge, 122, 130
Sheek Island, 161-62, 164, 179
shipbuilding, allowed inland by SLSPP,
  28, 199
shopping and retail: affected by SLSPP,
  156, 166(f); new shopping centres, 166,
  168, 171. *See also* Lost Villages and
  SLSPP; rehabilitation, connected to
  SLSPP; relocation, connected to SLSPP
shoreline: controlling of, 132, 156-57, 162-
  64, 226, 270n79, 273n132; democratiza-
  tion of, 177. *See also* environmental
  impact of SLSPP: on shoreline erosion
Simcoe (commercial ship), 192
single-stage dam. *See* engineering plans
  for SLSPP: evolution of
Skelton, O.D., 34, 40
Skibsrud, Johanna, xvi
smoke detectors, adoption of, 137
Soulanges: canal, 21; section, 7
Soviet Union, 11, 17, 48, 220-21, 227,
  282n25-26
Spalinski, Hugo, 257n50
Sparrowhawk Point, 132

special interests, ability to block legislation
  in US, 12, 28, 39, 87. *See also* Congress
  (US), and SLSPP
spoil, 132-33, 143, 226. *See also* environmental impact of SLSPP
St. Clair River, 5
St. Laurent, Liberal government of, 9, 10,
  54; approach to foreign policy and
  Canada-US relations, 11, 54, 77, 280n7;
  approach to SLSPP, 56-57, 67, 69, 71,
  74, 83, 86, 91, 214-15; January 9, 1953
  statement, 86
St. Laurent, Louis, 13, 48, 54, 58, 60-62,
  70, 80, 96, 208-9, 215
St. Lawrence Agreement of 1954, 101-8,
  209
St. Lawrence Board of Review (Ontario),
  153, 160-61, 164, 168. *See also* land acquisition and expropriations for SLSPP:
  complaints, appeals, and resistance
  regarding; Ontario, Government of,
  involvement in relocation and rehabilitation of dislocated communities
St. Lawrence Commission, 31
St. Lawrence County, New York, 161, 199
St. Lawrence Development Act (Ontario),
  142, 146, 148, 153
St. Lawrence Joint Board of Engineers,
  31, 36, 116, 118, 181, 186, 188. *See also*
  International Joint Commission (IJC)
St. Lawrence nationalism. *See* nationalism,
  Canadian
St. Lawrence Power Development Commission, 36
St. Lawrence River: as bridge and/or barrier, 15, 208; in Canadian historiography,
  xvii-xx, 12-15, 176, 237n15; flow volume
  and levels, 7; importance to Canadian
  history, xvii-xx, 3-4, 13-15, 21, 176; importance to United States, 27; Ontario
  vs Quebec nationalism and, 15; physio-
  graphic features, 5-7, 264n52; separation
  of power and navigation aspects, 10;
  as transnational space, 176-77. *See also*
  Laurentian thesis; nationalism,
  Canadian
St. Lawrence River Board of Engineers, 92

St. Lawrence River Power Company, and application for submerged weir, 27, 244n28

St. Lawrence Seaway and Power Project (SLSPP): as attractive to electorate, 88, 90-91, 217-18, 281n13; as Cold War symbol, 17, 184, 192-93, 220-21; cost, 32, 37, 43, 87, 93, 99, 107, 112-13, 179, 277n39; debates about depth, 23, 118; debates about lock size, 118, 198; dimensions and magnitude of, 5, 17, 98-107, 111-12, 128, 143-44, 179-80, 198, 225; as ecological imperialism, 178; as engineering achievement, xv, 7, 111, 130, 133; feasibility, 9, 34; generating capacity, 7, 111; historiography, 11-15; industrial and economic growth resulting from, 133-34, 175-76; as largest transborder water control project, 111; legacy of, 194-207; as megaproject, 7, 17, 211, 224; as mistake, 207; sections, 7, 235n2; separate Ontario/New York power project, 49, 52-53, 69-70, 76, 87; as social imperialism, 179; as spatial reorganization, 142, 170, 272n119; winter navigation on, 205. *See also* engineering plans for SLSPP: evolution of; joint seaway

St. Lawrence Seaway Authority (SLSA), 71, 113, 116, 119, 122, 130, 192, 195-97

St. Lawrence Seaway Authority Act, 71, 98

St. Lawrence Seaway Management Corporation (SLSMC), xv, 199-200

St. Lawrence Securities Company, 24

St. Lawrence Survey (Danielian), 42-43, 50

St. Lawrence Valley Union Cemetery, 162. *See also* cemeteries, affected by SLSPP

St. Marys River, 5

St. Regis (Akwesasne) First Nation and SLSPP, 122-28, 205-6, 279n78

Stacey, C.P., 29

Stanley, Meg, 227

staples, 14-15, 229. *See also* staples thesis

staples thesis, xix, 13-15, 218

state, nature of, 16

state building. *See* nation building

state formation. *See* nation building

Stewart, Gordon, 32

Stimson, Henry L., 35

stores, affected by SLSPP. *See* shopping and retail

strikes. *See* labour and working conditions on SLSPP

Strait, Russell, 163

Straits of Mackinac, 5

tar sands and pipelines. *See* oil sands and pipelines, parallels between SLSPP and

tariffs, 32, 34-35

Taschereau, Louis-Alexandre, 30, 36

technocratic. *See* expertise

technological imperative, 79. *See also* Francis, R. Douglas

technological nationalism. *See* nationalism, Canadian

technological projects, as compromises, xvi, 128, 183

technological sublime, 16

technology: borders between environment and, xvii; faith in, 62, 164, 187, 207, 219, 221; as nation building, xiii, 220. *See also* expertise; high modernism; models (hydraulic), use of; nationalism, Canadian; progress

Tennessee Valley Authority (TVA), dam projects, 65, 141, 220-21, 227, 282n23. *See also* New Deal, approach to water resources

third-party shipping, dispute over, 98-105

Thoreau, Henry David, xviii

Thousand Islands section, 7, 24, 33, 37, 114, 129

Three Gorges Dam, 221

tolls, and self-liquidation, for SLSPP, 50-52, 90-91, 100, 102, 104, 107, 121, 180, 189-91, 191(t), 195-97, 199, 225

tourism and public viewing of SLSPP, 25, 135, 155, 162, 265n67, 268n50

Toussaint Island, 132

traffic on SLSPP, 195-98, 225, 276n37-38

Trail Smelter dispute, 39, 234n24

TransCanada Air Lines, xiii

Trans-Canada Highway, xiv, 49

TransCanada Pipeline, 49, 218

transnationalism, approaches to, 10-11
transportation: Canadian vs American state support for, 28-29, 79, 176; and nation building, xii-xiv, 28, 78-80, 176, 222; SLSPP and reordering of networks of, 130, 222. *See also* mobility
Trent Canal, 25
Truman, Harry S.: SLSPP and administration of, 48, 50, 53-54, 58, 60, 66-76, 80-85, 249n18
tunnels, highway and railway, 117, 121-22, 130, 197(f). *See also* transportation
Two Solitudes (MacLennan), 3

Uhl, Hall and Rich, 116, 221
Underhill, Frank, xix
Ungava region, 49, 51, 59-60, 63. *See also* iron ore
unions. *See* labour and working conditions on SLSPP
United Empire Loyalists, 4, 127, 164
United States, alternate deep waterway routes, 22, 33, 35, 46
United States, foreign policy in early Cold War: creation of open markets, 48; nuclear weapons, 48; spreading liberal democracy, 48
United States, imperialism toward Canada: 11, 51, 78-80, 89, 93, 213, 220
United States, relationship to Canada. *See* Canada-US relationship
Upper Canada Migratory Bird Sanctuary, 164. *See also* environmental impact of SLSPP
Upper Canada Village, xv, 162, 164-65, 173
uranium, in Cold War period, 49, 64

Vandenburg, Arthur H., 50
Verdun, Quebec, 123
vernacular, knowledge. *See* knowledge, local
Veyret, Paul, xvii
Victoria Bridge, 130, 203(f)

Voices for the Watershed: Environmental Issues in the Great Lakes–St. Lawrence Drainage Basin (edited volume), 205

Waddington, New York, 133, 140, 162-63, 176, 270n77
Wales, Ontario, 139, 153
War of 1812, 131, 164-65, 176
water: commodification of, 49, 226; as culturally constructed, 16, 240n38; and identity, 14-15, 78; statist control of and power over, 17
water levels of SLSPP and Great Lakes. *See* engineering plans for SLSPP
Water Resources Policy Commission, US, 65
waterscape, xxii, 9, 111, 126. *See also* hybrid waterscape
Welland Canal, xviii, 6(m), 21, 26, 29, 33, 35, 37, 44, 53, 63, 107, 112, 118, 129, 136, 192, 193(f), 197, 205, 209, 245n38
wheat. *See* grain and SLSPP
Wiley Bill. *See* Congress (US), and SLSPP
Wiley-Dondero Act, 96-98, 103, 214
Wiley-Dondero Ship Canal, 114, 126, 129, 129(f), 133, 262n28
Williamsburg Canals, 21, 63
Willoughby, William, 11, 35
Wilson, Norman D., 45-56
Wilson Hill development, 163
Wilson Report. *See Rehabilitation of the St. Lawrence Communities* (Wilson)
Winter Vault (Michaels), xvi, xxiv-xxv
Woodlands, Ontario, 139
Wooten-Bowden Report, 29, 46, 112
workers, social and living conditions of, 134-35, 169, 271n99. *See also* labour and working conditions on SLSPP
Worster, Donald, 17, 284n51
Wrong, Hume, 65, 73, 81
Wynn, Graeme, 219

zebra mussels, 204, 232n7
Zeller, Thomas, 226

NATURE | HISTORY | SOCIETY

Claire Elizabeth Campbell, *Shaped by the West Wind: Nature and History in Georgian Bay*

Tina Loo, *States of Nature: Conserving Canada's Wildlife in the Twentieth Century*

Jamie Benidickson, *The Culture of Flushing: A Social and Legal History of Sewage*

William J. Turkel, *The Archive of Place: Unearthing the Pasts of the Chilcotin Plateau*

John Sandlos, *Hunters at the Margin: Native People and Wildlife Conservation in the Northwest Territories*

James Murton, *Creating a Modern Countryside: Liberalism and Land Resettlement in British Columbia*

Greg Gillespie, *Hunting for Empire: Narratives of Sport in Rupert's Land, 1840-70*

Stephen J. Pyne, *Awful Splendour: A Fire History of Canada*

Hans M. Carlson, *Home Is the Hunter, The James Bay Cree and Their Land*

Liza Piper, *The Industrial Transformation of Subarctic Canada*

Sharon Wall, *The Nurture of Nature: Childhood, Antimodernism, and Ontario Summer Camps, 1920-55*

Joy Parr, *Sensing Changes: Technologies, Environments, and the Everyday, 1953-2003*

Jamie Linton, *What Is Water? The History of a Modern Abstraction*

Dean Bavington, *Managed Annihilation: An Unnatural History of the Newfoundland Cod Collapse*

Shannon Stunden Bower, *Wet Prairie: People, Land, and Water in Agricultural Manitoba*

J. Keri Cronin, *Manufacturing National Park Nature: Photography, Ecology, and the Wilderness Industry of Jasper*

Jocelyn Thorpe, *Temagami's Tangled Wild: Race, Gender, and the Making of Canadian Nature*

Darcy Ingram, *Wildlife, Conservation, and Conflict in Quebec, 1840-1914*

Caroline Desbiens, *Power from the North: Territory, Identity, and the Culture of Hydroelectricity in Quebec*

Sean Kheraj, *Inventing Stanley Park: An Environmental History*

Justin Page, *Tracking the Great Bear: How Environmentalists Recreated British Columbia's Coastal Rainforest*